ALLOY STEEL – PROPERTIES AND USE

Edited by **Eduardo Valencia Morales**

Alloy Steel – Properties and Use
Edited by Eduardo Valencia Morales

Published by InTech

Janeza Trdine 9, 51000 Rijeka, Croatia

Publishing Process Manager Romana Vukelic

Technical Editor Teodora Smiljanic

Cover Designer InTech Design Team

Additional hard copies can be obtained from orders@intechopen.com

Alloy Steel – Properties and Use, Edited by Eduardo Valencia Morales
p. cm.
ISBN: 978-953-307-484-9

For EU product safety concerns:
IN TECH d.o.o., Prolaz Marije Krucifikse Kozulić 3, 51000 Rijeka, Croatia,
info@intechopen.com or visit our website at intechopen.com.

Meet the editor

Dr. Eduardo Valencia Morales is a Professor of Physics and Materials Science at the Central University "Marta Abreu" of Las Villas (UCLV) in Cuba. He has been a visiting Professor at several universities in America, where he has developed different international projects on physical metallurgy in subjects such as: Strengthening Mechanisms in Commercial Hot Strip Microalloyed Steels, Kinetics of the Phase Transformation in Low Alloy Steels, etc. He has trained several post-doctoral researchers and he has published numerous articles on materials in approximately 20 different international journals corresponding to the SCI. His works have been presented in multiple international symposia, and he is a reviewer of several international journals. Dr. Morales has received two prizes from the Cuban Academy of Sciences and the Special Award of the Ministry of Higher Education in Cuba for his scientific activity. He recently received the order Carlos J. Finlay, the main distinction that the Cuban State grants to persons by their recognized scientific work.

Contents

Preface

The main objective of this book is to show, in a logical sequence, a set of papers presenting new ideas that have been the result of the great advance in the knowledge of steel within the last few decades. Amongst the fundamental topics dealt with is the relationship between steel's properties in terms of its structure and composition and how this is affected by its surrounding environment (i.e. stresses, temperature, strains, etc.).

It is not possible to affirm that one knows all the links of this chain with details. But today, thanks to significant technological and theoretical advances, we know the enough to understand the most important phenomena that take place in these according to its uses. In this sense, a number of well-known specialists have consented to write on the various principal branches in this subject and the editor has been responsible for preserving a basic unit among the expert contributions.

Each section is devoted to new approaches and usages of the stainless steels, the influence of the environments on the behavior of some class of steels, new structural concepts to understand some fatigue processes and new insight on strengthening mechanisms and toughness in micro-alloyed steels. The kinetics during tempering in low-alloy steels is also discussed through a new set-up that uses a modified Avrami formalism. And each section contains an updated list of references which will allow to the readers to delve further into a particular subject.

I am grateful to the authors for their courtesy and patience, and hope that through their efforts this book earns its place alongside the more cited books about alloy steels. Finally, I should like here to acknowledge the sustained helpfulness and dedication of the publisher's staff, in particular of Ms. Romana Vukelic by its insistence for concluding this interesting project.

Dr. E. V. Morales
Department of Physics,
Central University of Las Villas, Villa Clara
Cuba

Part 1

Stainless Steels: New Approaches and Usages

1

Alloy Steel: Properties and Use First-Principles Quantum Mechanical Approach to Stainless Steel Alloys

L. Vitos[1,2,3], H.L. Zhang[1], S. Lu[1], N. Al-Zoubi[1], B. Johansson[1,2], E. Nurmi[4], M. Ropo[4], M. P. J. Punkkinen[4] and K. Kokko[4]

[1]*KTH Royal Institute of Technology,*
[2]*Research Institute for Solid State Physics and Optics,*
[3]*Uppsala University,*
[4]*University of Turku,*
[1,3]*Sweden*
[2]*Hungary*
[4]*Finland*

1. Introduction

Accurate description of materials requires the most advanced atomic-scale techniques from both experimental and theoretical areas. In spite of the vast number of available techniques, however, the experimental study of the atomic-scale properties and phenomena even in simple solids is rather difficult. In steels the challenges become more complex due to the interplay between the structural, chemical and magnetic effects. On the other hand, advanced computational methods based on density functional theory ensure a proper platform for studying the fundamental properties of steel materials from first-principles. In 1980's the first-principles description of the thermodynamic properties of elemental iron was still on the borderline of atomistic simulations. Today the numerous application-oriented activities at the industrial and academic sectors are paired by a rapidly increasing scientific interest. This is reflected by the number of publications on *ab initio steel research,* which has increased from null to about one thousand within the last two decades. Our research group has a well established position in developing and applying computational codes for steel related applications. Using our *ab initio* tools, we have presented an insight to the electronic and magnetic structure, and micromechanical properties of austenite and ferrite stainless steel alloys. In the present contribution, we review the most important developments within the *ab initio quantum mechanics aided steel design* with special emphasis on the role of magnetism on the fundamental properties of alloy steels.

Steels are mainly composed of iron and carbon and special properties are reached by introducing additional alloying elements. Stainless steels are among the most important engineering materials. They are alloy steels containing more than 12 percent Cr. Chromium forms a passive oxide film on the surface, which makes these alloys resistant against corrosion in various chemical environments (Wranglén, 1985). The main building block of ferrite stainless steels is the Fe-Cr alloy having the ferromagnetic α-Fe structure. Austenitic stainless steels form the largest sub-category of stainless steels and comprise a significant amount of Ni as well. At low temperature, these alloys exhibit a rich variety of magnetic structures as a function of chemical composition, ranging from ferromagnetic phase to spin-glass and antiferromagnetic alignments (Majumdar & Blanckenhagen, 1984). At ambient conditions, Ni changes the ferromagnetic α-Fe structure to the paramagnetic γ-Fe structure. Today austenitic stainless steels dominate the steels applications, where high corrosion resistance and excellent mechanical properties are required. The austenitic grades represent the primary choice also when nonmagnetic properties are concerned.

2. Fundamental properties

2.1 Mechanical properties

The behavior of materials under external load defines their mechanical properties. Deformations are usually described in terms of stress of force per unit area and strain or displacement per unit distance. Using the stress-strain relation one can distinguish elastic and plastic regimes (Aragon, Backer, McClintock, & al., 1966; Ghosh & Olson, 2002; Lung & March, 1999). At small stress, the displacement and applied force obey the Hooke's law and the specimen returns to its original shape upon unloading. Exceeding the so-called elastic limit, upon strain release the material is left with a permanent shape.

Within the elastic regime, the elastic constants play the primary role in describing the stress-strain relation, whereas in the plastic regime the mechanical hardness expresses the resistance of material to permanent deformations. Plastic deformations are facilitated by dislocation motion and can occur at stress levels far below those required for dislocation-free crystals. Mechanical hardness may be related to the yield stress separating the elastic and plastic regions, above which a substantial dislocation activity develops. In an ideal crystal dislocations can move easily because they only experience the weak periodic lattice potential. In real crystal, however, the movement of dislocation is impeded by obstacles, leading to an elevation of the yield strength. In particular, in solid solutions the yield stress is decomposed into the Peierls stress needed to move a dislocation in the crystal potential and the solid-solution strengthening contribution due to dislocation pinning by the randomly distributed solute atoms. The Peierls stress of pure metals is found to be approximately proportional to the shear modulus (Lung & March, 1999). Dislocation pinning by random obstacles is controlled by the size and elastic misfit parameters (Fleischer, 1963; Labusch, 1972; Nabarro, 1977). The misfit parameters, in turn, can be derived from the composition dependent elastic properties of bulk solids. The effect of alloying on the elastic moduli of Fe and Fe-based alloys was studied in several experiments (Ghosh & Olson, 2002; Speich, Schwoeble, & Leslie, 1972; Takeuchi, 1969). Many of those measurements, however, were performed on multiphase samples, and thus the obtained elastic parameters correspond to a mixed phase rather than to a well defined crystal structure and hence give no information about the solid-solution strengthening mechanism within a particular phase.

Besides the bulk parameters, the formation energies of two-dimensional defects are also important in describing the mechanical characteristics of solids. The surface energy, defined as the excess free energy of a free surface, is a key parameter in brittle fracture. According to Griffith theory (Lung & March, 1999), the fracture stress is proportional to the square root of the surface energy, that is, the larger the surface energy is the larger the load could be before the solid starts to break apart. Another important planar defect is the stacking fault in close-packed lattices. In these structures, the dislocations may split into energetically more favorable partial dislocations having Burgers vectors smaller than a unit lattice translation. The partial dislocations are bound together and move as a unit across the slip plane. In the ribbon connecting the partials the original ideal stacking of close-packed lattice is faulted. The energy associated with this miss-packing is the stacking-fault energy (SFE). The equilibrium separation of the partial dislocations is determined by the balance of the repulsive interaction and the stacking fault energy. Generally, larger stacking fault energy corresponds to smaller distance between the partials. During the dislocation movement, the partials must re-combine in order to overcome the obstacles (e.g. solute atoms). The resistance of materials to plastic deformation decreases with increasing SFE and hence in order to increase their strength the SFE should be lowered. In solid-solutions, the stacking fault energy may be varied, whereby wider or narrower dislocations can be produced and the mechanical properties can be altered accordingly. In practice, SFE is controlled by alloying elements towards desired properties such as strength or work hardening rate. Although, the stacking fault energy in austenitic steels has been determined from experiments (Rhodes & Thompson, 1977; Schramm & Reed, 1975), it should be mentioned that it is difficult to measure precisely and large inaccuracies are associated with the available experimental values (Vitos, Korzhavyi, Nilsson, & Johansson, 2008).

The immediate use of the stacking fault energies in steel design is beyond doubt. However, studying the stacking faults in steel alloys has some fundamental aspects as well. The stacking fault energy, to a good approximation, is proportional to the Gibbs energy difference between the hexagonal close packed (hcp) and face centered cubic (fcc) phases (Ishida, 1976). In contrast to the fcc phase, the magnetic free energy vanishes in the hcp Fe (ε-Fe) indicating that the local moments disappear in this phase (Grimvall, 1976). Therefore, the stacking fault energy appears to be a perfect candidate for detecting the footprint of room-temperature spin fluctuations on the mechanical properties of austenitic steels. In Sections 4.1, 4.2 and 4.4, we review some of our results on the elastic properties and stacking fault energies of Fe-based alloys.

2.2 Surface properties

A metallic solid solution, consisting of components which are immiscible at low temperature, is thermodynamically unstable when quenched from high temperature and phase separation occurs during aging or annealing. The phase separation may take place through two different paths: nucleation and growth (NG) and spinodal decomposition (SD). NG is initiated by small nuclei with large compositional fluctuations relative to the host and occurs within the metastable region of the miscibility gap. SD, on the other hand, is characterized by extended domains with fluctuating compositions which develop both in size and compositions toward their equilibrium states during aging.

For both paths, the interfacial energy (γ_i) between the decomposed phases plays an important role. According to Gibbs theory (Gibbs, 1948), the extremum of the work required to form a heterogeneous spherical grain of radius R determines the critical nucleus size R_{crit}, which in turn depends on γ_i. A nucleus will grow continuously with initial size $R \geq R_{crit}$ and disappear with $R < R_{crit}$. For SD, the interfacial energy corresponds to the gradient energy that determines the critical wave-length of the fluctuations as shown by the phenomenological models developed to describe the kinetics and thermodynamics of SD (Hillert, 1961).

Owing to the large miscibility gap below about 500 °C, Fe-Cr is a typical binary system showing phase decomposition. When aged at the temperature range of 300-500 °C, alloys with composition within the miscibility gap separates into α (Fe-rich) and α' (Cr-rich) phases, both having the body centered cubic (bcc) structure. The phenomenon is commonly known as the „475 °C embrittlement" and it degrades seriously the alloy properties. Although tremendous efforts have been made to investigate the phase decomposition of Fe-Cr alloys, due to the complexity of the interface the accurate determination of the composition dependent interfacial energy, either experimentally or theoretically, has been very limited. Recently, we evaluated the interfacial energy between the α and α' phases of the Fe-Cr alloys and investigated the effect of chemistry and magnetism (Lu, Hu, Yang, Johansson, & Vitos, 2010). Our study provides an insight into the fundamental physics behind the phase decomposition which is not accessible by the phenomenological theories.

The corrosion rate of Fe-Cr alloys decreases drastically within a narrow concentration interval (9-13 wt. % Cr), (Khanna, 2002; Ryan, Williams, Chater, Hutton, & McPhail, 2002; Wranglén, 1985) making the transition from iron-type to non-corrosive behavior quite abrupt. During oxidation, first a monolayer of oxide is formed instantly on the clean alloy surface exposed to oxidizing environment. The type of the initial oxide layer depends on the oxygen pressure, temperature and the actual alloy compositions within the first few surface layers. High surface density of the chemically less active atoms may also initiate the internal oxidation of the active alloy components. Focusing on the surface phenomena, further oxidation assumes transport of metal and oxygen ions through the initially formed oxide film. The ion transport is controlled by diffusion, which in turn is determined by the defect structure of the oxide layer. The high mobility of Fe in Fe oxides, especially in FeO, which is the dominant oxide component on pure iron above 570 °C, explains the corrosive nature of Fe. The passivity in Fe-Cr, on the other hand, is attributed to a stable Cr-rich oxide scale. Above the critical concentration a pure chromia layer is formed which effectively blocks the ion diffusion across the oxide scale.

Describing the oxide layer growth and ultimately the passivity of stainless steels is an enormous task as it requires the knowledge of the thermodynamic and kinetic properties of the oxide as well as oxide-metal and oxide-gas interfaces under oxidizing conditions. Many times the kinetics of the oxidation process is so slow that the real thermodynamic equilibrium is never reached during the active lifetime of the alloy product. Today massive information is available about the properties of the oxide scale on Fe-Cr, but the initial stage of the oxidation is still unclear. This is due to the experimental difficulties connected to the timescale of the initial oxidation of clean alloy surfaces. Large number of models were put forward for the kinetics of the thin layer oxidation and oxide scale formation (Khanna, 2002). Most of these theories, however, left in the shadow the active role of the metallic substrate in the oxidation process, simplifying it to a cation and electron reservoir. This might be a justified approximation after a monolayer or a few layers of oxide are built up. Nevertheless, the atomic level behavior of the metallic Fe-Cr surfaces is indispensable for understanding the oxygen chemisorption and the initial thin layer oxide formation on this class of materials.

Numerous former first-principles calculations focused on the properties of the Fe-rich Fe-Cr surfaces (Geng, 2003; Nonas, Wildberger, Zeller, Dederichs, & Gyorffy, 1998; Ponomareva, Isaev, Skorodumova, Vekilov, & Abrikosov, 2007; Ruban, Skriver, & Nørskov, 1999). However, due to the involved approximations and constrains, most of these studies failed to reproduce the experimentally observed Cr enrichment on the alloy surface (Dowben, Grunze, & Wright, 1983). A few years ago, we demonstrated that the Fe-Cr surfaces exhibit a compositional threshold behavior (Ropo et al., 2007). In particular, we showed that about 9 at. % chromium in Fe-Cr induces a sharp transition from Cr-free surfaces to Cr-containing surfaces. This surprising surface behavior was found to be a consequence of the complex bulk and surface magnetic interactions characteristic to the Fe-Cr system. The predicted surface chemical threshold has recently been confirmed by an independent theoretical study by Kiejna and Wachowicz (Kiejna & Wachowicz, 2008).

In Section 4.3, we review our theoretical study of the interfacial energy between the α and α' phases of Fe-Cr alloys, and in Section 4.5 we discuss the surface and magnetic properties of the iron-rich Fe-Cr alloys.

3. Computational approach

Today there is a large number of first-principles computational tools available which can in principle be employed to study the fundamental properties of Fe-based systems. When it comes to the Fe-based solid solutions and especially to paramagnetic austenitic stainless steel alloys, the number of suitable first-principles tools is very limited. Our ability to reach an *ab initio* atomistic level approach in the case of such complex systems has become possible by the Exact Muffin-Tin Orbitals (EMTO) method (Andersen, Jepsen, & Krier, 1994; Vitos, 2001, 2007). This *ab initio* computation tool is an improved screened Korringa-Kohn-Rostoker method for solving the one-electron equations within density functional method (Hohenberg & Kohn, 1964). It is based on the Green's functional and full charge density techniques (Kollár, Vitos, & Skriver, 2000). The problem of disorder is treated within the coherent-potential approximation (CPA) (Györffy, 1972; Soven, 1967; Vitos, Abrikosov, & Johansson, 2001). The total charge density is obtained from self-consistent calculations based on the local density approximation for the exchange-correlation potential and the total energy is evaluated within the Perdew-Burke-Ernzerhof (PBE) generalized gradient approximation for the exchange-correlation functional (Perdew, Burke, & Ernzerhof, 1996). The paramagnetic state of various Fe alloys is simulated by the so-called disordered local moment (DLM) model (Györffy, Pindor, Stocks, Staunton, & Winter, 1985). Within the DLM picture, a paramagnetic $Fe_{1-x}M_x$ binary alloy is described as a quaternary $(Fe^{\uparrow}Fe^{\downarrow})_{1-x}(M^{\uparrow}M^{\downarrow})_x$ alloy, with the equal amount of spin up (↑) and spin down (↓) atoms. Thereby, even though formally our calculations are performed at 0 K, the effect of the loss of the net magnetic moment above the Curie temperature on the total energy is correctly captured.

The EMTO approach in combination with the CPA is an efficient tool for describing alloying effects on the atomic-scale properties of random substitutional solid solutions. Its particular strength is that it is suitable to study properties and processes involving anisotropic lattice distortions or low symmetry structures. Due to the employment of optimized overlapping muffin-tin potential approach and the single-site approximation, this method has some limitations for systems with substantial charge transfer between alloy components or when the short range order and local lattice relaxation effects become important. To control the above problem, applications are often preceded by a series of test calculations to find the best numerical parameters for the problem in question.

4. Results and discussion

4.1 Elastic properties

Alloying plays a central role in designing advanced engineering materials with desired properties. Different solute atoms produce different effects on the fundamental properties of the host. In particular, the single-crystal and polycrystalline elastic parameters are amongst the basic intrinsic properties of materials. Understanding how the elasticity is affected by alloying provides important information on determining mechanical properties such as fracture, hardness, brittleness, plasticity *etc.* Due to the limited solubility, many of the experimental measurements of the elastic properties of Fe-based alloys (Ghosh & Olson, 2002; Speich et al., 1972; Takeuchi, 1969) were performed on multiphase samples, and thus the obtained elastic parameters correspond to a mixed phase rather than to a well-defined crystal structure and hence give no information about the solid-solution strengthening mechanism within a particular phase. The situation is even less satisfactory on the theoretical side.

4.1.1 Binary alloys

Iron is a major alloy component in modern industry due to its structural strength and high abundance. Bulk Fe is ferromagnetic (FM) with Curie temperature of 1043 K and crystallizes in the body-centered-cubic (bcc) structure. Above the Curie temperature, Fe adopts a paramagnetic (PM) bcc structure up to 1183 K. With increasing temperature, it transforms to face-centered-cubic phase (γ-Fe). At ambient pressure, paramagnetic γ-Fe is stable between 1183 K and 1667 K. Before the melting temperature of 1811 K, the paramagnetic bcc phase is stabilized again. Alloying makes the Fe phase diagram even more complex, since phase stability and magnetism are sensitive to alloying element, impurity concentration, and temperature. Different solute atoms may stabilize or destabilize the bcc or fcc phases. Here we quote some basic information regarding the binary Fe phase diagrams and the reader is referred to the multi-component phase diagrams for further details. Based on the experimental phase diagrams (Massalsk, 1986), the maximum solubility of Mn and Ni in α-Fe is 3 and 5.5 atomic percent (at.%), respectively. The bcc Fe-rich Fe-Al, Fe-Si, Fe-V, Fe-Cr, Fe-Co, and Fe-Rh alloys show solubility up to ~10 at. % impurity concentration. In the γ phase, the solubility limit of Al, Si, V, and Cr is about 1.3, 3.2, 1.4, and 11.9 at.%, respectively, whereas Fe-Mn, Fe-Co, Fe-Ni, and Fe-Rh form continuous solid solutions. Beyond these concentrations, the PM fcc Fe-Al, Fe-Si, Fe-V, and Fe-Cr alloys transform to PM bcc alloys (with relatively narrow fcc-bcc two phase fields).

The elastic properties of ferromagnetic bcc and paramagnetic fcc $Fe_{1-x}M_x$ (M=Al, Si, V, Cr, Mn, Co, Ni, and Rh; $0 \leq x \leq 0.1$) binary alloys have been investigated (H. L. Zhang, Johansson, & Vitos, 2009; H. L. Zhang, Punkkinen, Johansson, Hertzman, & Vitos, 2010a; H. L. Zhang, Punkkinen, Johansson, & Vitos, 2010) using the exact muffin-tin orbitals density functional method in combination with the coherent-potential approximation.

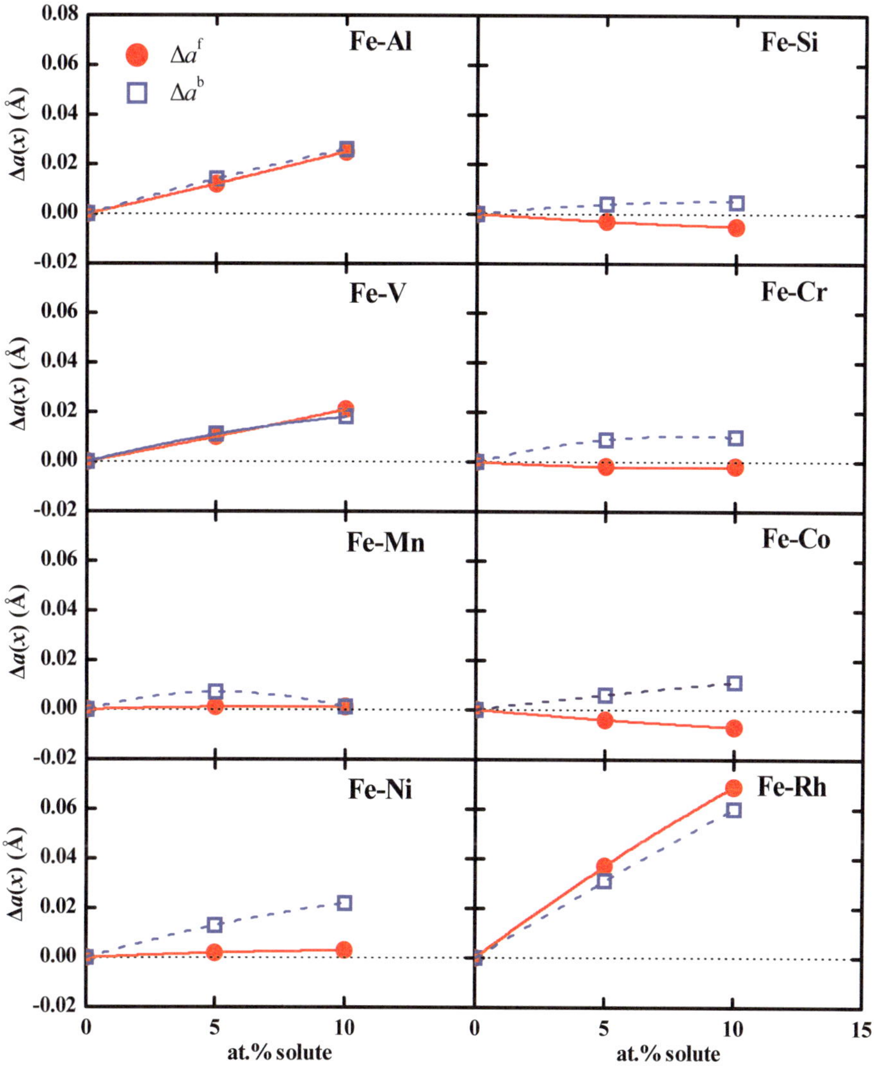

Fig. 1. Theoretical changes (relative to pure Fe) of the equilibrium lattice parameter $a(x)$ (in Å) for paramagnetic fcc (solid circles connected with solid lines) and ferromagnetic bcc (open squares connected with dashed lines) $Fe_{1-x}M_x$ (M=Al, Si, V, Cr, Mn, Co, Ni, and Rh; $0 \leq x \leq 0.1$) random alloys. The superscripts ^{f} and ^{b} denote the results for the fcc and bcc phases, respectively.

To calculate the elastic parameters of Fe-based alloys, first we computed the equilibrium lattice parameter for ferromagnetic bcc and paramagnetic fcc $Fe_{1-x}M_x$ (M=Al, Si, V, Cr, Mn, Co, Ni, and Rh; $0 \leq x \leq 0.1$) random alloys. To emphasize the alloying effect, in Fig. 1 we show the theoretical changes relative to pure Fe. We find that $a(x)$ of fcc Fe is strongly enlarged by Al, V, and Rh and slightly reduced by Si, Cr, and Co, while it remains nearly constant with Mn and Ni additions. Unlike the fcc phase, the calculated $a(x)$ of the bcc phase increases with alloying for all binaries considered here.

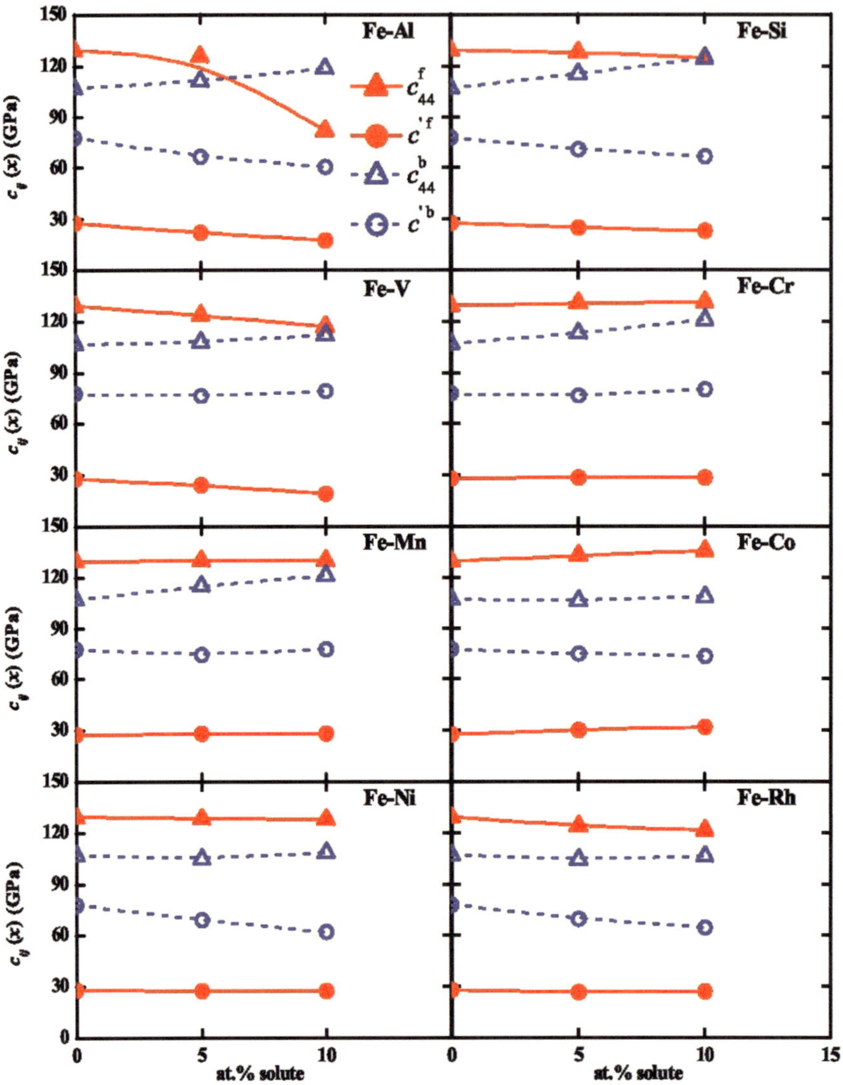

Fig. 2. Single-crystal shear elastic constants $c'(x)$ and $c_{44}(x)$ for paramagnetic fcc (solid symbols connected with solid lines) and ferromagnetic bcc (blue open symbols connected with dashed lines) $Fe_{1-x}M_x$ (M=Al, Si, V, Cr, Mn, Co, Ni, and Rh; $0 \leq x \leq 0.1$) random alloys. The superscripts ^{f} and ^{b} denote the fcc and bcc phases, respectively.

The elastic constants of paramagnetic fcc and ferromagnetic bcc $Fe_{1-x}M_x$ (M=Al, Si, V, Cr, Mn, Co, Ni, and Rh; $0 \leq x \leq 0.1$) random alloys are shown in Fig. 2. We find that the paramagnetic fcc alloys have smaller tetragonal elastic constant c' than the ferromagnetic bcc ones. Meantime, the c_{44} is larger for fcc alloys than for bcc ones, with exception of Fe-Al and Fe-Si at large impurity concentrations. On the average, the alloying effect on the c' and c_{44} is rather small for both fcc and bcc phases. The ferromagnetic bcc Fe-based alloys are more isotropic than the paramagnetic fcc counterparts, and this difference to a large extent is due to the soft tetragonal mode (c') in the fcc phase.

With some exceptions, alloying has much larger effects on ferromagnetic bcc alloys than on paramagnetic fcc ones. However, in order to see where this stronger effect comes from, one should carry out similar calculations for the paramagnetic bcc Fe-based alloys to be able to exclude the effect of crystal lattice from the above comparison.

4.1.2 Ternary alloys

The elastic constants are intrinsic properties of a particular crystal structure and thus their alloying and magnetic state dependence may be weaker than that experienced, e.g., in the case of stacking fault energies (Section 4.4). Indeed, our former calculation (Vitos, Korzhavyi, & Johansson, 2002) for the polycrystalline elastic moduli (derived from single crystal elastic constants) of paramagnetic fcc Fe-Cr-Ni alloys show weak composition dependence (Fig. 3). Nevertheless, as we will show below, these bulk parameters also exhibit magnetic state dependence and mechanisms changing the local magnetic moments within the paramagnetic phase are expected to alter the elastic properties of Fe-Cr-Ni alloys.

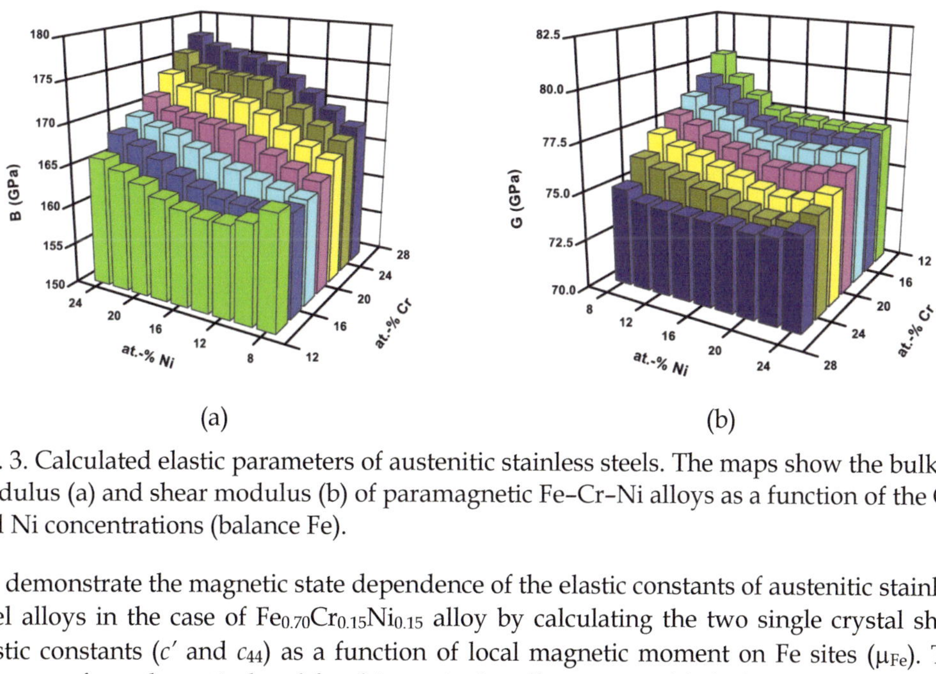

Fig. 3. Calculated elastic parameters of austenitic stainless steels. The maps show the bulk modulus (a) and shear modulus (b) of paramagnetic Fe–Cr–Ni alloys as a function of the Cr and Ni concentrations (balance Fe).

We demonstrate the magnetic state dependence of the elastic constants of austenitic stainless steel alloys in the case of $Fe_{0.70}Cr_{0.15}Ni_{0.15}$ alloy by calculating the two single crystal shear elastic constants (c' and c_{44}) as a function of local magnetic moment on Fe sites (μ_{Fe}). The accuracy of our theoretical tool for this particular alloy was established previously (Vitos & Johansson, 2009). The theoretical equilibrium bulk parameters and the elastic constants for $Fe_{0.70}Cr_{0.15}Ni_{0.15}$ are compared with the available experimental data in Table 1. We find ~3.0% mean absolute relative deviation between the theoretical and experimental (Teklu et al., 2004) single-crystal elastic constants. As a matter of fact, this error is much smaller than that obtained for ferromagnetic bcc Fe (H. L. Zhang, Punkkinen, Johansson, Hertzman, & Vitos, 2010a). The conspicuously better accuracy achieved for Fe-Cr-Ni compared to Fe may be ascribed to the fact that theory gives a highly accurate equation of state for paramagnetic Fe-Cr-Ni: the relative errors in the equilibrium atomic radius and bulk modulus being 0.4% and 1.1%, respectively.

	w_0	B	c_{11}	c_{12}	c'	c_{44}
theory	2.66	162.23	203.86	141.42	31.22	133.20
error	0.4	1.1	-2.5	4.0	-14.5	2.5
experiment	2.65	159-162	207-211	135-137	35-38	130

Table 1. Theoretical and experimental (Teklu et al., 2004) equilibrium Wigner-Seitz radius (w_0, in Bohr), bulk modulus (B, in GPa), and single-crystal elastic constants (c_{ij}, in GPa) of paramagnetic fcc $Fe_{0.70}Cr_{0.15}Ni_{0.15}$ alloy. The numbers from the second row are the relative deviations (in %) between the theoretical and the mean experimental values.

Figure 4a displays the c' and c_{44} elastic constants as a function of μ_{Fe} for spin-constrained calculations (solid lines) and for fully self-consistent calculations (single symbols at $\mu_{Fe} = 1.63$ μ_B corresponding to the self-consistent result). The fact that the fully self-consistent and the spin-constrained results are relatively close to each other is due to the fact that upon lattice distortion the local magnetic moments do not change significantly. This is illustrated in Fig. 4b, where we plotted the local magnetic moment on Fe atoms for the paramagnetic $Fe_{0.80}Cr_{0.15}Ni_{0.05}$ alloy as a function of volume (Wigner-Seitz radius, w) and tetragonal lattice constant ratio c/a. The theoretical equilibrium w for the fcc phase is 2.66 Bohr, and for the body centred cubic (bcc) phase 2.68 Bohr. We recall that the fcc structure has $c/a \approx 1.41$ and the c' elastic constant is proportional with the curvature of the total energy versus c/a (for fixed volume). It is interesting to note that the local magnetic moments slightly increase upon tetragonal lattice distortion in the fcc phase (c/a≈1.41) and decrease in the bcc phase (c/a=1.0).

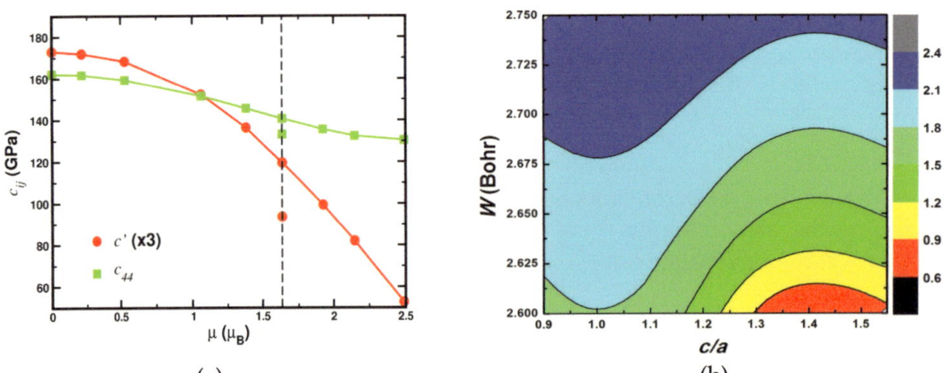

Fig. 4. Left panel: Single-crystal elastic constants of paramagnetic fcc $Fe_{0.70}Cr_{0.15}Ni_{0.15}$ alloy as a function of the local magnetic moment on the Fe atoms. Note that c' has been multiplied by three in order to match its scale to that of c_{44}. Shown are also the floating-spin results obtained at the equilibrium magnetic moment $\mu_{Fe} = 1.63$ μ_B (separate circle and square). All calculations were performed at paramagnetic volume (w=2.66 Bohr). Right panel: local magnetic moments on Fe atoms of paramagnetic $Fe_{0.80}Cr_{0.15}Ni_{0.05}$ alloy as a function of Wigner-Seitz radius w and tetragonal lattice parameter ratio c/a.

Returning to Fig. 4a, we observe that both elastic constants strongly depend on the local magnetic moment. The effect is somewhat more pronounced for the tetragonal elastic constant. From a polynomial fit to the data from Fig. 4a, for the slopes of the elastic constants versus magnetic moment we obtain $\delta c'/\delta\mu \approx -22.5$ GPa/μ_B and $\delta c_{44}/\delta\mu \approx -19.5$ GPa/μ_B. Hence, 0.1μ_B change in the local magnetic moment results in ~2 GPa change in the elastic constants, representing ~6% for c' and ~2% for c_{44}. This is an important effect, especially taking into account that we are dealing with a system well above its magnetic transition temperature. We suggest that by manipulating the magnetism, e.g., via chemical composition, chemical ordering, external field, or temperature, one is able to tailor the thermo-elastic properties of austenitic stainless steels. We have demonstrated the above effect in the case of the temperature dependence of the single crystal elastic constants of paramagnetic Fe-Cr-Ni alloys (Vitos & Johansson, 2009). In particular, we have shown that spin fluctuation in paramagnetic Fe-Cr-Ni alloys can account for 63% of $\delta c'/\delta T$ and 28% for $\delta c_{44}/\delta T$ as compared to the experimental measurements (Teklu et al., 2004).

4.2 Bain path

The transformation mechanism between the bcc and the fcc phases of Fe-based alloys is of key importance for the properties of alloy steels. This is a typical diffusionless structural change belonging to the group of the so called martensitic transformations. Several homogeneous paths have been suggested for describing the bcc-fcc transformation. In particular, the Bain path (Bain, 1924) is obtained by expanding the bcc lattice along one of the cubic axes (c) and contracting along the two others (a). Upon lattice deformation the crystal symmetry remains tetragonal and the unit cell is body centered tetragonal (bct). The tetragonality of the lattice is described by the c/a ratio. When c/a is 1 the bct lattice corresponds to the bcc one, whereas when c/a reaches $\sqrt{2}$ the bct lattice turns into the fcc one. The Bain path is an appropriate model for studying the energetics of the bcc-fcc martensitic transformation. Furthermore, monitoring the alloying induced softening or hardening of Fe-based alloys against tetragonal distortions (Al-Zoubi, Johansson, Nilson, & Vitos, 2011) is of key importance for understanding the interstitial driven martensitic transformations in alloy steels.

In Fig. 5, we present the calculated total energy maps for Fe-Cr and Fe-Cr-Ni alloys along with the Bain path. The energy map for $Fe_{0.90}Cr_{0.10}$ (Fig. 5a) shows that at the equilibrium volume the close-packed fcc structure is marginally more stable than the bcc modification, the energy differences between the fcc and the bcc structures being $\Delta E \equiv E_{fcc}-E_{bcc} = -0.021$ mRy. There is a clear energy barrier, a saddle point between the bcc and fcc local minima. Approximating the energy barrier by the total energy calculated for $c/a = 1.2$ and $w = 2.675$ Bohr, for $Fe_{0.9}Cr_{0.1}$ we obtain $\Delta E_f \equiv E_{1.2} - E_{fcc} = 1.019$ mRy (barrier relative to the fcc structure) or $\Delta E_b \equiv E_{1.2} - E_{bcc} = 0.998$ mRy (barrier relative to the bcc structure). Obviously $\Delta E_b - \Delta E_f = \Delta E$.

Since in the present calculations no temperature effects are taken into account (except the chemical and magnetic randomness in the total energy) the total energy difference ΔE between the fcc and bcc structures should be interpreted with precaution and should not be associated with the phase stability of Fe-Cr alloys. Nevertheless, we can make our conclusions more robust and identify the primary chemical effects on the phase stability by considering the $Fe_{0.90}Cr_{0.10}$ as reference and focusing on the total energy of $Fe_{0.85}Cr_{0.1}M_{0.05}$ (M stands for Cr and Ni) expressed relative to that calculated for $Fe_{0.90}Cr_{0.10}$. The corresponding relative fcc-bcc energy difference is denoted by $\Delta E(M)$ and the relative energy barriers by $\Delta E_f(M)$ or $\Delta E_b(M)$. According to this definition, for instance vanishing $\Delta E(M)$ and $\Delta E_{f/b}(M)$ mean that 5 % alloying addition M produces negligible effect on the corresponding energy differences of $Fe_{0.9}Cr_{0.1}$.

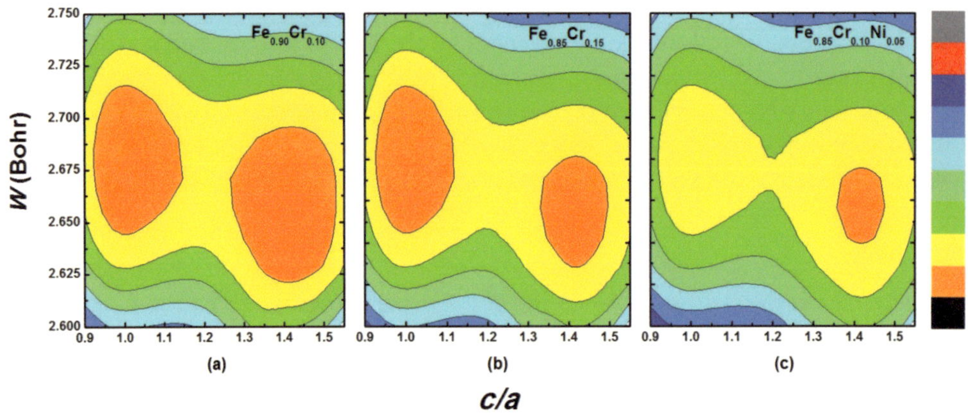

Fig. 5. Total energy contours (in mRy) for (a) $Fe_{0.9}Cr_{0.1}$, (b) $Fe_{0.85}Cr_{0.15}$ and (c) $Fe_{0.85}Cr_{0.1}Ni_{0.05}$ alloys as a function of the tetragonal lattice ratio (c/a) and the Wigner-Seitz radius (w). For each alloy, the energies are plotted relative to the minimum of the corresponding bcc ($c/a = (c/a)_{bcc} = 1$) total energy.

Next, we illustrate the effect of adding 5% Cr and Ni on the Bain path of the paramagnetic Fe-Cr alloy. We find that adding 5 % Cr to the $Fe_{0.9}Cr_{0.1}$ alloy increases the fcc-bcc total energy difference by $\Delta E(\text{Cr}) = 0.443$ mRy (Fig. 5b). In other words, the chemical effect of Cr is to stabilize the bcc phase relative to the fcc one. Alloying changes the energy barrier between the bcc and fcc structures as well. We obtain that 5 % Cr addition to $Fe_{0.9}Cr_{0.1}$ alters the barrier by $\Delta E_f(\text{Cr}) = -0.118$ mRy or $\Delta E_b(\text{Cr}) = 0.325$ mRy. That is, the energy minimum around the fcc phase becomes shallower and that around the bcc phase deeper by alloying with Cr. In terms of mechanical stability of alloys, the above alloying effect of Cr corresponds to mechanically less (more) stable fcc (bcc) phase. This is in line with the observation that Cr decreases the tetragonal elastic constant of paramagnetic fcc $Fe_{0.9}Cr_{0.1}$.

Nickel is calculated to have pronounced effects on the Bain path of paramagnetic Fe-Cr alloys (Fig. 5c). Adding 5 % Ni to $Fe_{0.9}Cr_{0.1}$ yields $\Delta E(\text{Ni}) = -0.300$ mRy and changes the energy barrier by $\Delta E_f(\text{Ni}) = 0.045$ mRy per atom or $\Delta E_b(\text{Ni}) = -0.255$ mRy per atom relative to that of $Fe_{0.9}Cr_{0.1}$. It is interesting to contrast the above trends for the energy barrier with those calculated for the elastic constants of paramagnetic fcc Fe alloys (Section 4.1.1). In particular, Ni is calculated to decrease slightly the tetragonal elastic constant of fcc Fe. Thus we may conclude that the trends in the elastic constants are not sufficient to predict the changes in the fcc-bcc energy barrier upon alloying.

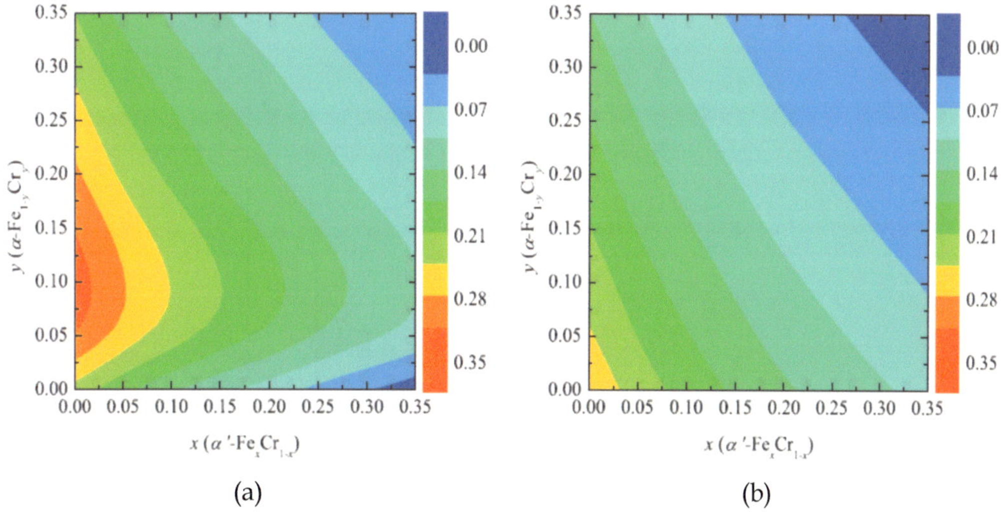

Fig. 6. The (001) interfacial energy (in J m^{-2}) between the Cr-rich α' phase (Fe_xCr_{1-x}) and the Fe-rich α phase ($Fe_{1-y}Cr_y$) for (a) ferromagnetic and (b) paramagnetic states.

4.3 Interfacial energies

The interfacial energy (γ_i) between the two phases (Fe-rich α and Cr-rich α') in Fe-Cr alloys is an important parameter when studying the phase separation by spinodal or nucleation and growth mechanisms. While the experimental determination of the interfacial energy is less feasible, it can easily be evaluated by first-principles calculations. The (110) and (100) interfacial energies between the Fe-rich α-$Fe_{1-y}Cr_y$ and Cr-rich α'-Fe_xCr_{1-x} phases have been calculated (Lu et al., 2010) as a function of composition ($0<x$, $y<0.35$). It is found that generally the (110) interface has lower energy than the (001) one as expected from a simple bond-cutting model, and the interface energies vary in a similar way with respect to the composition. The interfacial energy for the (001) interface varies between 0.02 and 0.33 Jm^{-2} for the ferromagnetic state and between 0.02 and 0.27 Jm^{-2} for the paramagnetic state with respect to composition (Fig. 6.). The paramagnetic γ_i shows a monotonous decreasing trend with increasing x and y, while the ferromagnetic γ_i decreases with increasing x, but increases with y for $0<y<0.1$ and then decreases for $0.1<y<0.35$. The nonlinear concentration dependence of the ferromagnetic interfacial energy has been attributed to the complex magnetic interaction near the interface (Lu et al., 2010).

Using a continuum model, the critical grain size (R_{crit}) for phase separation may be estimated from the calculated interfacial energies (Fig. 7.). It is shown that the theoretical critical radius exhibits a strong dependence on the composition of the initial homogenous alloy. The critical radius is very small inside the spinodal line and increases significantly between the spinodal and solubility lines, which agrees well with the fact that in these two composition regions different mechanisms, spinodal or nucleus and growth, control the phase separation process.

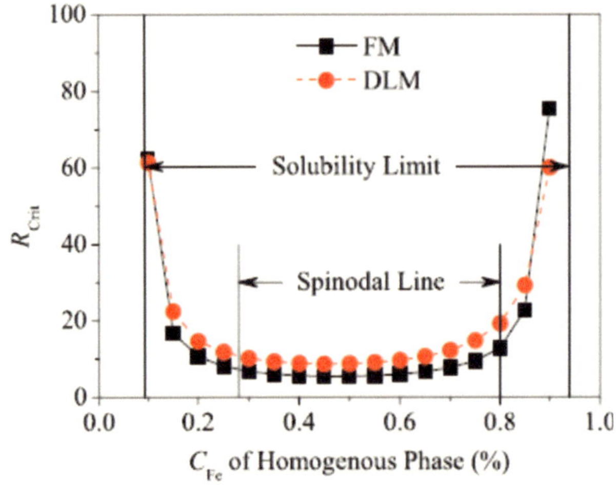

Fig. 7. Critical nucleus size (R_{crit} in Å) as a function of the composition of the homogenous phase Fe_cCr_{1-c} for ferromagnetic and paramagnetic states. The vertical lines show the experimental spinodal and solubility limits.

4.4 Stacking fault energies

The stacking fault energy of a material is an important characteristic since it is related to various mechanical properties such as strength, toughness and fracture *etc.* In austenitic stainless steels, SFE has been measured for various compositions (Rhodes & Thompson, 1977; Schramm & Reed, 1975). However, the experimental data show large scatter (Table 2) and thus the empirical relations between SFE and composition established based on such data cannot properly account for the correct alloying effects (Lo, Shek, & Lai, 2009).

Cr	Ni	theory	Experiment
15	14	40.1	46±7
17	13	30.9	23±5
17	20	38.8	31±5
19	10	11.7	7.2±1.5, 25±2.5, 16.4±1.1
22	13	18.4	18±4
26	20	42.0	40±5

Table 2. Comparison between the calculated and experimental (Vitos et al., 2008) stacking fault energies for six selected alloys. Compositions are given in atomic percent and SFE in mJ/m^2.

The SFE of Fe-Cr-Ni alloys have been calculated (Vitos, Korzhavyi, & Johansson, 2006; Vitos et al., 2008; Vitos, Nilsson, & Johansson, 2006) as a function of temperature and composition using the EMTO method (Fig. 8). These theoretical results reveal the underlying mechanisms determining the complicated relation between SFE and alloying elements. It is shown that the increasing effect of Ni on the SFE strongly depends on the content of Cr. In low Cr alloys, Ni has negligible effect on the SFE when the concentration of Ni is over ~12 at. %, while in high Cr alloys, the SFE is nearly linearly proportional to the amount of Ni. On the other hand, Cr decreases the SFE only in low Ni alloys and in high Ni alloys the slope of SFE vs. Cr content changes sign at ~20 at. % Cr.

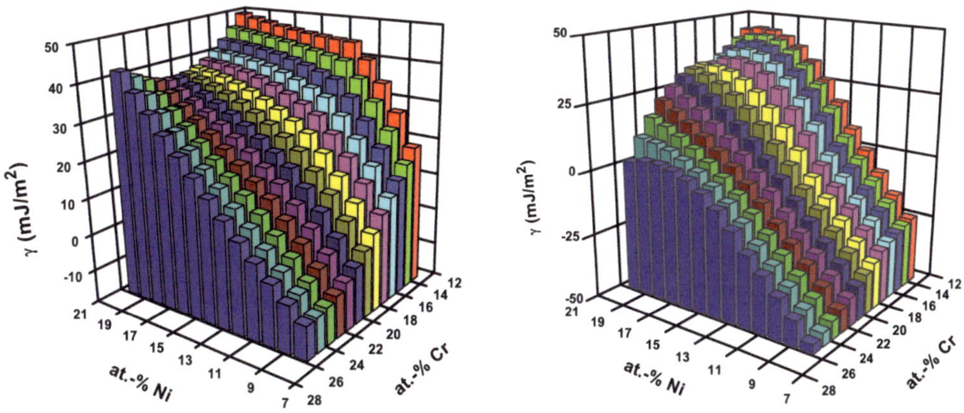

(a) (b)

Fig. 8. Calculated stacking fault energy (γ) of paramagnetic fcc Fe-Cr-Ni alloys plotted as a function of Ni and Cr contents for 300 K (left panel) and 0 K (right panel).

From Fig. 8a, one would conclude that in general with increasing Ni content in paramagnetic Fe-Cr-Ni alloys the width of the ribbon connecting the partial dislocations decreases so that the partials can more easily recombine and thus the resistance of the alloy against plastic deformation decreases. At the same time, Cr is predicted to enhance the strength of the alloy at low Ni content and have negligible effect at large Ni content. However, the above trends show strong temperature dependence. On the right panel of Fig. 8, we show the calculated SFE at 0 K. The overall effect of Ni at 0 K is similar to that from Fig. 8a, but Cr is found to decrease the SFE at any Ni content. We will show below that the reason behind this change is the behavior of the local magnetic moment with alloying.

We find that the above chemical effects of alloying additions are accompanied by major magnetic effects, which in fact stabilize the most common industrial alloy steels at normal service temperatures. Note that according to Fig. 8b, all Fe-Cr-Ni alloys encompassing less than ~11-17 % Ni (depending on Cr content) have negative SFE at 0 K. At 300 K, only alloys within a small compositional range have still negative SFE and they are located in the low-Ni-high-Cr part of the map from Fig. 8a. Within the present model, the temperature part of the SFE corresponds mainly to the magnetic entropy contribution to the SFE. Since the local magnetic moments in the double hexagonal structure are calculated to be close to those within the fcc structure, the magnetic fluctuation part of the SFE reduces to $\gamma^{mag} = -T[S^{hcp} - S^{fcc}]/A_{2D}$ (where S stands for the magnetic plus electronic entropy and A_{2D} is the interface area). γ^{mag} is plotted in Fig. 9a for 300 K. We can observe that γ^{mag} exhibits a strongly nonlinear composition dependence, especially for low-Ni alloys (as a function of Cr content) and for high-Cr alloys (as a function of Ni content). Because at 300 K the leading term in the entropy is the magnetic contribution, the above trends are direct consequences of the composition dependence of the magnetic moments for the fcc and hcp phases (Vitos, Nilsson et al., 2006).

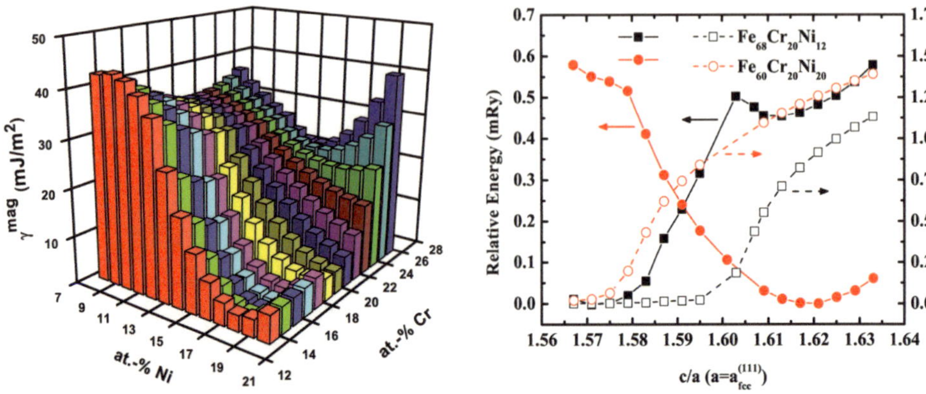

(a) (b)

Fig. 9. Left panel: magnetic fluctuation contribution to the stacking fault energy (γ^{mag}) of fcc Fe-Cr-Ni alloys calculated for 300 K. Right panel: total energy (left axis) and local magnetic moment (right axis) for the hcp Fe-Cr-Ni alloys as a function of hexagonal lattice parameter *c/a* keeping the in-plane lattice constant *a* fixed to that of the fcc lattice ($a_{fcc}^{(111)}$). Notice the different orientations of the Ni and Cr axes on Fig. 8 and Fig. 9a.

We illustrate the change of the local magnetic moments in the case of $Fe_{0.68}Cr_{0.20}Ni_{0.12}$ and $Fe_{0.60}Cr_{0.20}Ni_{0.20}$ alloys. According to the axial interaction model (Vitos, Nilsson et al., 2006), the stacking fault energy is computed from the total energies of the double hexagonal, hcp and fcc lattices. In these calculations, the atomic volume is assumed to be constant and equal to that of the parent fcc lattice. However, in real alloys due to the vanishing local magnetic moments in the hcp environment, the hcp lattice prefers a smaller equilibrium volume than that of the fcc lattice. Due to the in-plane lattice constraint volume relaxation can be realized only along the direction perpendicular to the stacking fault plane. To mimic this situation, in all our calculations we relaxed the *c* lattice constant of the hcp lattice while keeping the in-plane lattice constant *a* fixed to $a_{fcc}^{(111)}$ (Lu, Hu, Johansson, & Vitos, 2011). The calculated total energies are shown in Fig. 9b (left axis) as a function of *c/a*. In alloys containing 12 % Ni, the hexagonal lattice is nonmagnetic (the local magnetic moments vanish within the hcp phase, see Fig. 9b right axis) and thus there should be a large volume relaxation relative to the volume of the fcc lattice. This is reflected by the very small equilibrium $c/a\approx 1.57$ obtained for hcp $Fe_{0.68}Cr_{0.20}Ni_{0.12}$ and the large γ^{mag} calculated for this alloy (Fig. 9a). When the Ni content is increased to 20 %, the hcp lattice becomes weakly magnetic (small local magnetic moments appear on Fe sites, see Fig. 9b right axis). Therefore, the equilibrium volume of hcp $Fe_{0.60}Cr_{0.20}Ni_{0.20}$ should be close to that of the fcc phase. Indeed, the calculated equilibrium $c/a\approx 1.62$ for hcp $Fe_{0.60}Cr_{0.20}Ni_{0.20}$ is very close to the ideal one (~1.63), meaning that in this alloy no substantial volume relaxation takes place around the stacking fault. In consequence, the magnetic fluctuation contribution to the stacking fault energy of $Fe_{0.60}Cr_{0.20}Ni_{0.20}$ alloys becomes very small (Fig. 9a).

The results summarized in Figs. 8 and 9 clearly demonstrate the importance of the disordered local moments for the stacking fault energies of steels. For the magnetic contribution to the free energy one may employ models which are more advanced than the mean-field approximation used here. However, this will not change the general conclusion that local magnetic moments have a marked contribution to the energetic of the stacking faults. Any mechanism (alloying, temperature or strong magnetic field) that can alter the magnetic structure of these alloys is predicted to have large impact on the stacking fault energies and thus on the strength of the paramagnetic Fe-Cr-Ni alloys.

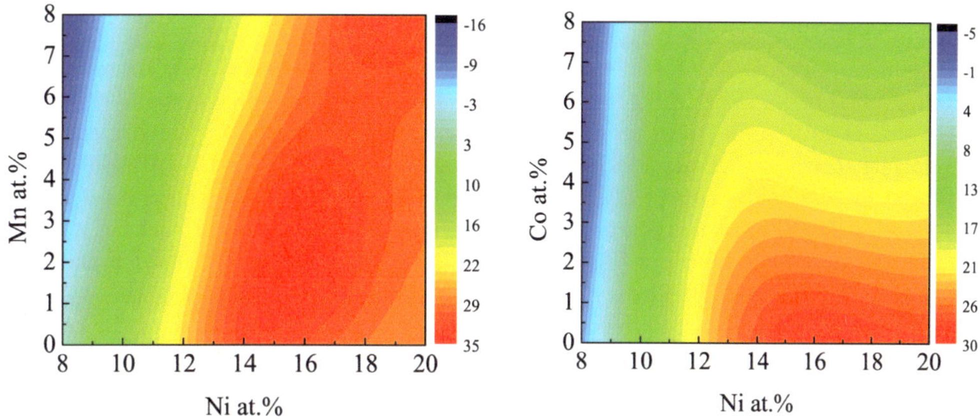

Fig. 10. Calculated stacking fault energy maps of Fe-Cr-Ni-Mn (left panel) and Fe-Cr-Ni-Co (right panel) alloys plotted as a function of composition for $T = 300$ K.

One important implication of the above finding is that the same alloying element can cause totally opposite changes in the SFE of alloys with different host composition, indicating that in practice no universal composition equations for the SFE can be established. To illustrate this effect, in Fig. 10 we show the theoretical room-temperature SFE maps for Fe-Cr-Ni-Mn and Fe-Cr-Ni-Co alloys as a function of composition (Lu et al., 2011). It is found that Mn decreases the SFE in alloys with less than 16 at.% Ni, beyond which the SFE slightly rises with Mn. On the other hand, Co always tends to decrease the SFE and the decreasing effect is enhanced in high-Ni alloys. Cobalt is known as a useful alloying element in improving the steel resistance against galling. Enhanced galling effect, in turn, is thought to be associated with enhanced ductility. According to our study (Fig. 10b), Co decreases the SFE and thus decreases the ductility of austenitic stainless steels. This might explain why Co acts as an efficient anti-galling alloying ingredient.

4.5 Surface properties

Stainless steels have versatile mechanical properties. However, the corrosion resistivity makes these materials unique among the engineering materials. For instance, the corrosion rate of ferritic stainless steels decreases drastically within a narrow concentration interval (9-13 wt. % Cr), (Khanna, 2002; Ryan et al., 2002; Wranglén, 1985) making the transition from iron type to non-corrosive behavior quite abrupt. The type of the oxide layer formed on the surface depends on the oxygen pressure, temperature and the alloy compositions in the vicinity of the surface. Further oxidation assumes transport of metal and oxygen ions through the initially formed oxide scale. The ion transport is controlled by diffusion, which in turn is determined by the defect structure of the oxide layer. The high mobility of Fe in Fe oxides, especially in FeO, which is the dominant oxide component on pure iron above 570°C, explains the corrosive nature of Fe. The passivity in Fe-Cr, on the other hand, is attributed to a stable Cr-rich oxide scale. Above the critical concentration a pure chromia layer is formed on the surface which effectively blocks the ion diffusion across the oxide scale.

Kinetics of the thin layer oxidation and oxide scale formation have been explained using various models (Khanna, 2002). Numerous first-principles calculations have also focused on the properties of the Fe-rich Fe-Cr surfaces (Geng, 2003; Nonas et al., 1998; Ponomareva et al., 2007; Ruban et al., 1999). However, due to the involved approximations and constrains, most of these studies failed to reproduce the experimentally observed Cr enrichment on the alloy surface (Dowben et al., 1983). Later, using the exact muffin-tin orbitals method (Andersen et al., 1994; Vitos, 2001, 2007; Vitos et al., 2001) in combination with the generalized gradient approximation, it was demonstrated that the Fe-Cr surfaces exhibit a compositional threshold behavior (Ropo et al., 2007). In particular, it was shown that about 9 at. % chromium in Fe-Cr induces a sharp transition from Cr-free surfaces to Cr-containing surfaces. This surprising surface behavior was found to be a consequence of the complex bulk and surface magnetic interactions characteristic to the Fe-Cr system. The predicted surface chemical threshold has recently been confirmed by an independent theoretical study by Kiejna and Wachowicz (Kiejna & Wachowicz, 2008). Using the present achievements of the first-principles quantum mechanical approach many previously controversial results can now be merged into a consistent model of Fe-rich Fe-Cr alloys.

Before going into the details of the calculated surface properties we discuss some common procedures related to the first-principles surface calculations. The thermodynamically stable surfaces of the $Fe_{1-c}Cr_c$ alloys can be modeled by using periodic slab geometry consisting of a certain number of atomic layers, with surface alignment, joined to a set of empty layers representing the vacuum region. The thickness of the metal and vacuum regions is optimized keeping in mind both the computational cost on the one hand and the calculational accuracy on the other. In practice, the most common procedure in calculating the surface concentrations is to optimize the chemical composition of the surface layer only and to keep the concentrations of the other atomic layers fixed to the bulk value.

In magnetic systems, surface magnetism is observed to reduce the surface energy of open surfaces to the extent that the usual anisotropy of the surface energy is reversed (Aldén, Skriver, Mirbt, & Johansson, 1992, 1994; Vitos, Ruban, Skriver, & Kollár, 1998). In particular, the magnetic contribution to the surface energy of the (100) facet of pure Cr (Fe) is about -50% (-41%) compared to -2% (-16%) obtained for the close-packed (110) facet (Aldén et al. 1994). Accordingly, the most stable surfaces for pure Cr and for Fe-rich Fe-Cr alloys are the (100) crystal facet of the B2 lattice and the (100) crystal facet of the bcc lattice, respectively.

The atomic origin of the chemical threshold behavior of Fe-Cr surfaces (Ropo et al., 2007), becomes evident by considering the effective chemical potentials (ECPs) of the bulk and the (100) surface. Figure 11 (left axis) shows the bulk and surface ECP plotted as a function of the bulk Cr concentration. Data is shown for surfaces containing 0, 10, 20, and 30 at. % Cr. Comparing these curves one can easily construct a clear picture of the driving forces behind the peculiar trend of the surface chemistry of Fe-Cr alloys. At low Cr concentrations in bulk ($c < 0.08$), the ECP in bulk is above the ECP at the pure Fe surface. As a consequence, for these alloys the Fe terminated surface is more favorable than the Cr containing surface. That is, Cr atoms are influenced by a large chemical driving force from the surface towards the bulk (Geng, 2003; Ropo et al., 2007). However, near 8 at. % Cr in the bulk alloy the ECP drops below the surface ECP leading to the transition from pure Fe terminated surfaces to Cr containing surfaces. The finding of the outburst of bulk Cr to the surface at about 9 at. % Cr in the bulk (Ropo et al., 2007) agrees well with the theoretical prediction based on Ising model (Ackland, 2009).

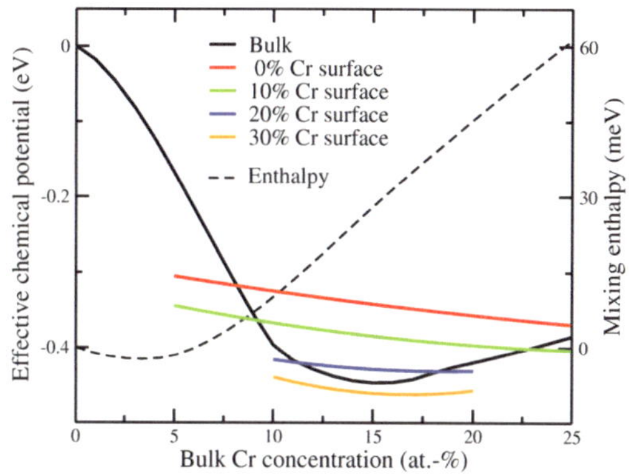

Fig. 11. Left axis (solid lines): Effective bulk (black) and surface (short coloured lines) chemical potentials (in eV) of ferromagnetic $Fe_{1-c}Cr_c$ alloys as a function of bulk Cr concentration (at. %) at $T = 0$ K. All curves are plotted relative to the bulk chemical potential for the dilute alloy. Left axis (dashed line): The mixing enthalpy (in meV) of disordered Fe-Cr alloy. The standard states are the ferromagnetic bcc Fe and antiferromagnetic B2 Cr. The inflection point in the mixing enthalpy around 15% Cr corresponds to the minimum of the bulk effective chemical potential (Ropo et al., 2007).

Figure 12 represents the calculated Cr concentration in the surface layer as a function of the Cr concentration in bulk, which describes well the characteristics of the experimentally observed compositional threshold (Wranglén, 1985). In particular, we emphasize that the calculated transition interval (8-12 % Cr) from Cr-free surfaces to surfaces with bulk-like composition is in excellent agreement with the concentration range within which the observed corrosion rate in Fe-Cr alloys drops from 0.1 mm per year near 9 % Cr to below the detectable limit at 13 % Cr (Wranglén, 1985). It should be noted that the sharp increase in the surface Cr content around the theoretical threshold in Fig. 12 can be traced back to the particular stability of pure Fe-terminated surfaces in low-Cr alloys rather than to a considerable surface segregation of Cr in high-Cr alloys.

In the following we analyze the data shown in Figs. 11 and 12 in more detail using the available theoretical data on bulk and surface Fe-Cr alloys. According to the surface segregation model (Ruban et al., 1999) of alloys with isostructural components the surface energy is an important driving force behind the segregation, namely, the alloy component with the lowest surface energy segregates toward the surface of the alloy. However, in Fe-Cr alloys the situation seems to be more complex. It turns out that, in spite of the large surface magnetic effects (Aldén et al., 1992, 1994; Vitos et al., 1998), the surface energy of pure Cr is significantly (~ 30%) larger than the one calculated for pure Fe (1.41 eV per surface Fe atom)

(Ropo et al., 2011; Ropo et al., 2007), thus preventing Cr atoms going to the surface. The surface energy difference between the pure Fe- and pure Cr-terminated surfaces is even larger (~36%) in Fe-rich Fe-Cr alloys. Thus, in contrast to Fig. 12 and experimental observations (Lince, Didziulis, Shuh, Durbin, & Yarmoff, 1992; Suzuki, Kosaka, Inone, Isshiki, & Waseda, 1996), from standard surface energy considerations the Cr-containing surfaces should always be energetically less favorable compared to the Cr-free surfaces. As a matter of fact, this contradictory picture is in line with several theoretical investigations carried out on diluted Fe-Cr alloys (Geng, 2003; Nonas et al., 1998; Ponomareva et al., 2007; Ruban et al., 1999).

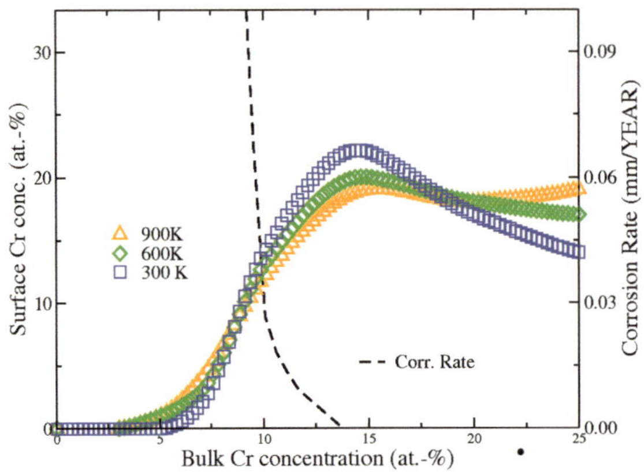

Fig. 12. The calculated Cr concentration in the surface layer (left axis) and the experimental corrosion rate (right axis), as a function of Cr concentration of bulk alloy in 300 K (blue square), 600 K (green diamond), and 900 K (yellow triangle).

The obvious failure of the surface energy considerations to explain the composition of the Fe-Cr surfaces indicates that the bulk part of the alloy could play a key role in the stability of Cr-containing surfaces. Bulk Fe-Cr alloys have a broad and slightly skewed miscibility gap, allowing the solubility of a small amount of Cr in Fe but not vice versa (Hultgren, Desai, Hawkins, Gleiser, & Kelley, 1973). The slightly negative mixing enthalpies at low Cr concentrations of the ferromagnetic solid solutions have been predicted theoretically (Klaver, Drautz, & Finnis, 2006; Olsson, Abrikosov, Vitos, & Wallenius, 2003; Olsson, Abrikosov, & Wallenius, 2006). It has been demonstrated (Ackland, 2006, 2009; Klaver et al., 2006; Nonas et al., 1998; Olsson et al., 2006; Olsson, Domain, & Wallenius, 2007) that the limited solubility of chromium in iron is connected to the complex magnetic interactions characteristic of solid solutions between antiferromagnetic (Cr) and ferromagnetic (Fe) species. These interactions originate from magnetic frustrations due to the strong antiparallel coupling between Cr and the Fe matrix and also between different Cr atoms (Nonas et al., 1998).

The energetically unfavorable magnetic interactions in Fe-Cr can be avoided or minimized by forming Cr-rich clusters (Klaver et al., 2006) and simultaneously moving some of the Cr atoms to the alloy surface. The latter phenomenon becomes clear if one compares the bulk ECP and the mixing enthalpy (Fig. 11). At low temperatures, apart from a constant shift and sign, the slope of the mixing enthalpy gives to a good approximation the value of the bulk effective chemical potential (Nonas et al., 1998). Similarly, the second order concentration derivative (curvature) of the mixing enthalpy gives the slope of the bulk ECP. In particular, the large negative slope of the bulk ECP for Fe-Cr (Fig. 11, left axis) is related to the positive curvature of the mixing enthalpy of alloys with Cr content below ~ 15% (Fig. 11, right axis). When compared to the surface ECP, one can see that the crossover between the bulk and surface chemical potentials is indeed a consequence of the rapidly rising (convex) mixing enthalpy. On these grounds (Badini & Laurella, 2001; Götlind, Liu, Svensson, Halvarsson, & Johansson, 2007; Klaver et al., 2006), we can conclude that the magnetism-driven solubility of Cr in Fe is in fact the main factor responsible for the increasing stability of Cr containing surfaces compared to Fe-terminated surfaces (Ropo et al., 2011).

The decisive role of magnetic interactions in the bulk properties of Fe-Cr alloys has been proposed in several theoretical investigations (Ackland, 2006; Klaver et al., 2006; Nonas et al., 1998; Olsson et al., 2006; Olsson et al., 2007), but their impact on the surface chemical composition was revealed only recently. Figures 11 and 12, and the above arguments give clear evidence for the magnetic origin of the stability of Cr enriched surfaces for bulk concentrations beyond the (9%) threshold. It should be pointed out that the strongly nonlinear change of the surface Cr content versus bulk composition is due to the delicate balance between bulk and surface effects. In particular, the lack of Cr at the surface of Fe-rich alloys is a direct consequence of the anomalous mixing of Fe and Cr at low Cr concentrations, which in its turn has a magnetic origin. This finding has important implication in modern materials science as it offers additional rich perspectives in the optimization of high-performance steel grades.

To close our discussion of surface properties we consider an example where Cr enhances corrosion resistance indirectly. Chromium oxide gives good corrosion protection at usual operating temperatures but since Cr forms volatile compounds at high temperature the corrosion protection at elevated temperatures requires, for instance, the more stable Al oxide scales on the alloy surface. Cr_2O_3 scale is protective up to 1000-1100 °C whereas Al_2O_3 scales up to 1400 °C (Brady, Gleeson, & Wright, 2000; Ebbinghaus, 1993). Unfortunately, for most of the Fe alloy applications the straightforward procedure to improve high temperature corrosion resistance by increasing the Al content in bulk, is not an acceptable solution. This is because the high Al content makes Fe-Al alloys brittle which poses a natural upper bound for the Al content in these alloys regarding to most of the applications (Palm, 2005). Fortunately, the additional alloying of Fe-Al with Cr boosts the formation of the Al oxide scale on the surface up to such a level that the Al content in bulk can be kept within the acceptable limits regarding to the required mechanical properties of the alloy. This phenomenon, called the third element effect, is still considered a phenomenon without generally accepted explanation (Badini & Laurella, 2001; Götlind et al., 2007; Niu, Wang, Gao, Zhang, & Gesmundo, 2008; Stott, Wood, & Stringer, 1995; Z. G. Zhang, Gesmundo, Hou, & Niu, 2006). In Fig. 13 the calculated effective chemical potentials of Fe and Al (μ_{Fe} - μ_{Al}) in Fe-Cr-Al are shown as a function of Cr content. As Fig. 13 shows, Al surfaces are favoured in all cases, but Cr addition up to 10 at% decreases μ_{Fe} - μ_{Al} in bulk whereas this quantity at the surface is almost constant. This builds up an increased driving force for the Al diffusion from bulk to the surface resulting in better corrosion resistance with less Al in bulk alloy.

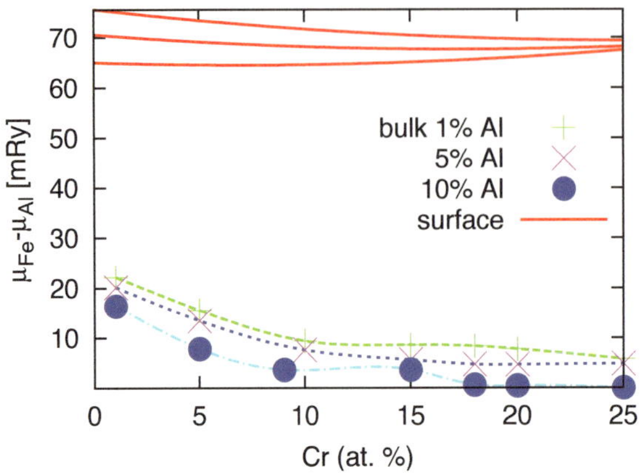

Fig. 13. Bulk and surface chemical potential differences (μ_{Fe} - μ_{Al}) of Fe-Cr-Al, (Al percentages for surface potentials from top to bottom: 1 at.%, 5 at.% and 10 at.%). Surface data is taken from two-dimensional polynomial fit and the calculated bulk values (shown by symbols) are connected by spline curves. The surface has the same composition as the bulk.

5. Conclusion

Magnetoelastic phenomena in magnetic materials and, in particular, in alloy steels have been known for a long time. However, the magnetic effects on the stacking fault energies and elastic constants of magnetic materials in their paramagnetic state have been less well documented. Here, using first-principles computational methods, we have investigated the atomic-scale chemical, magnetic and structural effects behind the elastic properties and stacking fault energies of paramagnetic Fe-Cr-Ni alloys. We have demonstrated that the presence of large disordered magnetic moments in the paramagnetic state can explain a wide diversity of properties that the austenitic stainless steels exhibit. Therefore, in this important class of „nonmagnetic" engineering materials, the „hidden" magnetism gives a major contribution to the fundamental bulk properties.

The interfacial energies between the Cr-rich Fe-Cr and Fe-rich Fe-Cr alloy phases have been calculated to be between ~0.02 and ~0.33 Jm^{-2} for the ferromagnetic state and between ~0.02 and ~0.27 Jm^{-2} for the paramagnetic state. The ferromagnetic interfacial energy exhibits strong nonlinear concentration dependence, whereas the paramagnetic interfacial energy follows smooth composition dependence. As an immediate application of the computed interfacial energies, we have estimated the critical grain size for phase separation using a continuum model. The theoretical critical radii depend very strongly on the composition of the initial homogeneous alloy. The rapidly increasing R_{crit} between the spinodal and solubility lines is in good agreement with the observation that in this region decomposition happens via the nucleation and growth mechanism. At a given volumetric driving force, it is the energy cost to form an interface which presents the main obstacle for decomposition. Therefore, shedding light on the atomic-level mechanisms behind the composition and magnetic state dependence of the interfacial energy opens perspectives for the quantum engineering of the Fe-Cr-based alloys. Our results present a step in this direction and give guidance for experimental and further theoretical investigations of the interfaces in Fe-Cr based alloys.

Investigating the surfaces of Fe-Cr alloys, we have demonstrated that the surface chemistry follows the peculiar threshold behavior characteristic of ferritic stainless steels. We find that in dilute alloys the surfaces are covered exclusively by Fe, whereas for bulk Cr concentration above ~10% the Cr-containing surfaces become favorable. The two distinctly dissimilar surface regimes appear as a consequence of two competing magnetic effects: the magnetically induced immiscibility in bulk Fe-Cr alloys and the stability of magnetic surfaces.

The above examples of *ab initio* study of steel materials have important message for modern materials science: they clearly show that a consistent approach to materials design must be based on first-principles quantum theory and thermodynamics. This combination offers a unique and probably the only possibility for a thorough control of the balance between competing atomic-level effects in steels.

6. Acknowledgment

The Swedish Research Council, the Swedish Steel Producers' Association (Jernkontoret), the Swedish Foundation for Strategic Research, the China Scholarship Council, the Erasmus Mundus External Cooperation Lot3, the Academy of Finland (Grant No. 116317) and Outokumpu Foundation and the Hungarian Academy of Sciences (research project OTKA 84078) are acknowledged for financial support.

7. References

Ackland, G. J. (2006). Magnetically Induced Immiscibility in the Ising Model of FeCr Stainless Steel. *Phys. Rev. Lett., 97*, 015502.

Ackland, G. J. (2009). Ordered sigma-type phase in the Ising model of Fe-Cr stainless steels. *Phys. Rev. B, 79*, 094202.

Airiskallio, E., Nurmi, E., Heinonen, M. H., Väyrynen, I. J., Kokko, K., Ropo, M., et al. (2010). High temperature oxidation of Fe-Al and Fe-Cr-Al alloys: The role of Cr as a chemically active element. *Corros. Sci. 52*, 3394.

Al-Zoubi, N., Johansson, B., Nilson, G., & Vitos, L. (2011). The Bain path of paramagnetic Fe-Cr based alloys. *J. Appl. Phys., 110*, 013708.

Aldén, M., Skriver, H. L., Mirbt, S., & Johansson, B. (1992). Calculated surface-energy anomaly in the 3d metals. *Phys. Rev. Lett., 69*, 2296.

Aldén, M., Skriver, H. L., Mirbt, S., & Johansson, B. (1994). Surface energy and magnetism of the 3d metals. *Surf. Sci., 315*, 157.

Andersen, O. K., Jepsen, O., & Krier, G. (1994). *Lectures on Methods of Electronic Structure Calculations.* edited by Kumar, V., Andersen, O. K., & Mookerjee, A. Singapore: World Scientific Publishing, pp. 63–124.

Argon, A. S., Backer, S., McClintock, F. A., et al., (1966). *Metallurgy and Materials.* Ontario: Addison-Wesley Publishing Company.

Badini, C., & Laurella, F. (2001). Oxidation of FeCrAl alloy: influence of temperature and atmosphere on scale growth rate and mechanism. *Surf. Coat. Techn., 135*, 291.

Bain, E. C. (1924). The nature of martensite. *Trans. Am. Inst. Min. Metal. Eng., 70*, 25.

Brady, M. P., Gleeson, B., & Wright, I. G. (2000). Alloy design strategies for promoting protective oxide-scale formation. *J. Min. Met. Mat. Soc., 52*, 16.

Dowben, P. A., Grunze, M., & Wright, D. (1983). Surface segregation of chromium in $Fe_{72}Cr_{28}$(110) crystal. *Surf. Sci., 134*, L524.
Ebbinghaus, B. B. (1993). Thermodynamics of gas-phase chromium species - the chromium oxides, the chromium oxyhydroxides, and volatility calculations in waste incineration processes. *Combustion and Flame, 93*, 119.
Fleischer, R. L. (1963). Substitutional solution hardening. *Acta Metall., 11*, 203.
Geng, W. T. (2003). Cr segregation at the Fe-Cr surface: A first-principles GGA investigation. *Phys. Rev. B, 68*, 233402.
Ghosh, G., & Olson, G. B. (2002). The isotropic shear modulus of multicomponent Fe-base solid solutions. *Acta Mater., 50*, 2655.
Gibbs, J. W. (1948). *Collected Works*. New Haven Yale University Press.
Götlind, H., Liu, F., Svensson, J.-E., Halvarsson, M., & Johansson, L.-G. (2007). The effect of water vapor on the initial stages of oxidation of the FeCrAl alloy Kanthal AF at 900 °C. *Oxid. Met., 67*, 251.
Grimvall, G. (1976). Polymorphism of metals 3. Theory of temperature-pressure phase diagram of iron. *Phys. Scr., 13*, 59.
Györffy, B. L. (1972). Coherent-Potential Approximation for a Nonoverlapping-Muffin-Tin-Potential Model of Random Substitutional Alloys. *Phys. Rev. B, 5*(6), 2382.
Györffy, B. L., Pindor, A. J., Stocks, G. M., Staunton, J., & Winter, H. (1985). A first-principles theory of ferromagnetic phase transitions in metals. *J. Phys. F: Met. Phys., 15*, 1337.
Heinonen, M. H., Kokko; K., Punkkinen M. P. J., Nurmi, E., Kollár, J., & Vitos, L. (2011). Initial oxidation of Fe-Al and Fe-Cr-Al Alloys: Cr as an alumina booster. *Oxid. Met. 76*, 331.
Hillert, M. (1961). A solid-solution model for inhomogeneous systems. *Acta Metall., 9*, 525.
Hohenberg, P., & Kohn, W. (1964). Inhomogeneous Electron Gas. *Phys. Rev. , 136*, B864.
Hultgren, R., Desai, P. D., Hawkins, D. T., Gleiser, M., & Kelley, K. K. (1973). *Selected Values of Thermodynamic Properties of Binary Alloys*. Metals Park OH.
Ishida, K. (1976). Direct estimation of stacking fault energy by thermodynamical analysis. *Phys. Stat. Sol. A, 36*, 717.
Khanna, A. S. (2002). *Introduction to High Temperature Oxidation and Corrosion*: ASM International, Materials Park OH.
Kiejna, A., & Wachowicz, E. (2008). Segregation of Cr impurities at bcc iron surfaces: First-principles calculations. *Phys. Rev. B, 78*, 113403.
Klaver, T. P. C., Drautz, R., & Finnis, M. W. (2006). Magnetism and thermodynamics of defect-free Fe-Cr alloys. *Phys. Rev. B, 74*, 094435.
Kollár, J., Vitos, L., & Skriver, H. L. (2000). *in Electronic Structure and Physical Properties of Solids: the Uses of the LMTO Method, Lectures Notes in Physics,* . Berlin: Springer-Verlag.
Labusch, R. (1972). Statistical theories of solid solution hardening. *Acta Metall., 20*, 917.
Lince, J. R., Didziulis, S. V., Shuh, D. K., Durbin, T. D., & Yarmoff, J. A. (1992). Interaction of O_2 with the $Fe_{0.84}Cr_{0.16}$(001) surface studied by photoelectron spectroscopy. *Surf. Sci., 277*, 43.
Lo, K. H., Shek, C. H., & Lai, J. K. L. (2009). Recent developments in stainless steels. *Mat. Sci. Eng. R, 65*, 39.
Lu, S., Hu, Q. M., Johansson, B., & Vitos, L. (2011). Stacking fault energies of Mn, Co and Nb alloyed austenitic stainless steels. *Acta Mater., 59*, 5728.

Lu, S., Hu, Q. M., Yang, R., Johansson, B., & Vitos, L. (2010). First-principles determination of the α-α' interfacial energy in Fe-Cr alloys. *Phys. Rev. B, 82*, 195103.

Lung, C. W., & March, N. H. (1999). *Mechanical properties of metals: atomistic and fractal continuum approaches*: World Scientific Publishing Co. Pte. Ltd.

Majumdar, A. K., & Blanckenhagen, P. v. (1984). Magnetic phase diagram of $Fe_{80-x}Ni_xCr_{20}$ ($10 < x < 30$) alloys. *Phys. Rev., 29*, 4079.

Massalski, T. B. (1986). *Binary Alloy Phase Diagrams* (Vol. 1). Ohio: American Society for Metals, Metals park.

Nabarro, F. R. N. (1977). The theory of solution hardening. *Phil. Mag., 35*, 613.

Niu, Y., Wang, S., Gao, F., Zhang, Z. G., & Gesmundo, F. (2008). The nature of the third-element effect in the oxidation of Fe-xCr-3 at.% Al alloys in 1 atm O_2 at 1000 °C. *Corros. Sci., 50*, 345.

Nonas, B., Wildberger, K., Zeller, R., Dederichs, P. H., & Gyorffy, B. L. (1998). Magnetic properties of 4d impurities on the (001) surfaces of nickel and iron. *Phys. Rev. B, 57*, 84.

Olsson, P., Abrikosov, I. A., Vitos, L., & Wallenius, J. (2003). Ab initio formation energies of Fe-Cr alloys. *J. Nucl. Mater., 321*, 84.

Olsson, P., Abrikosov, I. A., & Wallenius, J. (2006). Electronic origin of the anomalous stability of Fe-rich bcc Fe-Cr alloys. *Phys. Rev. B, 73*, 104416.

Olsson, P., Domain, C., & Wallenius, J. (2007). Ab initio study of Cr interactions with point defects in bcc Fe *Phys. Rev. B, 75*, 014110.

Palm, M. (2005). Concepts derived from phase diagram studies for the strengthening of Fe-Al-based alloys. *Intermetallics, 13*, 1286.

Perdew, J. P., Burke, K., & Ernzerhof, M. (1996). Generalized gradient approximation made simple. *Phys. Rev. Lett., 77*, 3865.

Ponomareva, A. V., Isaev, E. I., Skorodumova, N. V., Vekilov, Y. K., & Abrikosov, I. A. (2007). Surface segregation energy in bcc Fe-rich Fe-Cr alloys. *Phys. Rev. B, 75*, 245406.

Rhodes, C. G., & Thompson, A. W. (1977). Composition dependence of stacking-fault energy in austenitic stainless steels. *Metall. Trans. A, 8A*, 1901.

Ropo, M., Kokko, K., Airiskallio, E., Punkkinen, M. P. J., Hogmark, S., Kollár, J., et al. (2011). First-principles atomistic study of surfaces of Fe-rich Fe-Cr. *J. Phys.: Cond. Mat., 23*, 265004.

Ropo, M., Kokko, K., Punkkinen, M. P. J., Hogmark, S., Kollár, J., Johansson, B., et al. (2007). Theoretical evidence of the compositional threshold behavior of FeCr surfaces. *Phys. Rev. B, 76*, 220401.

Ruban, A. V., Skriver, H. L., & Nørskov, J. K. (1999). Surface segregation energies in transition-metal alloys. *Phys. Rev. B, 59*, 15990.

Ryan, M. P., Williams, D. E., Chater, R. J., Hutton, B. M., & McPhail, D. S. (2002). Why stainless steel corrodes. *Nature, 415*, 770.

Schramm, R. E., & Reed, R. P. (1975). Stacking-fault energies of 7 commercial austenitic stainless steels. *Metall. Trans. A, 6A*, 1345.

Soven, P. (1967). Coherent-Potential Model of Substitutional Disordered Alloys. *Phys. Rev., 156*, 809.

Speich, G. R., Schwoeble, A. J., & Leslie, W. C. (1972). Elastic constants of binary iron-base alloys. *Metall. Trans., 3*, 2031.

Stott, F. H., Wood, G. C., & Stringer, J. (1995). The influence of alloying elements on the development and maintenance of protective scales. *Oxid. Met., 44*, 113.

Suzuki, S., Kosaka, T., Inoue, H., Isshiki, M., & Waseda, Y. (1996). Effect of the surface segregation of chromium on oxidation of high-purity Fe-Cr alloys at room temperature. *Appl. Surf. Sci., 103*, 495.

Takeuchi, S. (1969). Solid-solution strenthending in single crystals of iron alloys. *J. Phys. Soc. Japan, 27*, 929.

Teklu, A., Ledbetter, H., Kim, S., Boatner, L. A., McGuire, M., & Keppens, V. (2004). Single-crystal elastic constants of Fe-15Ni-15Cr alloy. *Metall. Mater. Trans. A, 35*, 024415.

Vitos, L. (2001). Total-energy method based on the exact muffin-tin orbitals theory. *Phys. Rev. B, 64*, 014107.

Vitos, L. (2007). *The EMTO Method and Applications, in Computational Quantum Mechanicals for Materials Engineers* London: Springer-Verlag.

Vitos, L., Abrikosov, I. A., & Johansson, B. (2001). Anisotropic Lattice Distortions in Random Alloys from First-Principles Theory. *Phys. Rev. Lett., 87*, 156401.

Vitos, L., & Johansson, B. (2009). Large magnetoelastic effects in paramagnetic stainless steels from first principles. *Phys. Rev. B, 79*, 024415.

Vitos, L., Korzhavyi, P. A., & Johansson, B. (2002). Elastic property maps of austenitic stainless steels. *Phys. Rev. Lett., 88*, 155501.

Vitos, L., Korzhavyi, P. A., & Johansson, B. (2006). Evidence of large magneto-structural effects in austenitic stainless steels. *Phys. Rev. Lett., 96*, 117210.

Vitos, L., Korzhavyi, P. A., Nilsson, J.-O., & Johansson, B. (2008). Stacking fault energy and magnetism in austenitic stainless steels. *Phys. Scr., 77*, 065703

Vitos, L., Nilsson, J.-O., & Johansson, B. (2006). Alloying effects on the stacking fault energy in austenitic stainless steels from first-principles theory. *Acta Mater., 54*, 3821.

Vitos, L., Ruban, A. V., Skriver, H. L., & Kollár, J. (1998). The surface energy of metals. *Surf. Sci., 411*, 186.

Wranglén, G. (1985). *Introduction to Corrosion and Protection of Metals.* New York: Chapman and Hall.

Zhang, H. L., Johansson, B., & Vitos, L. (2009). Ab initio calculations of elastic properties of bcc Fe-Mg and Fe-Cr random alloys. *Phys. Rev. B, 79*, 224201.

Zhang, H. L., Punkkinen, M. P. J., Johansson, B., Hertzman, S., & Vitos, L. (2010a). Single-crystal elastic constants of ferromagnetic bcc Fe-based random alloys from first-principles theory. *Phys. Rev. B, 81*, 184105.

Zhang, H. L., Punkkinen, M. P. J., Johansson, B., & Vitos, L. (2010). Theoretical elastic moduli of ferromagnetic bcc Fe alloys. *J. Phys.: Cond. Mat. 22*, 275402.

Zhang, Z. G., Gesmundo, F., Hou, P. Y., & Niu, Y. (2006). Criteria for the formation of protective Al_2O_3 scales on Fe-Al and Fe-Cr-Al alloys. *Corros. Sci., 48*, 741.

2

Review – Metallic Bipolar Plates and Their Usage in Energy Conversion Systems

Justin Richards and Kerstin Schmidt
Fraunhofer Institute for Chemical Technology, Project Group Sustainable Mobility, Wolfsburg, Germany

1. Introduction

"Fuel cells, like batteries, are electrochemical galvanic cells that convert chemical energy directly into electrical energy and are not subject to the Carnot cycle limitations of heat engines." [48] Unlike batteries the active material for fuel cells is externally stored which allows capacity and power to be scaled independently.

1.1 History

The first primary battery was invented by Alessandro Volta in 1800, the "Volta Pile" [70]. The first secondary battery which gave the basis for the lead-acid batteries found in most of the automotive applications was developed in 1859 by Raymond Gaston Planté [55].

The idea of a fuel cell was first discovered in 1839 by Christian Friedrich Schönbein [64] and William Grove [27]. Independently from each other they provided the foundation for the development of many different kinds of fuel cells. Today's fuel cells are mostly classified by the type of electrolyte used in the cells. The most common types are polymer electrolyte fuel cells PEFC (developed by William Grubb in 1959), the alkaline fuel cell AFC (developed for the Apollo space Program in the 1960's), the phosphoric acid fuel cell PAFC (from 1965), the molten carbonate fuel cell MCFC and the solid oxide fuel cells SOFC. The beginnings of SOFC and MCFC can be dated back to the mid 1960's [23].

1.2 Importance of fuel cells

Global warming and the political situation pushed the focus further to the renewable energy sources such as wind and solar energy. Their discontinuous availability conflicts with the required energy need. To compensate the increasingly temporary imbalance between generation and demand of energy, innovation solutions must be found. A better adjustment of the reserves to meet the changing demands can be achieved by using decentralized storage devices. The electrolysis of water and the storing of hydrogen in tanks is a promising solution. At times where the demand for energy is high, the hydrogen can be used to supply fuel cells where it will be recombined with oxygen from the air generating electricity. Besides the stationary applications in combination with electrolyzer, fuel cells operating on hydrogen are a promising option for the electrical energy supply for passenger cars with electrical drive trains and medium operating ranges. Due to their high power density and fast start-up time, proton exchange membrane fuel cells show highest potential for automotive applications [23].

1.3 Assembly

Fuel cells are generally assembled according to the stack method (shown in Fig.1).

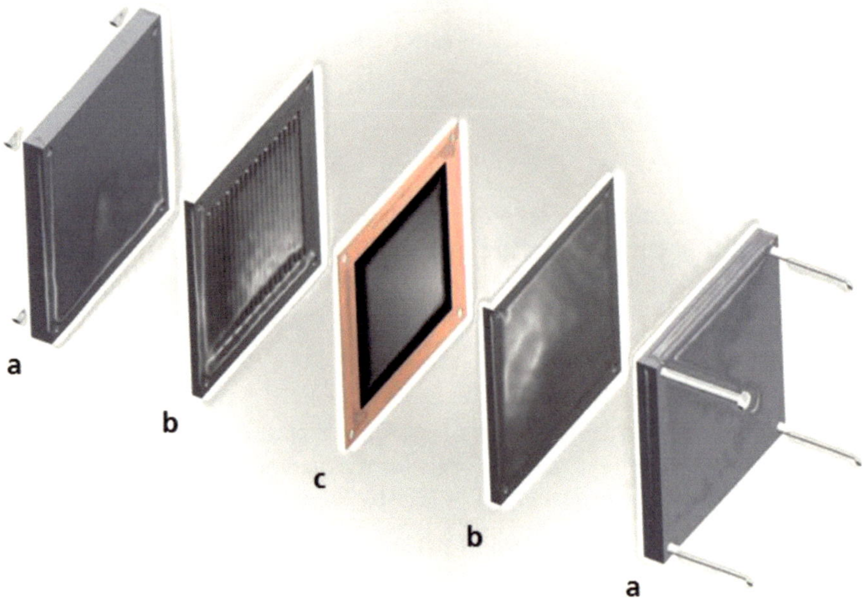

Fig. 1. Schematic design of a one cell stack

The endplates apply the necessary pressure on the stack (a).The bipolar plates (b) define about 60 % of the weight and about 30 % of the cost of one cell. They also provide conduits for the gas and fluid flows of reactants and products of a cell. They remove heat from the active areas and carry it current from cell to cell. The plates also constitute the backbone thus the mechanical stability of a stack. The two half cells are separated by an ionic conductor (c). Depending on the application it mostly is coated with a carbon supported catalyst.

To gain high voltages from a fuel cell the current collector plates of a cell, sometimes known as interconnectors (in SOFCs), are designed to be used as bipolar plates. One side supports the anode and the other side the cathode for the next cell. They are electrically connected in series (Fig. 2). The power (P) output of a fuel cell stack can be calculated by multiplication of the sum of the voltage differences (ΔU) and the current (I)

$$P[W]=I[A]\cdot\sum\Delta U[V] \tag{1}$$

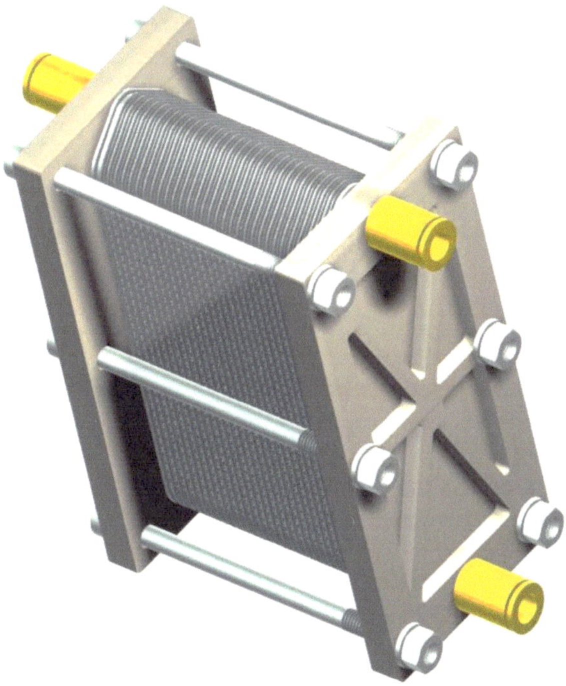

Fig. 2. Schematic design of an air-breathing PEMFC stack

1.4 Graphite – The state of the art material for bipolar plates

Currently graphite and graphite composites are considered the standard material for bipolar plates in e.g. PEMFC. This material provides a good electrical conductivity, a low contact resistance and an outstanding corrosion resistance [2]. However the mechanical properties of graphite limit the design and the dimensions of a stack. With a conductive electrolyte present, electro migration of ions into the graphite structure can take place which results in an expansion of the plate thus lowering the mechanical strength of the graphite. Other disadvantages of graphite are its permeability (e.g. for hydrogen) and its poor cost effectiveness for high volume manufacturing processes compared to metals such as stainless steels.

Recent fuel cell developments comply with the volumetric and gravimetric power density criteria (>1 kW/l and >1 kW/kg) [2]. The two major challenges hindering the technology to gain a firm footing in the energy market, the costs of a fuel cell and its durability.

1.5 Targets of the Department of Energy (DOE) for metallic bipolar plates

"The Department of Energy sets goals for all new technologies. Through R&D and technology validation programs, DOE gathers and reports progress towards the goals". [9]

Metals both treated and untreated are promising candidates for the usage as bipolar plates. However their low corrosion resistance and their high contact resistance resulting from the oxide layer formed on its surface hinder the general use in PEMFCs [37].

Parameter	Unit	DOE Targets *2005 Status*	*2010*	*2015*
Plate Cost	$*kW^{-1}	10	5	3
Plate Weight	Kg*kW^{-1}	0,36	<0,4	<0,4
H_2 Permeation rate	cm^{3}*s^{-1}*cm^{-2}	<2*10^{-6}	<2*10^{-6}	<2*10^{-6}
Corrosion[1]	µA*cm^{-2}	<1	<1	<1
Resistance[2]	Ohm*cm^{2}	<0,01	<0,02	<0,02
Resistivity [72]	Ohm*cm	-	<0,01	<0,01
Flexural Strength	MPa	>34	>25	>25
Flexibility	%	1.5 - 3.5	3 - 5	3 - 5
Durability with cycling [74]	h	-	5000[3]	-

[1]Electrolyte consist of pH 3 H_2SO_4 + 0.1 ppm hydrofluorhydric acid (HF) solution under 0.8 V (NHE) at 80 °C (Pontentiostatic Corrosion Current)
[2]Resistance including the contact resistance at 140 N/cm^{2}[74]; [3]<10% drop in power

Table 1. DOE Metal Plate Status and DOE's Targets

The corrosion of the metallic plates can result in a degradation of the membrane due to the affinity of the proton conducting groups in the membrane to adsorb the leached ions form the metal surface [69]. Furthermore the passive layers formed on the metal bipolar plates guard them from most corrosion attacks but increase the electrical resistivity. Consequently the fuel cell's efficiency declines as the oxide layers grow [3]. Therefore the department of energy states target values for various metallic bipolar plate parameters which are listed in Table 1 [75].

2. Material requirements

The material challenges are different for each energy conversion system. Operating temperatures, leached ions or different fuels have a great effect on the used components. The following chapter will give a brief overview on some of the types of energy conversion systems and on the theoretically used environment displaying them.

Low temperature proton exchange membrane fuel cell (LT-PEMFC)

Operating temperature	60 °C - 80 °C
Reacting agents	Air, oxygen , hydrogen
Operating pressure	1 bar to 3 bar absolute

Electrolyte solutions used for corrosion tests displaying PEMFC environments contain sulfate and mostly fluoride ions. The ions are the result of the degradation of the membrane and can be found in the effluent water [86]. The electrolyte is also purged with air displaying the cathode side or purged with H_2 displaying the anode side. The concentrations of sulfate-ions in the solutions vary from 0,001 mol*l^{-1} up to 1 mol*l^{-1}. Most electrolytes also exhibit fluoride ions at concentration up to 2 ppm. The corrosion tests are realized at different temperatures from ambient temperature up to 80 °C [3].

High temperature proton exchange membrane fuel cell (HT-PEMFC)

Operating temperature	120 °C - 180 °C
Reacting agents	Air, oxygen , hydrogen
Operating pressure	1 bar to 3 bar absolute

To overcome the challenges of the water and thermal management for LT-PEMFC several attempts have been made to develop so called high-temperature PEMFCs, which would operate in the temperature range 120 °C to 160 °C [23], [65].The elevated temperatures cause great challenges for the applied materials. Additionally the proton conductivity of the membrane used in HT-PEMFCs has to be realized differently unlike using H_2O in LT-PEMFC membranes. Different approaches are known in the literature [89]:

- Modified PFSA membranes
- Sulfonated polyaromatic polymers and composite membranes (PEEK, sPEEK)
- Acid-based polymer membranes (phosphoric acid-doped PBI).

For each approach the simulated corrosion environment changes. For example bipolar plate material tests for doped PBI are pickled in 85 % phosphoric acid at 160 °C for 24 hours. [89]

Solid oxide fuel cell (SOFC)

Operating temperature	600 °C - 1000 °C
Reacting agents	Air, oxygen, hydrogen
Operating pressure	~10 bar

The basis for the SOFC was the discovery of Nernst in 1890. He realized that some pervoskites were able to conduct ions in certain temperature ranges [23].

The interconnectors used in SOFCs can be divided into two categories, ceramics and metal alloys. The Ceramics are used at temperatures around 1000 °C whereas the metals can be found in applications with temperatures around 650 °C to 800 °C. The metals have to withstand the thermal cycling while still providing adequate contact resistances [19]. To investigate the durability of the metals Ziomek-Moroz [91] treated metal tubes in 99 % N_2 and 1 % H_2 for one year at temperatures of 600 °C, 700 °C and 800 °C analyzing the effects on the metal specimen afterwards.

Direct methanol fuel cell (DMFC)

Operating temperature	40 °C - 80 °C
Reacting agents	Air, methanol
Operating pressure	Ambient

The DMFCs are mostly used for portable applications [23]. Corrosion properties of materials for bipolar plates can be analyzed in a theoretical environment consisting of a 0.5 M H_2SO_4 solution with 10 % methanol [47].

3. Precious metals

Precious metals such as platinum and gold provide excellent corrosion resistance as well as good electrical conductivity [37], [84]. Nevertheless the costs for gold tripled in the last decade and is now as high as ~1065 €/ounce [80]. Platinum had an intermediate peak in 2008 with around 2200 €/ounce. Today's price for one ounce is around 1800 € [79]. These high prices prohibit the utilization of theses metals as bipolar plate material for commercial use. Table 2 compares the costs of gold and other usable materials.

Material	Material cost (US $/g)	Density (g/cm³)
Graphite	0.105	1.79
Aluminum	0.0088	2.7
Gold	9.97	19.32
Electroless nickel	0.034	8.19

Table 2. Bipolar plate material and high-volume material costs [37]

4. Stainless steel alloys

To further increase the power density of fuel cells and to decrease the costs of the stacks, the use of very thin steel bipolar plates which can be formed in a low cost hydroforming process would be beneficial.

4.1 Untreated

The major concerns for bulk metal alloys are the corrosion resistance against the harsh environment and the resulting increase in contact resistance once the passive layer forms on the surface.

Many different alloys have been tested for the use as bipolar plate materials in fuel cells. Herman et al. [31] reports that aluminum, titanium, stainless steels and nickel exposed to a PEMFC-like environment exhibit corrosion and dissolution. While a protective layer forms on the surface of the metal specimen to provide protection against the corrosion attacks, the contact resistance increases thus lowering the overall performance of the stack.

Davies et al. [14] indicates that compared to tested metal alloys the graphite material exhibits the lowest interfacial resistance. The resistance values (measured with a compaction force of 220 N/cm² similar to the force imposed in a fuel cell) of the metal alloys decrease in the order 321[1] > 304[1] > 347[1] > 316[1] > Ti > 310[1] > 904[1] >Incoloy 800[1] > Inconel 601[1] > graphite. The influence on the surface resistivity depends on the compression force seen in Fig. 3. Kraytsberg [39] suggests that texturing the bipolar plate surface would improve contact points thus reducing the resistivity.

4.2 Surface treatments

To improve corrosion resistance, steels can be provided with some kind of corrosion protection [13], [28], [56], [58], and [60]. Richards et al. [61] reported corrosion-test results of stainless steel specimen treated with a commercial coating solution (CCS) in combination with physicochemical surface treatments, electro-polishing and thermal annealing. Fig. 4 displays an improvement for the electro-polished and thermal annealed sample at high potentials. The other treated test specimen showed even higher corrosion currents than the bulk material.

[1]Metall alloys are named with their symbol or material number by European standards DIN EN 10027 [90]

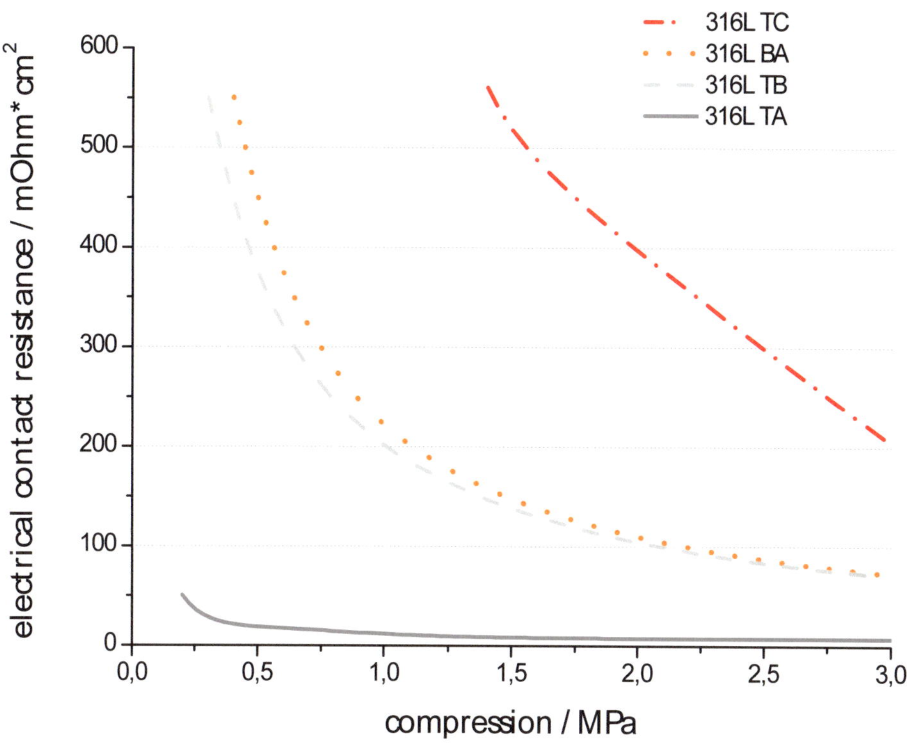

Fig. 3. Qualitative diagram of the electrical contact resistance of 316L samples for different pressure values [1]

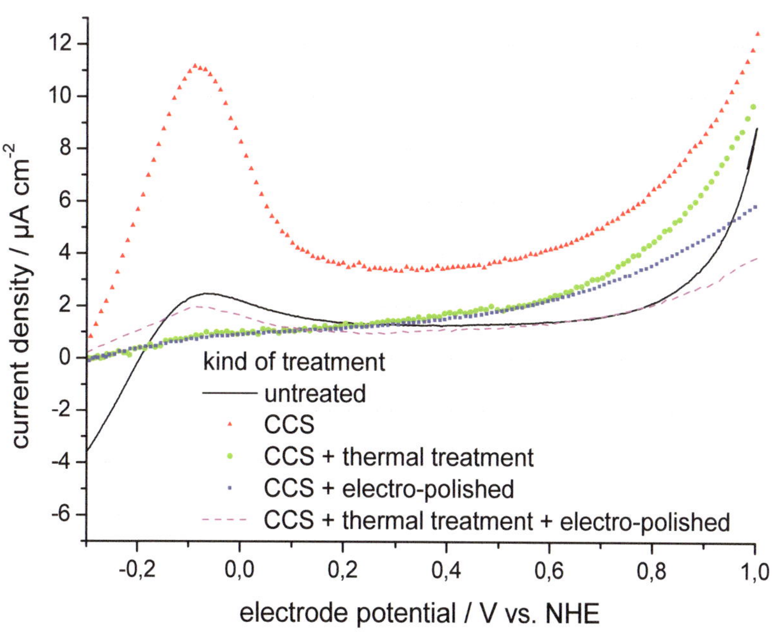

Fig. 4. LSV of alloy 2.4605 after different surface treatments at 50 °C [61]

The electro-polishing of stainless steels is an electrochemical surface treatment which is widely used for decorated utilization as well as for applications in medical and food science. The work done by Lee [44] provided results to the application of electro-polishing on stainless steels for bipolar plates. He used electrochemical surface treatments on 316L and reported that the content of chrome increased significantly in the passive layer on the specimen thus decrease the corrosion currents.

4.3 Amorphous metal alloys

Metal alloys or metallic glasses were first discovered in 1960 where Klement Willens and Duwes [67] reported the synthesis of amorphous metals by rapidly quenching (10^6 K) an Au-Si alloy from ~1300 °C to room temperature. To reduce the cooling rate and the material costs new alloys were developed. In 1995 the first iron based bulk metallic glasses was invented. One of the crucial aspects to design new amorphous alloys is that the difference in atomic size must be greater than 12 % [50].

Jayaraj [34], [33] compared the properties of the alloys $Fe_{48}Cr_{15}Mo_{14}Y_2C_{15}B_6$, $Fe_{44}Cr_{16}Mo_{16}C_{18}B_6$ and $Fe_{50}Cr_{18}Mo_8Al_2Y_2C_{14}B_6$ to a standard alloy 316L. Unfortunately he discovered that despite their unique properties, such as high strength, good hardness, good wear resistance, and high corrosion resistance, the amorphous alloy did not meet the DOE targets. In 2006 Jin, S. et al [35] detected a four times lower corrosion current of $Zr_{75}Ti_{25}$ (0.021 μA/cm²) compared to 316L after a period of 5 hours at 80 °C in simulated anodic PEMFC environment. He also reported that 316L and $Zr_{75}Ti_{25}$ showed a considerable difference on forming an oxide film on their surface (Fig. 5).

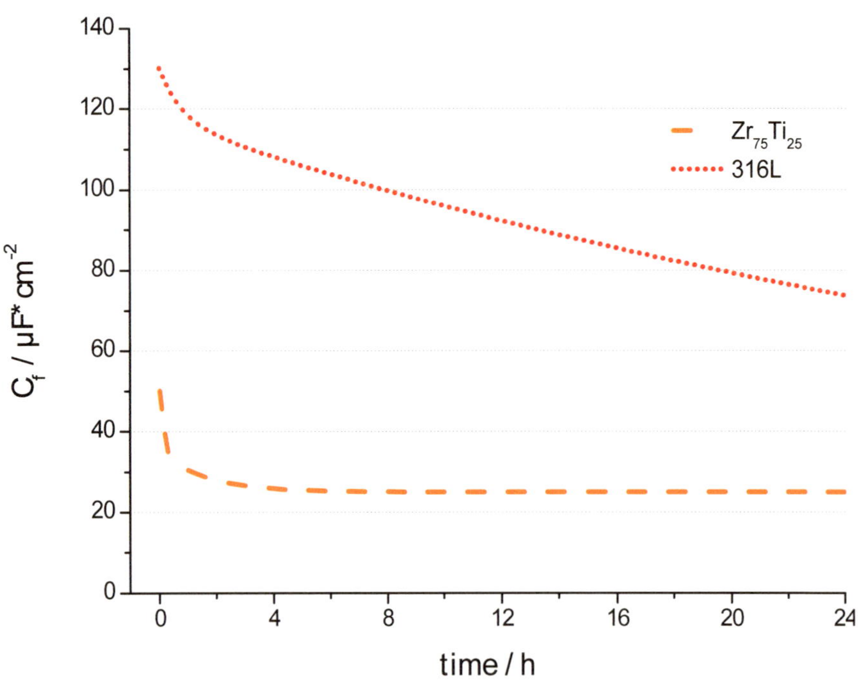

Fig. 5. Qualitative diagram of the oxide film capacity of 316L and Zr75Ti25 in the acid solution bubbled with H2 at 25 °C [35]

5. Coated metals

Coatings for metallic bipolar plates should be conductive and adhere to the substrate material with little to no porosity [37].

5.1 Electroplating

The conductive substrate material is dipped into an electrolyte bath containing dissolved metal ions. By connecting the substrate as cathode the dissolved ions are deposited on the surface.

Walsh et al. [71] investigated the influence of the porosity of electrochemically deposited metal coatings on the corrosion properties of the specimen. Fig. 6 indicates a rapid decrease of the corrosion rate of nickel coatings on aluminum with increasing thickness of the layer. From a thickness of ~2.5 µm a horizontal region indicates a pore free passive film [71].

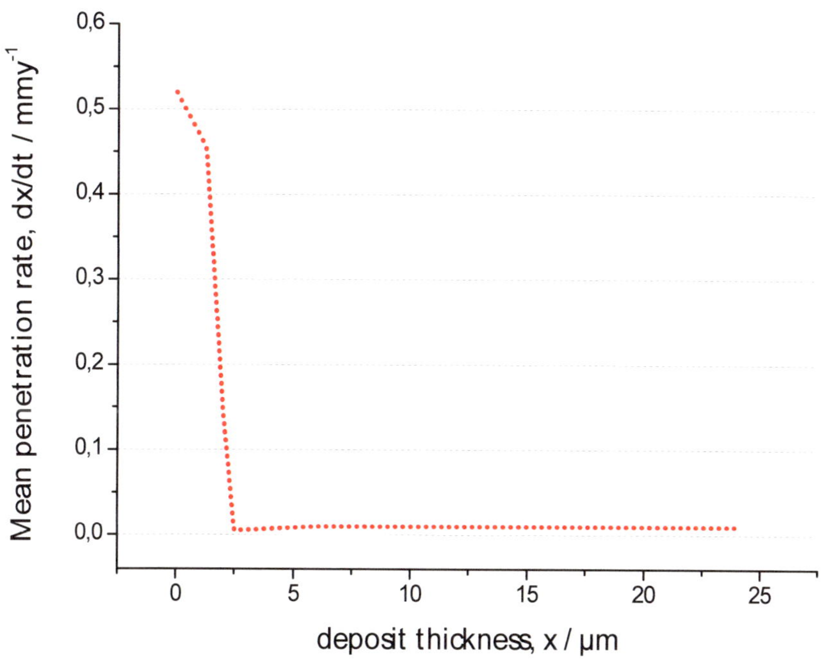

Fig. 6. Qualitative corrosion rate vs. electroless nickel coating thickness on aluminium in 0.85 mol/dm³ (5%) NaCl at 295 K [71].

Hental et al. [29] coated an aluminum substrate material with gold. First tests showed an almost equal performance to graphite (1.2 A/cm² at 0.7 V). But very quickly the performance dropped to 60 mA/cm² at 0.5 V. Subsequent analysis revealed that parts of the gold layer had been lifted off the surface and became embedded in the membrane. Other tests by Yoon et al. [88] indicated very good performance of gold coated 316L. The DOE targets for corrosion currents in an anodic environment and for the contact resistance were met with 316L coated with 10 nm gold. But only the ZrNAu coating on 316L showed satisfying results in the cathode environment.

El-Enin et al. [18] investigated aluminum as substrate material with different electro-deposited coatings. His results show that the Ni-Co coating exhibits the best corrosion properties due to its structure. Showing good performances in corrosion resistance as well the author recommends Ni-Mo-Fe-Cr coating with further annealing to be applied in a fuel cell instead of graphite.

5.2 Chemical vapor deposition

"Chemical vapor deposition may be defined as the deposition of a solid on a heated surface from a chemical reaction in the vapor phase." [54]

Another approach for stabilizing metallic bipolar plates in energy conversion environments is to deposit good conducting and inert material compositions onto the metal surfaces. In 2009 Chung and colleagues [12] investigated coatings of carbon, formed from a C_2H_2/H_2 gas mixture at 690 °C to 930 °C on SUS304 metal substrates.

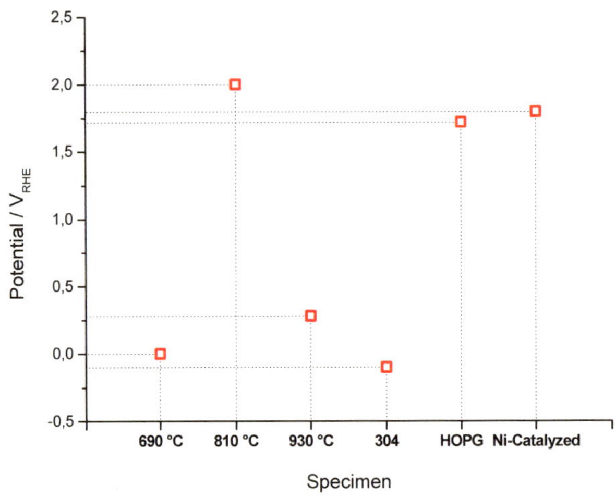

Fig. 7. Corrosion potential of specimen tested in 0.5 M H2SO4 solution at 25 °C [11]

All reported carbon films in this study showed an improved resistivity for the metal plates to a level comparable to the highly oriented pyrolytic graphite (HOPG). The results show a significant influence of the temperature range of the treatment on the corrosion potentials of the specimen. The specimen prepared at a temperature of 810 °C exhibits the highest corrosion potential followed by the Ni-catalyzed, the HOPG and the specimen prepared at 930 °C. The sample coated at 690° C showed a little higher corrosion potential than the bulk specimen. The results for the corrosion potentials for the tested samples are displayed in Figure 7.

Good properties for the use in energy conversion systems can also be provided by chromizing metal surfaces. Lee et al. [42] investigated a chromizing process for 316L by CVD enhanced pack cementation. They used two different powder mixtures consisting of 50 % Cr, 43 % Al_2O_3 (A) and 7% NH_4C respectively 25 % Cr, 72 % Al_2O_3 and 3 % NH_4C (B). The corrosion resistance improved compared to the bulk material. The interfacial contact resistance was significantly influenced by the composition and procedure of the chromizing process. Specimen A exhibits the highest resistance in the test. It was up to three times higher (~50 Ohm*cm^2) than the resistance of the bulk specimen and the specimen B resulting from the absence of the chrome carbide in the outer surface layer. Wen et al. [81] reported that modified SS420 chromized by a low temperature process possessed a reduction of four to five orders of magnitude when compared with its bulk substrate (100 µA/cm² - 10 µA/cm²). The contact resistances value at 12-18 mOhm*cm² was below those of Lee [42] and colleagues released in 2009. Other experiments by Bai et al. [5] showed an influence on the pretreatment of AISI1020 as the substrate material which was also chromized by pack cementation. With electrical discharge machining (EDM) as pretreatment theses specimen showed corrosion potentials little higher (-0,45 V and 0.37 V vs. SCE[2]) than that of the bulk substrate (~4.5 V vs. SCE) but with little contact resistances (11.8 mOhm*cm² and 17.7 mOhm*cm²).

5.3 Physical vapor deposition

In the physical vapor deposition processes (PVD) or thin film processes a solid or liquid material is vaporized by e.g. electron beams and is deposited onto the surface of a substrate. Thin films of a few nanometers to a thousand nanometers can be achieved. Multilayers and free standing structures like ribs are also possible [49].

Stainless steel specimen used as substrates in PVD processes need to be pretreated by sputtering the surface with e.g. Ar-ions which remove the passive film on the material. Fu et al. [21] investigated three different carbon based films for the use as bipolar plate materials. The C-Film and the C-Cr-N-film exhibited relatively high contact resistances. The C-Cr deposited film provided a good corrosion resistance little higher than the DOE-targets and much lower contact resistance (~10 mOhm/cm²) than the other tested samples. Dur et al. [17] showed the effect of the manufacturing process on the corrosion resistance of modified stainless steel samples. The specimen were formed by stamping or hydroforming into bipolar plates and subsequently coated with TiN, CrN and ZrN. Among the coated samples the corrosion resistance was best for ZrN followed by CrN and TiN. Surprisingly TiN coated samples showed even lower resistances than their bulk samples. Regardless of the coating type the blank unformed samples provided the highest corrosion resistance. Wang, Y. [77] states that besides a corrosion resistant coating the substrate itself should have a high corrosion resistance. Those results were gained by tests with 316L and SS410 coated with TiN.

Investigations on a Ni-Cr enriched layer by Feng K. [19] with a thickness of 60 nm showed an improvement in corrosion resistance and a decrease in interfacial resistance. However the results are not at the desired level for PEMFC applications. A different solution was provided by Cha [11] who reported that a multiphase NbN/NbCrN film performed better than a single phased NbN film as a coating for metallic bipolar plates. The specimens were fabricated by sputtering Nb, Cr and using N_2 as reaction gas. Cha states that concerning the interfacial resistance the chromium amount in the layer does not make much of a difference but the influence in the gas ratio is noticeable. The corrosion resistance improved significantly compared to the bare material but pitting corrosion appears with certain coating process parameters. In the work done by Zhang et al. [90] it is shown that the corrosion resistance as well as the electrical conductivity and also the mechanical properties (hardness) are increased by arc ion plating. Zhang and colleagues coated 316L with a multilayer consisting of Cr/CrN/Cr.

[2]SCE – saturated calomel electrode

Lee et al. [43] coated aluminum (PVD coating of diamond-like film on 5052 Al) with the bulk 316L, bulk Al and graphite. The results were displayed in a tafel plot. The corrosion resistance of the coated samples is higher than its uncoated substrate but not as good as the 316L specimen. But the treated Al samples exhibits lower contact resistance than the 316L thus increasing the power of the cell.

5.4 Diffusion coating with Nitride

Different approaches by nitridation or nitrating steel samples can be found in the literature. Yang et al. [87] modified the surface of Fe-27Cr-6V and Fe-27Cr-2V alloy with high temperature nitridation. Their published results showed good corrosion and low interfacial resistances for the prepared samples. They also discovered that the addition of vanadium in the specimen influence the external nitride layer positively. A TEM cross section image of the nitride surface structure can be seen in Fig. 7. Improvements of contact resistances for a thermally nitride AISI446 stainless steel alloy at 1100 °C for 24 h was shown in the reported work of Wang H. et al. [73]. An increase in corrosion resistance was also observed for the cathodic environment. Unfortunately it exhibited poor corrosion resistance properties in the anodic conditions.

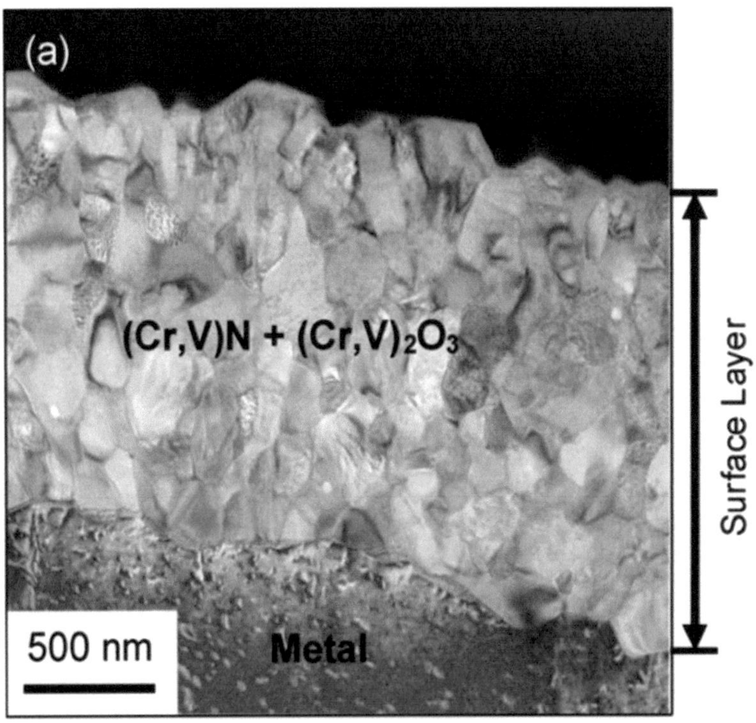

Fig. 8. Microstructures of nitrided Fe–27Cr–6V (850 °C, 24 h, N_2–$4H_2$) bright-field cross-section TEM [87][3]

[3]This article was published in Journal of Power Sources, 174, Yang, B. et al., Protective nitride formation on stainless steel alloys for proton exchange membrane fuel cell bipolar plates, 228–236, Copyright Elsevier (2007)

Feng et al. [20] and colleagues investigated the effect of high and low temperature nitrogen plasma immersion ion implantation (PIII) of titanium. No significant effects were discovered for the low temperature PIII compared to the bulk material. Although the high temperature PIII treated specimen showed better results for the corrosion as well as for the interfacial resistance analysis.

Brady et al. [8] and Toops et al. [68] investigated thermal nitridation of stainless steel alloys. They showed that the modified Fe-20Cr-4V possessed the best potential for the usage as bipolar plate materials compared to its untreated material and 904L alloy. In the work done by Hong et al. [32] austenitic steel (316L) was nitrided by inductively coupled plasma using a gas mixture of N_2 and H_2 at relatively low temperatures (~270 °C - 380 °C). The specimen prepared at ~316 °C showed good electrical and electrochemical properties.

5.5 Conductive polymer coating

Extrinsic

An extrinsically conductive material is composed of at least two different components, the binding agent, which is a polymer in general, and the electrical conductive filler. As filler mostly a metallic or carbon powder is used. The amount of the electrical conductive material needs to be high enough so the distance between the particles undermatches 10 nm to bring the composite above the percolation threshold where the conductivity of the composite increases excursively [63].

Carbon nanotubes (CNT) as a layer between steel plate and coating

Yang-Bok Lee and his colleagues [45] showed, that a CNT layer between a stainless steel plate and a coating of a pressure molded polypropylene-graphite-carbon fiber composite decreases the interfacial contact resistance. The best results were reached by direct deposition of the CNT on the plates with a decrease of 90%. The corrosion resistance test results showed that the chosen polymer-composite is a good protection for the stainless steel.

PTFE[4]-Composites with CNT and Ag as electro-conductive filler

Show and Takahashi [66] reported on a CNT/PTFE composite which was coated on a stainless steel bipolar plate in dispersed form. With an amount of 75% of CNT a conductivity of 12 S/cm has been measured. Besides a better corrosion resistance and a decrease in contact resistance between the surface of the modified bipolar plate and the carbon paper 46 mOhm*cm^2 for bare plates to 12 mOhm*cm^2 for the coated samples was achieved. A PEMFC equipped with these coated plates reached a 1.6 times higher power output in comparison to a system with uncoated separators.

Another PTFE-composite used as coating material is shown by Yu Fu et al. [22]. In this case an Ag-PTFE composite was electro-deposited on the plate surface. The coated plates showed a similar performance to those of the uncoated ones for the interfacial contact resistance and

[4] Polytetrafluoroethylen one of the known trademarks is Teflon®

the corrosion stability. An improvement was reported for the contact angle being an important factor in the water management of a fuel cell.

Resin-Composites

A resin-composite-coating was tested by Kitta et al.[38]. They coated a stainless steel plate with a thin composite film of bisphenol A-type epoxy resin and 60 vol.% of graphite. Additionally to the coating they brought a rib structure on the plate using a different resin graphite system. In this case the researchers used a cresol-novolak-type epoxy resin with 70 vol.% of graphite as electrical conductor. The phenol-type hardener showed a high stability at 90 °C. Only a small amount of ionic impurities (<20 µS/cm) was found. The area specific resistance of the described plate was reported as 13.8 mOhm*cm at a tightening force of 1 MPa. Epoxy resins are also found in other applications such as shielding electromagnetic/radio frequency interferences and in dissipating static charges. Azim et al. [4] reported about a composite of polysulfide modified epoxy resin cured with a polyethylene polyamine and carbon based electro-conducting fillers. They showed that at an amount of 55 vol.% of conductive filler in the composite, consisting of 85 % graphite and 15% carbon black, the material had a specific volume resistance $2*10^{-5}$ Ohm*m.

Rubber-Titanium nitride-Composite

Another approach to reduce the interfacial contact resistance of stainless steel bipolar plates is shown by Kumagai, M. et al. [40]. The stainless steel plates were coated with a suspension of TiN and styrene butadiene rubber (SBR) by using electro-phoretical deposition. Fig. 8 compares the degradation by the help of cell voltages over 300 hours in a single cell of treated plates. The diagram shows that the plates coated with TiN/SBR provided voltage losses which were comparable to graphite. As shown in Fig. 7, the TiN/SBR coating enlarges the contact area of the bipolar plate and the gas diffusion layer. This effect is due to the elasticity of the SBR.

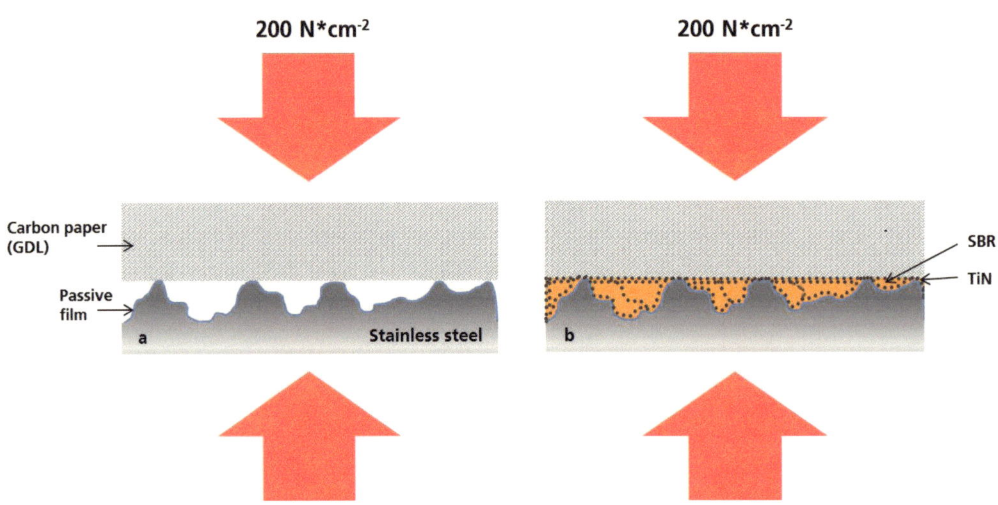

Fig. 9. Effect of TiN/SBR coating (b) on the contact between BPP and GDL in comparison to an as-polished stainless steel (a) [40]

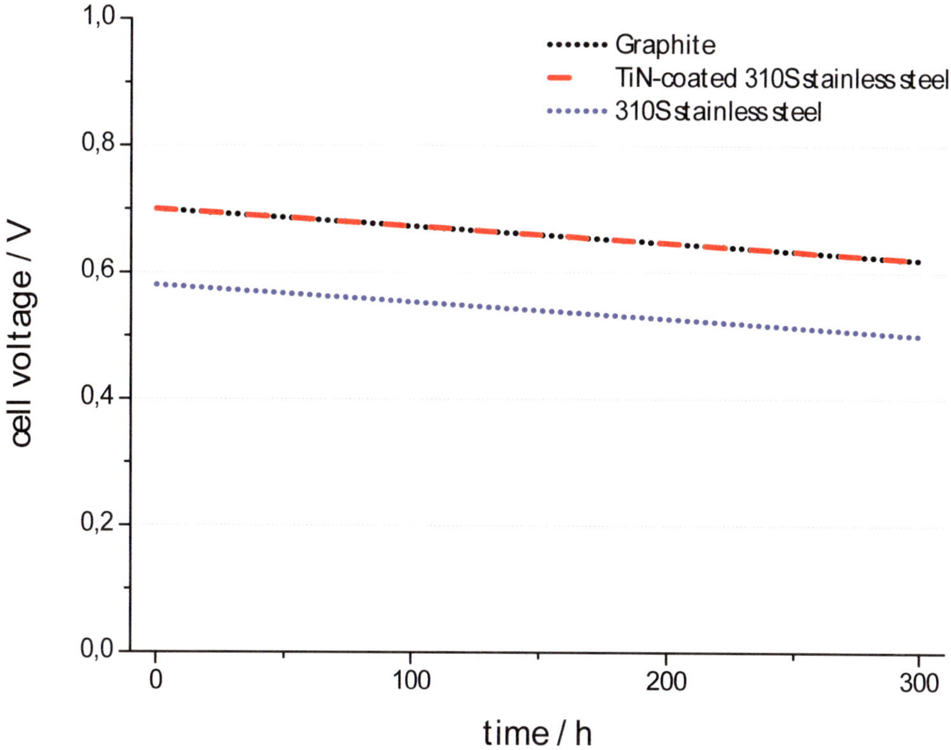

Fig. 10. Qualitative diagram of the cell voltage loss of the treated specimen [40]

Intrinsic

Intrinsically conductive polymers were first discovered by Letheby in 1862. In 2000 the Nobel Prize in Chemistry was given to three scientists who have revolutionized the development of electrically conductive polymers. The conductivity is realized through conjugated double bonds along the backbone of the polymer (seen in Fig. 9, Fig. 10) unlike fillers in extrinsically conductive materials [51]. A very low electrical resistance of these polymers can be reached by oxidation. This procedure is known as "doping" and the effects defect electrons on the conjugated polymer chains.

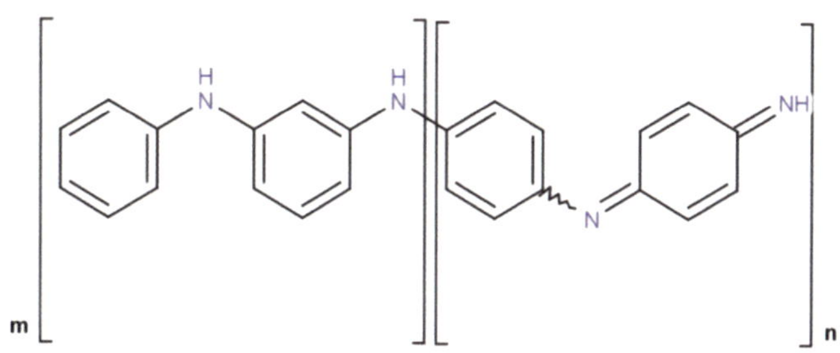

Fig. 11. Chemical structure of Polyaniline (PAni) [7]

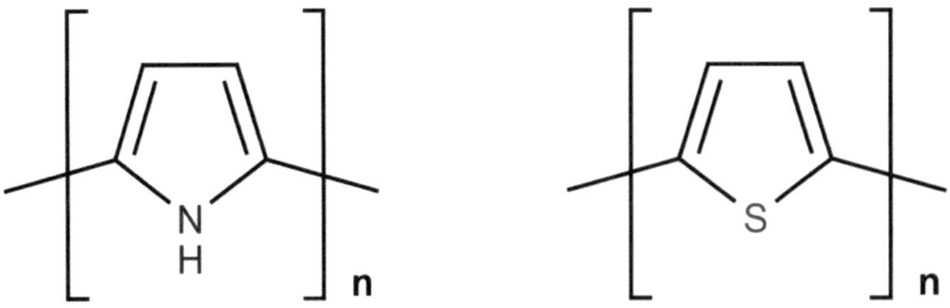

Fig. 12. Chemical structure of Poly(pyrrole) PPy (left), Polythiophene (PTh) (right) [7]

Besides the electrical conductivity and light emission of the polymer the corrosion protection properties were also investigated. Different polymers are qualified to be used as corrosion improvements for steels [46]:

- Poly(pyrrole) PPy
- Polythiophene (PTh)
- Polyaniline (PAni)

Ren et al. [59] reported about the coating of stainless steel plates with polypyrrole and polyaniline. The coatings were brought onto the steel surface by galvanostatic deposition for polypyrrole and by cyclic voltametrie deposition for polyaniline. The PPy were placed first as an inner layer, following covered by the PAni. Electrochemical measurements showed a better corrosion resistance in 0.3 M HCl for the composite coating compared to a single PPy layer. The coatings increased the potential for corrosion (~400 mV) and for pitting corrosion (600 mV) as compared to the bare stainless steel sample. Though the corrosion current was similar to that of the bulk specimen it was rather an indication of the oxidation reduction reaction than the corrosion of the substrate material. The chemical stability for the composite coating was higher than that of the single layer PPy after 36 days exposure to the electrolyte. The contact resistances measurement results for PPy/PAni coating on steel as substrate compared to graphite and uncoated steel is shown in Fig. 11.

Joseph et al. [36] published results comparing PPy and PAni coated stainless steel plates. The coated plates showed improved corrosion resistances in fuel cell environment. Best results for corrosion resistance (~350 mV vs. SCE) were achieved after the third deposition cycle. However with increasing of the thickness of the coating the corrosion potential decreased to ~120 mV vs. SCE. The contact resistance was higher for the PPy and PAni coated steel samples compared to the bulk substrate.

PPy coatings were electro-polymerized by García et al. [25] and the corrosion properties under fuel cell conditions were investigated. The results show that the coatings are capable of reducing the corrosion current. For room temperature tests the results show a decrease of up to two orders of magnitude and for 60 °C operation temperature of even up to four orders of magnitude. Nevertheless the protective properties did not last for longer times of immersion in an electrolyte of 0.1 M H_2SO_4. Subsequent tests showed the incorporation of polymer particles of DBSA doped PPy into the metal oxide film on the substrate surface.

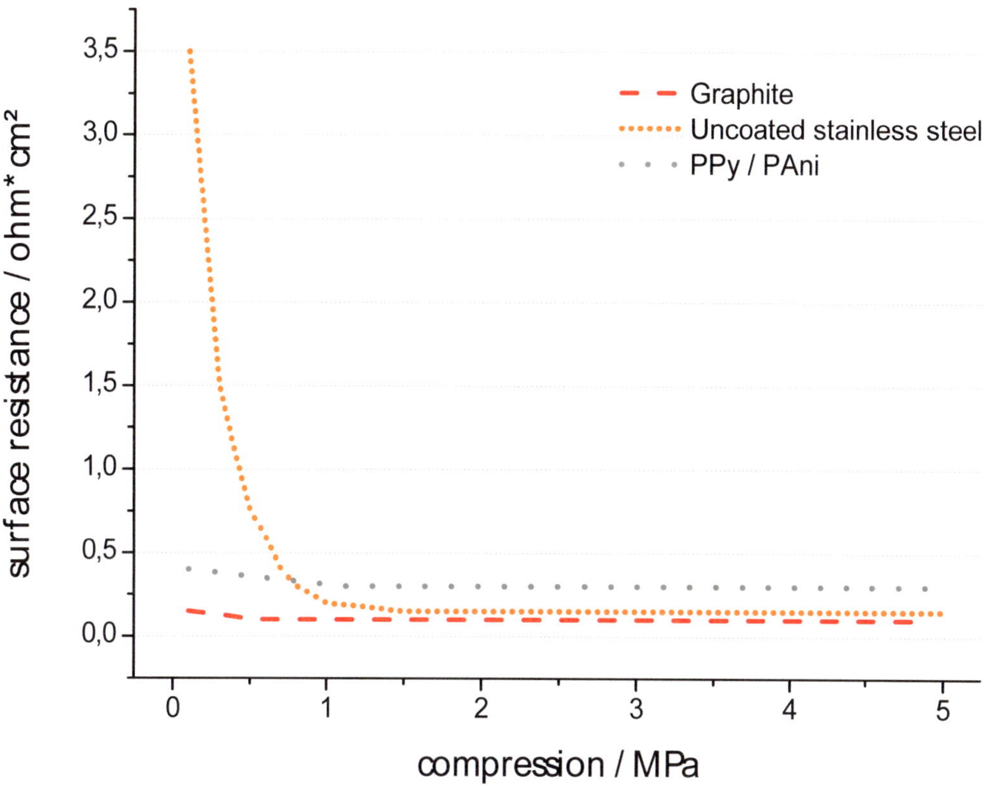

Fig. 13. Contact resistance of the PPy/PAni on stainless steel compared to graphite and uncoated steel [59]

6. Defect detection for coatings

Defects of coatings on metallic bipolar plates can be detected in many different ways. Invasive analytical methods are e.g. the Solid Particle Impact Method, the Colloidal Probe Technique or the Vibration Method [52], [53], [62], [78]. More elegant results are provided by the non-invasive methods such as localized electrochemical impedance spectroscopy (LEIS), the ultrasonic pulse-echo scheme or the scanning kelvin probe. Defects like pinholes as well as high porous coatings are examples for high corrosion currents so they might be detectable directly through an electrochemical corrosion test.

Wooh and Wei reported in 1999 [85] a high-fidelity ultrasonic pulse-echo scheme for detecting delamination in thin films. They stated that the influence of surrounding noises can interfere with the results of the analytical method. Another noninvasive method is shown in the work done by Fürbeth et al. [24] and Wielant et al.[82] - the scanning kelvin probe. This technique has been established to detect electrode potentials at buried polymer/oxide/substrate interfaces. The results by Wielant showed a possible correlation between the polymer/iron-oxide peel-off force and the polar surface energy component [82].

Localized EIS, another approach of detecting defects in coatings, is displayed in the work by Dong et al. [16].The results indicated that the corrosion of coated steel is significantly influenced by cathodic protection potential and the defect geometry. The work also shows a purely capacitive behavior with a phase angle of 90° for a faultless coated steel electrode over most of the measurement frequencies [15]. An LEIS diagram around a defect in a coating after one day of immersion in an alkaline test solution consisting of 0.05 M Na_2CO_3, 0.1 $NaHCO_3$ and 0.1 M NaCl und cathodic potential of -850 mV vs. SCE is shown in Fig. 12.

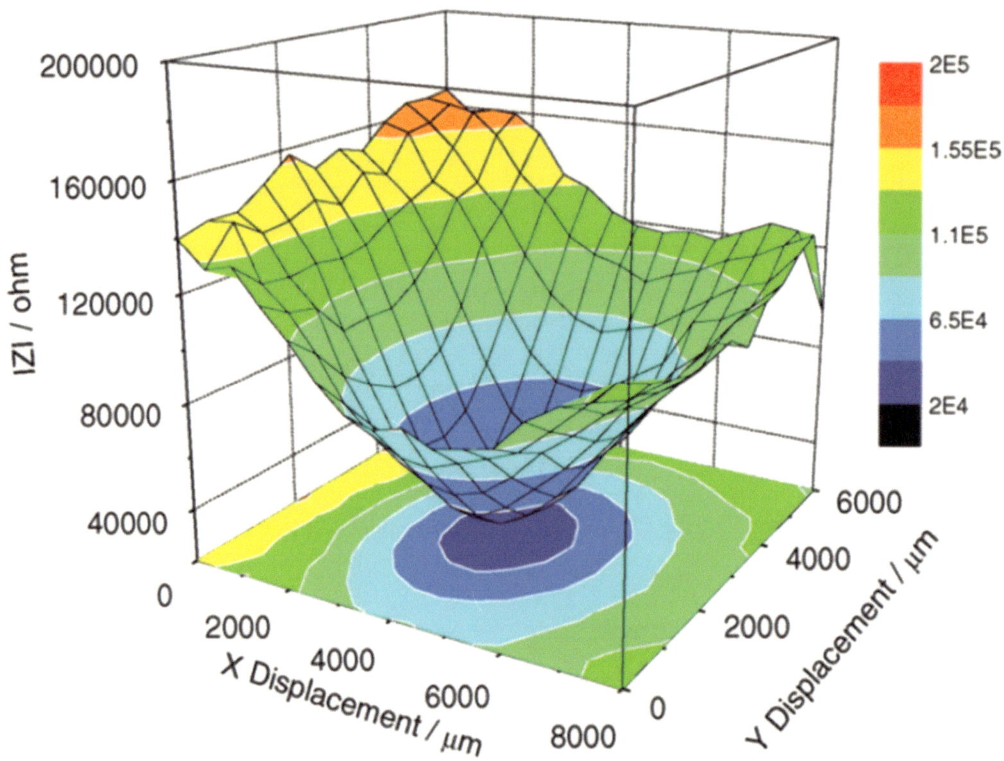

Fig. 14. LEIS maps around the defect [15][5]

7. Summary and conclusion

Metals as bipolar plates in energy conversion systems offer great opportunities. The high corrosion and the interfacial contact resistance hinder their commercialized usage as substrate materials. During the last decades different approaches were reported by researchers worldwide. To this day it is not clear which the optimal treatment for metals is

[5]This article was published in Electrochemica Acta, 54, Dong, C.F. et al., Localized EIS characterization of corrosion of steel at coating defect under cathodic protection, 628-633, Copyright Elsevier (2008)

to resist the harsh environments in an energy conversion system. Many modifications on metals provide good potentials.

Solutions like electro-plating metals or coating them with conductive polymer exhibit similar difficulties by delamination of the coatings. Those defects are mostly noticed through high corrosion currents and degradation of the cells. Modifications on the metal surface through CVD or PVD sometimes show similar challenges for delamination as polymer coatings but they also provide very high corrosion potentials and low contact resistances. Some researcher groups report that the best results are achieved by combining methods like electro-plating and nitration by Wang [74].

For further improvements in surface treatments and for designing new of coatings for metals, noninvasive defect detection methods are a essential research topic. These methods will be able to provide in-situ results for metallic bipolar plates in e.g. fuel cells.

Amorphous metals are possibly the most interesting approach. These bulk metallic glasses provide a good prospect for the use as bipolar plates in energy conversion systems due to their corrosion resistance and manufacturing method. But no long term studies have been made so far.

In former times treating metal alloys to overcome environmental challenges was little more than a niche market. This evolved into a field of greater importance, so that many researchers develop, advance and combine ways to coat metals or even design new metal alloys that exhibit the necessary properties.

8. References

[1] André, J. et al. International Journal of Hydrogen Energy 34 (2009) 3125-3133
[2] André, J. et al., International Journal of Hydrogen Energy 35 (2010) 3684-3697
[3] Antunes, R. A. et al. International Journal of Hydrogen Energy 35 (2010) 3632-3647
[4] Azim, S. et al., Progress in Organic Coatings 55 (2006) 1-4
[5] Bai, C.-Y. et al., International Journal of Hydrogen Energy 34 (2009)
[6] Berns, H; Theisen, W.; Eisenwerkstoffe - Stahl und Gusseisen, 4. Bearbeitete Auflage (2008)
[7] Bhadra, S. et al. Progress in Polymer Science 34 (2009) 783–810
[8] Brady, M. P. et al. Journal of Power Sources 195 (2010) 5610-5618
[9] California Fuel Cell Partnership, DOE Targets, Webpage:
[10] http://www.fuelcellpartnership.org/progress/technology/doetargets
[11] Cha, B.-C. et al. International Journal of hydrogen Energy 36 (2011) 4565-4572
[12] Chung, C.-Y., Journal of Power Sources 186 (2009) 393–398
[13] Datz, A. et al., DE 10 2006 017 604 A1 (2007).
[14] Davies, D.P et al., Journal of Applied Electrochemistry 30 (2000) 101-105
[15] Dong, C.F. et al. Electrochemica Acta 54 (2008) 628-633
[16] Dong, C.F. et al. Progress in Organic Coatings 67 (2010) 269-273
[17] Dur, E. et al. International Journal of hydrogen Energy 36 (2011) 7162 - 7173
[18] El-Enin, S.-A. A. et al. Journal of Power Sources 177 (2008) 131-136
[19] Feng, K. et al. International Journal of hydrogen Energy 35 (2010) 690 - 700

[20] Feng, K. et al., Journal of Power Sources 195 (2010) 6798-6804
[21] Fu, Y. et al. International Journal of hydrogen energy 34 (2009) 405-409
[22] Fu, Y. et al., Journal of Power Sources 182 (2008) 580-584
[23] Fuel Cell Handbook - Seventh Edition (2004)
[24] Fürbeth, W. et al. Corrosion Science 43 (2001) 229-241
[25] García, L. et al. Journal of Power Sources 158 (2006) 397-402
[26] Georges Leclanché, French Patents, No. 69,980 and 71,865, 1866
[27] Grove, W. R.; The London and Edinburgh Philosophical Magazine and Journal of Science - Series 3, 14 (1839) 127-130
[28] Henkel, G., Hinweise zum Passivschichtpänomen bei austensitischen Edelstahllegierungen (2003)
[29] Hental, P. L. et al., Journal of Power Sources 80 (1999) 235-241
[30] Heras, N. et al., Energy & Environmental Science 2 (2009) 206-214
[31] Hermann, A. et al., International Journal of Energy 30 (2005) 1297-1302
[32] Hong, W. et al. International Journal of hydrogen Energy 36 (2011) 2207-2212
[33] Jayaraj, J et al. Material Science and Engineering A 449-451 (2007) 30-33
[34] Jayaraj, J et al., Science and Technology of Advanced Materials 6 (2005) 282-289
[35] Jin, S. et al., Journal of Power Sources 162 (2006) 294-301
[36] Joseph, S. et al. International Journal of Hydrogen Energy 30 (2005) 1339-1344
[37] Journal of Power Sources 163 (2007) 755–767
[38] Kitta, S. et al., Electrochemica Acta 53 (2007) 2025-2033
[39] Kraytsberg, A. et al. Journal of Power Sources 164 (2007) 697-703
[40] Kumagai, M. et al., Electrochemica Acta 54 (2008) 574-581
[41] Kumagai, M. et al., Journal of Power Sources 185 (2008) 815–821
[42] Lee, S.B. et al. Journal of Power Sources 187 (2009) 318-323
[43] Lee, S.-J. et al. Journal of Material Processing Technology 140 (2003) 688-693
[44] Lee, S.-J. et al., Journal of Power Sources 131 (2004) 162-168
[45] Lee, Y. et al., International Journal of Hydrogen Energy 34 (2009) 9781-9787
[46] Léon C.P. de et al. Transaction of the Institute of Metal Finishing 86 (2008) 34-40
[47] Lin, J.-Y. et al., Surface Coatings Technology 205 (2010) 2251-2255
[48] Linden, D.; Reddy, T.; Handbook of Batteries - Third Edition (2002) 1.6
[49] Mattox, D. M., Handbook of Physical Vapor Deposition (PVD) Processing - second edition (2010)
[50] Miller, M; Liaw, P.; Bulk metallic glasses - An overview (2008)
[51] Nobel Prize in Chemistry - Conductive Polymers (2000)
[52] Oka, Y. et al. Second International Conference on Erosive and Abrasive Wear. 258 (2005) 92–99.
[53] Pashley, D. H. Dental Materials 11 (1995) 117–125.
[54] Pierson, H. O; Handbook of chemical vapor deposition (CVD) - Principles, Technology, and Application - second edition (1999)
[55] Planté, R.G., Comptes rendus hebdomadaires des séances de l'Académie des sciences 49 (1859) 402-405
[56] Poligrat GMBH, Webpage:
[57] http://www.poligrat.de/de/Aktuelles/index.htm (2011)

[58] Preißlinger-Schweiger, S. et al. EP 1 923 490 A2 (2007).
[59] Ren, Y.J. et al. Journal of Power Sources 182 (2008) 524-530
[60] Reuter, M.. Nichtrostende Stähle: Arten, Eigenschaften, Wärmebehandlung, Schädigung. Teil 12 (2001).
[61] Richards, J. et al. ECS Transactions 25 (2009) 747-755
[62] Ripperger, S. et al. China Particuology 3 (2005) 3–9
[63] Rüdiger, D et al. Carbon Leitlacke 148 (2004)
[64] Schoenbein, C. F., The London and Edinburgh Philosophical Magazine and Journal of Science - Series. 3, 14 (1839) 43-45
[65] Shao, Y. et al. Journal of Power Sources 167 (2005) 235-242
[66] Show, Y et al., Journal of Power Sources 190 (2009) 322-325
[67] Suryanarayana, C.; Inoue, A.; Bulk Metallic Glasses (2011)
[68] Toops, J. T. et al. Jouranl of Power Sources 195 (2010) 5619-5627
[69] Trappmann, C.; Metallische Bipolarplatten für Direkt-Methanol Brennstoffzellen (2010)
[70] Volta, A; Philosophical Transactions of the Royal Society of London, Vol. 90, pp 403-431, 1800
[71] Walsh, F.C. et al. Surface & Coatings Technology 202 (2008) 5092-5102
[72] Wang, Fuel Cell Project Kickoff Meeting – Presentation (2009)
[73] Wang, H. et al. Journal of Power Sources 138 (2004) 79-85
[74] Wang, H.; Turner, J.; Corrosion Protection of Metallic Bipolar Plates for Fuel Cells (2006)
[75] Wang, H.; Turner, J.; Reviewing Metallic PEMFC Bipolar Plates, (2010)
[76] Wang, J. et al. Surface diffusion modification AISI 304SS stainless steel as bipolar plate material for proton exchange membrane fuel cell, International Journal of hydrogen Energy (2011)
[77] Wang, Y. et al. Journal of Power Sources 191 (2009) 483-488
[78] Wang, Y. M. et al. Applied Surface Science 255 (2009) 6875–6880.
[79] Webpage: http://www.finanzen.net/rohstoffe (2011)
[80] Webpage: http://www.gold-goldbarren.com/goldpreise/goldpreisentwicklung/ (2011)
[81] Wen, T.M. et al., Corrosion Science 52 (2010) 3599-3608
[82] Wielant, J. et al. Corrosion Science 51 (2009) 1664-1670
[83] Wind, J.; Metallic bipolar plates for PEM fuel cells, Journal of Power Sources 105 (2002) 256–260
[84] Wonseok, Y.; Evaluation of coated metallic bipolar plates for PEMFC; Journal of Power Sources 179 (2008) 265–273
[85] Wooh et al. Composites. Part B 30 (1999) 433-441
[86] Worcester Polytechnic Institute, Project Report: Observation of a Polymer Electrolyte Membrane Fuel Cell Degradation Under Dynamic Load Cycling (2009)
[87] Yang, B. et al., Journal of Power Sources 174 (2007) 228-236
[88] Yoon, W. et al., Journal of Power Sources 179 (2008) 265-273
[89] Zhang, J. et al. Journal of Power Sources 160 (2006) 872-891
[90] Zhang, M. et al. Journal of Power Sources 196 (2011) 3249-3254

[91] Ziomek-Moroz, M.; Corrosion behavior of stainless steel in solid oxide fuel cell simulated gaseous environments DOE (2003)

Part 2

New Insight in Microalloyed and Low-Alloy Steels

3

Comments About the Strengthening Mechanisms in Commercial Microalloyed Steels and Reaction Kinetics on Tempering in Low-Alloy Steels

Eduardo Valencia Morales
Department of Physics, Central University of Las Villas, Villa Clara
Cuba

1. Introduction

It is well known that hot-rolled microalloyed steels derive their overall strength from different strengthening mechanisms that simultaneously operate, such as: solid solution strengthening, hardening by the grain size refinement, precipitation strengthening and transformation induced dislocation strengthening [1]. Precipitation of fine carbonitride particles during thermomechanical processing has been used for many years to improve the mechanical properties of the microalloyed steels, where very small amounts (usually below 0.1 wt%) of strong carbide and nitride forming elements such as niobium, titanium and/or vanadium are added for grain refinement and precipitation strengthening. Both grain refinement and precipitation strengthening in microalloyed steels depend upon the formation of fine carbonitride particles, of about 10 nm or less in diameter, which may form in austenite during hot rolling, along the γ/α interface during the austenite to ferrite transformation (interphase precipitation), or as semicoherent particles in ferrite during final cooling. Each one of these basic precipitation modes will lead to its own characteristic particle distribution, and to generally different effects on steel properties [2]. First systematic investigations on microalloyed steels were carried out in the early sixties at the University of Sheffield [3,4], including initial observations of carbonitride particles by transmission electron microscopy (TEM). According to the early literature on niobium steels yield strength contribution of about 100 MNm^{-2}could be obtained in the as rolled condition due to the presence of fine carbonitride particles, which were observable in the TEM [5]. Even larger contributions of up to 200 MNm^{-2} were reported for niobium/vanadium [6] and titanium steels [7]. In principle, these experimental results appeared to be in good agreement with theoretical predictions, based upon the Orowan-Ashby model of precipitation strengthening with carbonitride particles of about 3 nm in diameter [7,8].

The most of the early results cited above were obtained by the observation in TEM of the carbonitride particles but did not determine the origin of the observed carbonitrides [9]. It was only later that electron diffraction methods were employed to distinguish unequivocally between the three modes of carbonitride precipitation [10], and the importance of carbonitride formation in ferrite for the effectiveness of the precipitation strengthening mechanism was generally realized [11]. Today, most authors agree that a significant strengthening effect can only be obtained when carbonitride particles precipitate semicoherently in the ferrite phase [12], and that such precipitation will be particularly effective in the case of hot strip products where a combination of shorter rolling times, higher finishing temperatures, and rapid cooling rates after rolling should cause a larger amount of microalloying elements to remain in solution before coiling [13]. Then, a larger volume fraction of very fine particles will thus be available for a more efficient precipitation strengthening during final cooling of the coil. However, no ferrite-nucleated carbonitride particles were found in commercial Nb, NbTi and NbTiV microalloyed steels processed under industrial conditions on a hot strip mill [9, 14-16]. Besides, it was demonstrated that all precipitation strengthening will be provided by carbonitrides particles which have nucleated in austenite during finish rolling, or by interphase precipitation nucleated during the $\gamma \rightarrow \alpha$ transformation. The contribution of dislocation strengthening has been usually neglected in these hot strip steels because of their polygonal ferrite microstructure. In this sense, relatively high dislocation densities were found in hot strip microalloyed steels with higher carbon and manganese contents, although the microstructure had remained polygonal ferrite + pearlite [15].

It should be realized that many of the results which are presented in the literature have been derived from laboratory tests and processing. The characteristic hot strip processing conditions during finish rolling (high strain rates and short interpass times), however, are difficult to simulate in the laboratory [16]. It is therefore important to study the effects of hot strip rolling in industrially processed materials in order to verify whether real results conform to generally accepted expectations.

2. Origin of carbonitrides and strengthening mechanisms in commercial hot strip microalloyed steels

2.1 Nb microalloyed steel

A first study was carried out on commercial hot strip steel where niobium was the only microalloy element with the following chemical composition: 0.07% C, 0.014% Si, 0.68% Mn, 0.035% Al, 0.04% Nb, 0.0096% N and rest of Fe [14]. The processing parameters of industrial hot strip rolling were: Soaking temperature-1150 °C, finish rolling start temperature-1080 °C, finish rolling end temperature-890 °C, cooling rate-10 °C/s, coiling temperature-650 °C and final thickness 10 mm.

In austenite, at roughing temperature, the carbonitrides nucleate preferentially on the grain boundaries where simultaneously are occurring the recrystallization processes. During the finish rolling at low temperatures also occurs an extensive precipitation of fine carbonitrides on subgrain boundaries, suggesting that they have nucleated in deformed (unrecrystallised) austenite, although the carbonitrides can also choose the γ-γ boundaries as suitable places for nucleation showed in the TEM images by its aligned distribution. Figure 1a shows the extensive precipitation in austenite during the finish rolling (TEM dark field image). The diffraction pattern in this Figure indicates the position of the objective aperture which was used for the dark field illumination of carbonitrides. As can be appreciated, the position of the carbonitride reflection (showed by the aperture objective position) not obeys the Baker-Nutting orientation relationship with respect to the surrounding ferrite, Figure 1b. Thus, the distribution of fine carbonitrides (~10 nm in diameter) decorating the previous subgrain boundaries and not obeying the Baker-Nutting orientation relationship with ferrite suggests that the above precipitation occurred in the austenite phase at finish rolling temperatures where a high plastic deformation has taken place in the microalloyed steel.

(a)

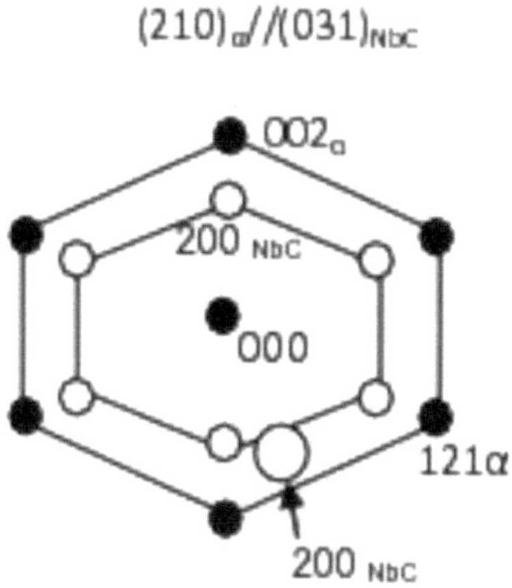

(b)

Fig. 1. (a) Nb(CN) precipitation at the austenite boundary cells during last stages of the hot rolling. (b) Composite diffraction pattern of carbonitride precipitation in ferrite phase showing the nearest Baker-Nutting orientation relationship, indicating (by arrow) objetive aperture position [14].

No carbonitrides were found that could have formed from supersaturated ferrite after the phase transformation. On the other hand, clear evidence for the presence of interphase precipitation in the form of row formation (obeying only one variant of the Baker-Nutting orientation relationship) was detected on a coarse scale (not very different from the precipitation in austenite) in only two grains of the twenty grains carefully observed at TEM, Figure 2. Interphase precipitation has been associated previously with a very high strengthening potential [17,18], generating yield strengths of more than 600 MPa in a high-titanium steels after isothermal transformations [17].

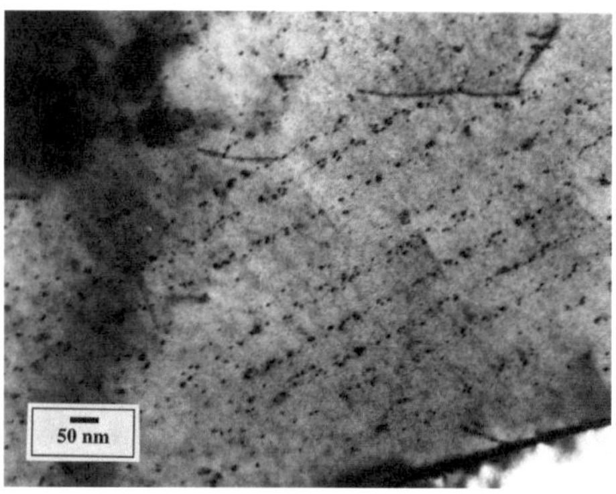

Fig. 2. Interphase precipitation in Nb microalloyed steel[15].

A normalising treatment at 900 °C during 30 minutes was conducted in the as coiled samples in order to verify if very fine precipitation had occurred in ferrite during coiling. In this sense, when very fine carbonitrides precipitate semicoherently in ferrite, a higher yield strength in the as rolled and coiled product is manifested, while particle coarsening and loss of particle coherence should lead to a lower yield strength after normalising. Test results indicated yield strength of 310 MPa and 312 MPa before and after the normalising treatment respectively, confirming the absence of fine scale carbonitride precipitation in ferrite during the final cooling and coiling.

According to the literature, a base value for the yield strength which includes the effects of solid solution and grain size hardening can be determined from the well-known structure-property relationship for low carbon steels, originally developed by Pickering and Gladman [19]:

$$\sigma_y(\mathrm{MPa}) = 15.4\left[3.5 + 2.1(\%\mathrm{Mn}) + 5.4(\%\mathrm{Si}) + 23(\%\mathrm{N_f}) + 1.13d^{-1/2}\right] \quad (1)$$

where the (%Mn), (%Si) and (%N_f) are the weight percentages of manganese, silicon and free nitrogen dissolved in ferrite, and d is the ferrite grain size in millimeters. The weight percent of free nitrogen that remain in solution was calculated for this microalloyed steel. It resulted to be: 0.0025% [14].

Thus, the additional contributions from dislocations and precipitation strengthening can conveniently be estimated by subtracting the base value from the total yield strength as determined by tensile testing. The results are shown in Table 1 for this Nb microalloyed steel.

Treatment of the Nb Steel	Ferrite Grain Size (μm)	Yield Strength Calculated (MPa)	Yield Strength Measured (MPa)	Additional Strengthening (MPa)
Coiled	10.0	252	310	58
Normalized	10.0	252	312	60

Table 1. Comparison between yield strength predictions from equation (1) and the results of tensile testing [14].

The results showed in Table 1 give an additional strengthening contribution of about 60 MPa. According to Gladman et al. [7], the Orowan-Ashby model of precipitation strengthening can be expressed quantitatively (in MPa) as:

$$\Delta s = \left[\left(10.8f^{1/2}\right)/D\right]\ln(1630D) \quad (2)$$

where $\Delta\sigma$ represent the precipitation strengthening increment in MPa, f is the precipitate volume fraction and D the mean particle diameter in micrometers. The application of equation (2) to the austenite precipitation gave as result an additional strengthening increment of about 65 MPa. This value seems to agree with the above difference between the measured and calculated yield strength. Quantitative estimates from several grains where the foil thickness has been measured by counting the number of grain boundary fringes under two-beam contrast conditions indicated an average dislocation density of about 10^8 cm^{-2} for this Nb microalloyed steel. According to the early literature [20], dislocation strengthening can be quantified by:

$$\Delta\sigma = m\alpha\mu b\rho^{1/2} \quad (3)$$

where $\Delta\sigma$ is the dislocation contribution to yield strength, m the appropriate Taylor factor for polycrystals, α a geometrical factor that depends upon the type of dislocation interaction, μ the shear modulus (82,300 MPa for ferrite), **b** the dislocation Burgers vector (0.25 nm in ferrite), and ρ the measured dislocation density. A value of $m\alpha$= 0.38 has been determined experimentally for pure iron [21]. Alternatively, theoretical values of m= 2.733 for bcc crystals [22] and of $\alpha=1/2\pi$ for dislocation forest cutting [23] would give a slightly higher estimate of $m\alpha$=0.435. Selecting $m\alpha$=0.4 as an intermediate value [15], a dislocation density of 10^8 cm^{-2} would contribute with 8 MPa to the strength of the Nb microalloyed steel, which would be considered negligible. As the interphase precipitation has only occurred in a very small fraction of the grains (two of the twenty) and in a coarse scale, their influence on the yield strength is also negligible [14].

2.2 NbTi microalloyed steel

New results were obtained from another commercial NbTi microalloyed hot strip steel [15], which reached a yield strength of 534 MPa and, according to expectations, lost part of that strength during normalizing. A Nb steel, (above referred) which only reached 310 MPa and maintained that strength after normalizing, was used as a reference material. Chemical compositions and industrial processing conditions of this NbTi steel are shown in Tables 2 and 3. Part of the material was normalized at 900 °C for 30 minutes. Optical and electron microscopy were used to study the microstructure, and yield strength values before and after normalizing were determined as the average of five tensile tests.

Steel	C	Mn	Si	P	S	Al	Nb	Ti	N
NbTi	0.12	1.21	0.33	0.023	0.008	0.048	0.057	0.059	0.008

Table 2. Chemical Composition of Hot Strip NbTi Steel in Weight Percent [15]

Steel	Soaking Temperature	Finish Rolling Start	Finish Rolling End	Cooling Rate	Coiling Temperature	Final Thicknees
NbTi	1150 °C	1079 °C	870 °C	10 °C/s	650 °C	7 mm

Table 3. Processing Parameters of Industrial Hot Strip Rolling [15]

The NbTi steel exhibited the smaller ferrite grain size than the Nb steel, presumably due to its higher carbon and manganese contents, which should have decreased the transformation temperature. Quantitative metallography, yield strength measurements, and structure-property relationships were used for a quantitative estimate of different strengthening contributions [15]. To begin with, the well-known empirical equation (1) served to calculate the contributions from chemical composition and ferrite grain size. Yield strength predictions from Eq. (1) are compared to tensile test results in Table 4. The difference between calculated and measured strength is usually attributed to some additional strengthening mechanism such as carbonitride precipitation or substructure strengthening. The important point in Table 4 is the very large additional strengthening of 177 MPa in the case of the NbTi steel, which was reduced to 69 MPa after normalizing [15]. As mentioned previously, normalizing did not reduce the yield strength of the Nb steel, and the additional strengthening contribution in this steel remained at around 60 MPa, a level very close to the 69 MPa exhibited by the NbTi steel after normalizing.

Treatment of the NbTi Steel	Ferrite Grain Size (μm)	Yield Strength Calculated (MPa)	Yield Strength Measured (MPa)	Additional Strengthening (MPa)
Coiled	5.0	367	534	177
Normalized	5.5	357	426	69

Table 4. Comparison between yield strength predictions from equation (1) and the results of tensile testing, [15].

Fine carbonitride precipitation was identified in all of the observed grains (twenty) at TEM, but orientation relationships determined from electron diffraction showed that these particles had nucleated in austenite [15]. In addition, carbonitride distributions appeared to be very similar to the distributions observed in a previous investigated steel [9]. In that case, quantitative metallography and the application of the Orowan-Ashby model of precipitation strengthening had indicated a strengthening contribution of about 60 to 80 MPa for carbonitride particles formed in austenite, in good agreement with the additional strengthening shown in Table 1 for the Nb steel and also for the NbTi steel after normalizing, as it is shown in Table 4.

As in the previous investigations,[9,14] no carbonitrides were found that could have formed from supersaturated ferrite after the phase transformation. On the other hand, clear evidence for the presence of interphase precipitation in the form of row formation [15] was detected in this steel (Figure 3). It is apparent; from a comparison between Figures 2 and 3 that interphase precipitation occurred on a much finer scale and thus should have contributed to strength in the case of the NbTi steel. However, interphase precipitation seemed to occur not very frequently because it was also encountered in only two grains in this steel. On the other hand, the visibility of TEM diffraction contrast from very fine carbonitride particles may require closely controlled sample orientations, which may not have been established in all the ferrite grains under observation for this NbTi microalloyed steel [15].

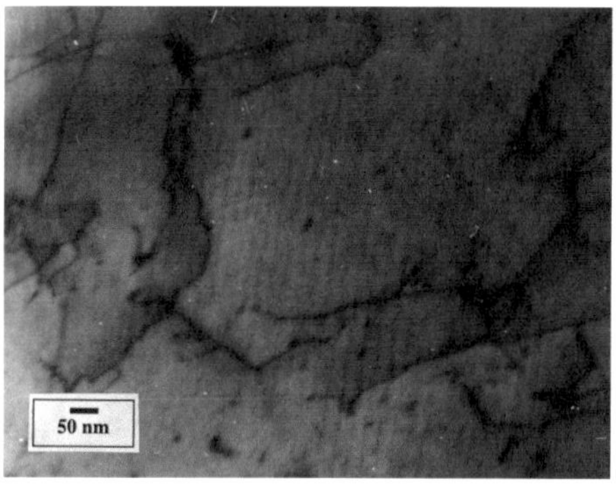

Fig. 3. Interphase precipitation in microalloyed steel[15].

As a first approximation, if it is assumed that about one-half of the total microalloy addition would be available for fine-scale carbonitride precipitation during thermo-mechanical processing [9,15] and applying the Orowan–Ashby model, a maximum strengthening contribution of 195 MPa could be predicted for the interphase precipitation of the NbTi steel shown in Figure 3, based upon a particle size of 2.0 nm and a volume fraction of $8x10^{-4}$. If interphase precipitation had occurred in only 25 pct of the ferrite grains, a simple rule of mixtures would thus suggest a strengthening contribution of about 50 MPa in the case of the NbTi steel [15].

Another strengthening contribution could come from the presence of dislocations. In fact, dislocation densities were always higher in the NbTi steel, which again can be explained by its lower transformation temperature in comparison with the Nb steel. Quantitative estimates from several grains where the foil thickness had been measured by counting the number of grain boundary fringes under two-beam contrast conditions indicated an average dislocation density of $5x10^9$ cm^{-2} for the NbTi steel [15]. Such numbers are in reasonable agreement with previous measurements of dislocation densities in microalloyed steels [24] and confirm the possibility of a transformation-induced dislocation substructure even in the case of polygonal ferrite grains.

According to Eq. (3), a strengthening contribution of 58 MPa due to the dislocation substructure of the NbTi microalloyed steel was obtained. Thus, in the case of the NbTi steel, the individual strengthening contributions from general precipitation in austenite (~70 MPa), localized interphase precipitation (~50 MPa), and dislocation substructure (~60 MPa) would add to a total of 180 MPa, a value which compares very favorably with the additional strengthening contribution found for the NbTi steel [15] in Table 4. During normalising, coarsening of fine interphase precipitate distributions and elimination of the dislocation substructure (which would not form again during air cooling due to a higher transformation temperature) can be expected to reduce the strengthening level to the contribution of austenite precipitation alone (69 MPa according to Table 4). This later strengthening contribution should survive the effect of normalizing because the formation of carbonitride particles during finish rolling occurs within the range of typical normalizing temperatures [15]. In the case of the previous Nb steel, a higher transformation temperature before coiling would leave the carbonitride precipitation in austenite as the only strengthening mechanism before and after normalizing.

2.3 Another NbTi and NbTiV microalloyed steels

Strength and microstructures of three new commercial microalloyed steels were investigated as a function of their chemical compositions. They were compared with the above two microalloyed steels [16]. As a common feature, all five steels had been hot rolled under similar thermomechanical processing conditions on an industrial hot strip mill, and each of them exhibited a polygonal ferrite+ pearlite microstructure. In contrast, carbon and manganese contents ranged from 0.05 wt.% to 0.14 wt.% C and from 0.5 wt % to 1.5 wt.% Mn, respectively, and microalloyed additions included pure Nb, Nb-Ti, and Nb-Ti-V combinations. Chemical compositions are shown in Table 5, together with maximum total volume fractions ($V_{f\ max}$) for carbonitride precipitation which were calculated assuming appropriated lattice parameters of 0.445, 0.430 and 0.415 nm for the fcc unit cell of niobium, titanium and vanadium carbonitrides, respectively [16].

Steel	C	Mn	Si	P	S	Al	Nb	Ti	V	N	$V_{f\ max}$
Nb	0.07	0.68	0.01	0.012	0.009	0.04	0.04	-	-	0.009	0.00045
NbTi-1	0.05	0.55	0.02	n.d.	n.d.	0.02	0.02	0.06	-	0.006	0.00131
NbTi-2	0.12	1.21	0.33	0.023	0.008	0.048	0.057	0.059	-	0.008	0.00166
NbTi-3	0.11	1.54	0.28	0.026	0.007	0.01	0.04	0.11	-	n.d.	0.00261
NbTiV	0.14	1.38	0.25	0.018	0.007	0.07	0.04	0.04	0.03	0.008	0.00174

Table 5. Steel compositions (wt %) and maximum total volume fraction for carbonitride precipitation [16].

Thermomechanical processing conditions are given in Table 6, confirming rather similar processing parameters for all the steels, with the exception of steel NbTiV which was rolled to smaller thickness of 3 mm. The last column in Table 6 shows the yield strength after coiling. Two distinct strength levels can be recognized: Low yield strength values in the range of 300 MPa for steels Nb and NbTi-1, and significantly higher yield strength values in the range of 500 to 600 MPa for steels NbTi-2, NbTi-3 and NbTiV [16].

Steel	Soaking /°C	Roughing /°C	Finishing /°C	Cooling /°C*min^{-1}	Coiling /°C	Thickness /mm	Y. S. MPa
Nb	1150	≥1080	890	10	650	10	310
NbTi-1	1230	≥1100	870	20	630	8	332
NbTi-2	1150	≥1070	870	10	650	7	534
NbTi-3	1225	≥1100	895	10	650	8	638
NbTiV	1225	≥1100	895	10	670	3	599

Table 6. Thermomechanical processing conditions and yield strength after coiling [16].

As it is shown in [16], all steels had transformed to polygonal ferrite+ pearlite. The ferrite grain size decreased from about 10 μm (average diameter) for low strength alloys Nb and NbTi-1, to 5 μm and below for high strength steels NbTi-2, NbTi-3 and NbTiV. Such grain refinement can be related to lower transformation temperatures caused by larger carbon and manganese additions to higher strength materials. Table 7 shows the additional contributions from dislocations and precipitation strengthening by subtracting the base value obtained by Eq. (1) from the total yield strength as determined by tensile testing[16]. The results showed in Table 7, ranging from low additional strengthening contributions below 100 MPa for steels Nb and NbTi-1 to much larger additional strengthening contributions between 150 and 250 MPa for steels NbTi-2, NbTi-3 and NbTiV.

Very low dislocations densities were found in the low strength alloy, with quantitative estimates remaining at about 10^8 cm^{-2} which would be typical value for well annealed ferrite steel. On the other hand, distinctly higher dislocation densities in the range of 10^9 to 10^{10} cm^{-2} were encountered in the high strength steels [16]. Such an increase in dislocation density may also be related to lower transformation temperatures, and a sizeable contribution to yield strength may thus be expected to come from transformation-induced dislocations even in the case of polygonal ferrite microstructures. According to the Keh equation [15, 25], this contribution could reach about 50 MPa for dislocation densities in the range of $5x10^9$ cm^{-2}.

In the above steels, two different modes of fine carbonitride precipitation were detected in the as-coiled samples: Precipitation on the deformation-induced dislocation substructure in austenite, and interphase precipitation where carbonitrides had nucleated on the $\gamma \rightarrow \alpha$ interface during transformation. Carbonitride precipitation in austenite was identified by electron diffraction and was found to be present in all the grains investigated. Mean particle diameters were observed to increase in proportion with the maximum theoretical precipitate volume fraction [16].

Steel	Grain Size (μm)	σ (MPa)	Δσ(MPa
Nb	10.0	252	58
NbTi-1	9.4	254	78
NbTi-2	5.0	367	167
NbTi-3	4.2	393	245
NbTiV	3.3	421	178

Table 7. Calculation of base yield strength, σ, and of additional strengthening from dislocations and precipitation, Δσ [16].

Interphase precipitation was detected in only a small number of grains, but in most of these observations was recognized through row formation. This mode of carbonitride precipitation may have occurred in other grains as well. Preliminary measurements indicated that mean particle diameters of about 2 nm were associated with the smaller sheet spacing, but reached 5 nm in other samples where the sheet spacing were larger. On one occasion, both larger and smaller sheet spacings were present in the same ferrite grain which probably had transformed during cooling through an extended temperature interval [16]. Quantitative estimates of interphase particles volume fraction gave $3.5x10^{-4}$ in NbTi-2 steel, $4.9x10^{-4}$ in Nb steel, and $7.8x10^{-4}$ in NbTi-3 steel [16].

An Orowan-Ashby analysis showed strengthening contributions of about 60 to 100 MPa from particle volume fractions in austenite in the range of 10^{-4} as a function of the particle diameter [16]. Local strengthening contributions from practical interphase precipitation phenomena would reach 110 to 180 MPa. But it must be remembered that interphase precipitation does not seem to occur in all the ferrite grains, so that its effective contribution would be reduced through some sort of rule of mixture. It thus appears that carbonitrides nucleated in austenite do make a sizeable contribution to the steel's yield strength [16].

For a more realistic estimate of the strengthening potential of interphase precipitation in commercial microalloyed steels, it is therefore important to find out more about its heterogeneous particle distributions. This has been the principal objective of the investigation showed in [26]. In order to investigate the influence on the overall strengthening in hot strip microalloyed steels due to the interphase precipitation, three commercial hot strip steels (Nb steel, NbTi-2 and NbTi-3 steels) containing different additions of niobium and titanium were selected [26].

Steel selection was based on the following arguments:

1. Different levels of microalloy additions were expected to vary the total amount of carbonitride precipitation. In particular, real volume fractions and the average particle size of interphase carbonitrides were expected to increase for larger values of V_{fmax}.
2. Different base compositions, with particular attention to carbon and manganese contents, were supposed to modify the transformation temperature and, as a consequence, to change mean spacings of the interphase precipitation sheets [27].
3. Thermomechanical processing conditions were desired to be similar, as it is shown in Table 6.

The difference between predicted values, σ_y, using Eq. (1), and experimental data derived from tensile testing can then be related quantitatively to the presence of additional strengthening mechanisms, as shown in Table 8.

Processing Conditions	Steel	Ferrite Grain Size (μm)	Yield Strength Calculated (MPa)	Yield Strength Measured (MPa)	Additional Strengthening (MPa)
As coiled	Nb	10.0	252	310	58
	NbTi-2	5.0	367	534	167
	NbTi-3	4.2	393	603	210
Normalised	Nb	10.0	252	312	60
	NbTi-2	5.5	357	426	69
	NbTi-3	4.5	386	462	76

Table 8. Yield strength and additional strengthening contributions from equation (1) in the as coiled and after normalising conditions [26].

Several points should be emphasized:

1. A significant part of the differences in yield strength for the as rolled condition was caused by additional strengthening mechanisms (see last column in Table 8).
2. The normalising treatment drastically reduced the additional strengthening contributions in steels NbTi-2 and NbTi-3, but not in steel Nb.
3. After normalising, additional strengthening contributions were similar for all three steels. As expected from previous studies[15,16], additional strengthening in the present

case should have come from carbonitride particles nucleated in austenite, carbonitride particles formed by interphase precipitation, and from dislocations introduced by the $\gamma \rightarrow \alpha$ transformation [26].

The microhardness measurements carried out on individual ferrite grains with the aim of determining the percentage of grains with and without interphase precipitation [26] show some aspects that should be emphasized, Figure 4:

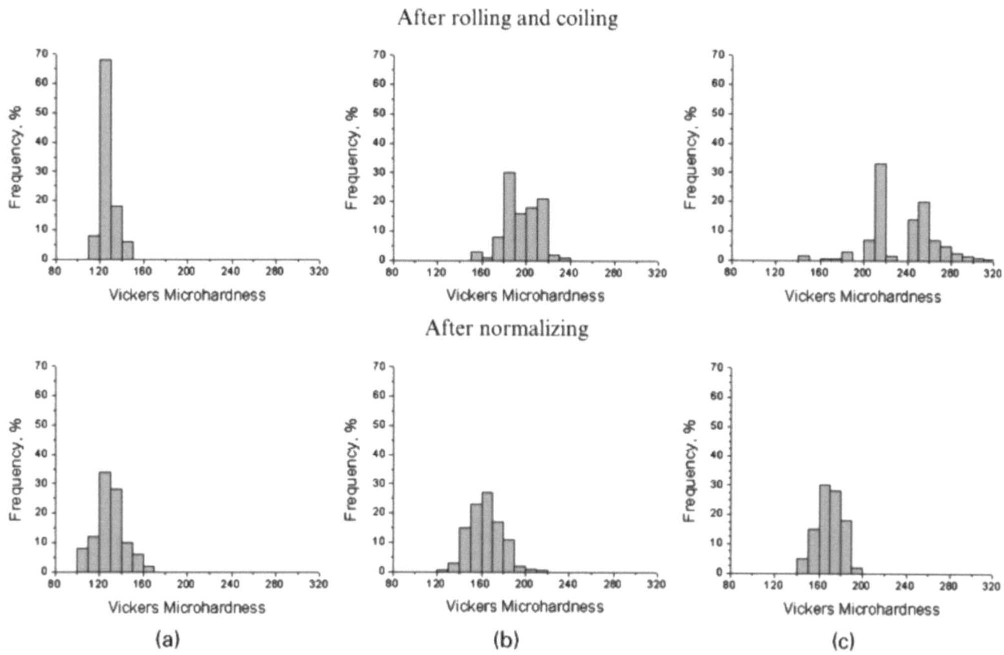

Fig. 4. Vickers microhardness from 200 individual ferrite grains for the as coiled condition and after normalising. a) Nb steel, b) NbTi-2 steel, c) NbTi-3 steel [26].

1. Two separate peaks appeared for both NbTi steels in the as coiled condition, with a very distinct peak separation in the case of steel NbTi-3, (about of 40 MPa).
2. For both NbTi steels, the second hardness peak disappeared after normalising, meaning that the regions of higher hardness lost their additional strengthening after spending 30 min at 900° C and being retransformed to ferrite at a lower cooling rate (in comparison with water spray cooling after rolling).
3. Ferrite grains in steel Nb did not show a second hardness peak in the as coiled condition, and their medium hardness values were not affected by normalising.
4. In case of the NbTi steels, normalising removed not only the second (higher) hardness peak but also reduced the level of the lower hardness peak.

As shown in [26], these findings are consistent with different degrees of precipitation and dislocation strengthening in the as rolled condition, and with the effects of particle coarsening and dislocation removal due to normalising.

A detailed TEM investigation was carried out on steel NbTi-3 in order to evaluate the role of precipitation hardening in both as rolled and normalised conditions [26]. Confirming the results of our previous studies on commercial hot strip steels [9, 15, 16], all fine carbonitride particles had either formed in austenite during rolling or on γ/α phase boundaries during transformation. No additional carbonitride populations were found that could have formed in supersaturated ferrite, despite the unusually large microalloy addition of 0.04%Nb + 0.11%Ti. On the other hand, interphase carbonitrides were relatively large and observed frequently, indicating that a substantial fraction of the microalloy addition had remained in solution at the time of transformation. Furthermore, some form of austenite precipitation was encountered in all the ferrite grains that were investigated. Interphase precipitation was present in only some of the ferrite grains. Occasionally, grains were found to be covered completely by carbonitrides in a row formation. On other occasions, interphase precipitation would occupy only parts of a particular ferrite grain. The presence of interphase precipitation in only part of a given ferrite grain must therefore be accepted as a real phenomenon [26].

During a TEM study of a large number of grains, of which more than a hundred exhibited interphase precipitation, it was found that random particle distributions were dominant only when thin foils had been prepared parallel to the rolling plane, while row formation was encountered very frequently in longitudinal and transverse sections. Such observations can only be explained by a preferred alignment of the interphase precipitation sheets parallel to the rolling plane.

As a result of this detailed analysis, interphase precipitation was identified in 27 out of a total of 51 ferrite grains that were investigated. Thus, about one half of the grains in steel NbTi-3 should have been strengthened by interphase precipitation [26].

The effects of normalising on the mechanical properties as shown in Table 8, suggest that important changes occurred during this heat treatment with respect to precipitation and/or dislocation hardening. The first important observation, therefore, was that many carbonitrides continued to decorate typical deformation subgrain structures. It can thus be concluded that the normalising treatment did not have a major effect on carbonitride distributions that had been formed in austenite, although the average particle size was increased. The second important observation was that row formation could no longer be detected after normalising. This means that extensive particle coarsening must have occurred during normalising, including the transfer of microalloy atoms from dissolving particles in one of the original interphase precipitation sheets to growing particles located in another sheet [26]. Another important result of normalising was the reduction in dislocation density.

Quantification of local strengthening contributions in the NbTi-3 steel showed two aspects that should be emphasised in those grains strengthened by both austenite and interphase precipitation, Table 9. First, the total carbonitride volume fraction after normalising ($13.0\text{x}10^{-4}$) was not very far from the combined volume fraction of austenite +interphase precipitation after coiling ($6.2 + 5.0 = 11.2\text{x}10^{-4}$), confirming the previous interpretation that, after normalising should have included the coarsened interphase particles. Second, the total level of precipitation strengthening for the as rolled and coiled condition was calculated by using a new average particle size (5.2 nm) determined for both the austenite and interphase precipitate populations [26].

Sample condition	Precipitation strengthening from Eq. (2)				Dislocation hardening from Eq(3)	
	Particle origin	D, nm	$fx10^{-4}$	$\Delta\sigma$, MPa	ρ, cm^{-2}	$\Delta\sigma$, MPa
As rolled after coiling	Austenite	7.2	6.2	92	$5.2x10^{9}$	64
	Interphase	4.0	5.0	112		
	Total	5.2	11.2	145		
After normalising	Austenite	12.0	6.9	70	10^{8}	<10
	Total	11.0	13.0	102		

Table 9. Substructural strengthening contributions in steel NbTi-3 [26].

It can be seen from Table 9, that the largest individual contribution was associated with the interphase precipitation mode, yielding average values of 112 MPa, against 92 MPa for carbonitride precipitation in austenite and 64 MPa for dislocation hardening. For

the overall strength of the steel, it is claimed in [26] that the interphase precipitation should be less effective, because it occurs only in some fraction of the ferrite grains. In addition, moving dislocations do not distinguish between the origin of the carbonitride obstacles that have to be overcome by Orowan bowing. In the presence of other carbonitride particles nucleated in austenite, therefore, the total contribution of precipitation strengthening will not amount to 92+112=204 MPa but to only 145 MPa as shown in Table 9. As a result, the local contribution of interphase precipitation to the strength of those particular ferrite grains (about 50% in steel NbTi-3) would only be 112(145/204)=79.5 MPa, against a local contribution of 92 (145/204)=65.5 MPa from austenite precipitation in the same ferrite grains [26].

Accepting the idea that austenite precipitation during rolling and dislocation generation during transformation occurred throughout the material whereas interphase precipitation was present in only 50% of the ferrite grains, considering, in addition, that precipitation and dislocation hardening would act independently and adopting a simple rule of mixture for the effects of grains with and without interphase precipitation, the final contributions of the additional 'substructural' mechanisms to the yield strength of steel NbTi-3 may be written (see Table 9) as $\Delta\sigma_y$=0.5(92+64)+ 0.5(145+64)=182.5 MPa, in reasonable agreement with the additional strengthening of 210 MPa derived from tensile testing and the generally accepted structure-property relationship (Eq. (1)) [26] (see Table 8). Following the same lines of argument [26], we would expect steel NbTi-3 to reach an additional 'substructural' strengthening contribution of $\Delta\sigma_y$=92+64=156 MPa without interphase precipitation. The difference, of only 182.5-156=26.5 MPa, would indicate a rather modest contribution of the interphase precipitation mode to the strength of steel NbTi-3.

Microhardness measurements can be used in principle to detect additional strengthening mechanisms which operate in only part of the ferrite grains. In this sense, it is shown in [26] that for an estimated peak separation of 40 HV (Figure 4), the additional strengthening mechanism in 53% of the ferrite grains would have contributed with an average of 99 MPa, a number that is not very far from the 79.5 MPa contribution of interphase precipitation strengthening derived from the TEM observations.

3. New kinetic approaches applied to reactions during tempering in low-alloy steels

3.1 Isothermal tempering

3.1.1 Introduction

The precipitation reactions which occur on tempering of low-alloy steels may all be classified as nucleation and growth transformations [28]. Extensive studies have been carried out to understand and to model the mechanisms that take place during the tempering of steels. Although it is well accepted that models based on physical principles rather than empirical data fitting give a better understanding of the individual mechanisms which occur on tempering, these models do not contemplate the complexity with which the reactions proceed in each situation T-t. In this sense, many models do not consider the overlapping of precipitation processes of different chemical natures [29, 30], and when they are taken into account, specific nucleation rates are assessed to fit the entire experimental data without considering the change that the nucleation rate could have during the progress of the reaction [31]. Another situation that commonly occurs when fitting the models, such as Johnson-Mehl-Avrami- Kolmogorov (JMAK)-like models, to the entire curve of the fraction transformed ($\xi(t)$) vs. lnt, is to discard the experimental uncertainties in the determination of the fraction transformed [31, 32]. Fitting such models to the entire experimental curve of the fraction transformed could therefore, in certain circumstances result in the prediction of unrealistic kinetic parameters [33].

In other works some attempts have been made [34-37] to deconvolute such experimental master curves into components due to individual processes, but it is difficult to see whether some of these fitting parameters have any physical meaning.

Recently, a general modular model for both isothermal and isochronal kinetics of phase transformations in solid state has been published [38, 39]. This model incorporates a choice of nucleation (nucleation of mixed nature) and growth mechanisms, as well as impingement. Also, the JMAK formulation has been deeply modified to suit isochronal case [40-42], but these analytical approaches need of the nucleation protocols in order to provide a suitable description of phase transformation kinetics during both isothermal and isochronal heat treatments.

In the following, an overview is given about the kinetic theory of overlapping phase transformations (KTOPT) [43] which is based on the Avrami model. This new approach permits the determination of the kinetic parameters (n, k) for simultaneous diffusion-controlled precipitation reactions based on the knowledge of a specific macroscopic parameter P(t), chosen to study the ongoing reaction. The present approach does not need to assume nucleation and growth protocols in its formulation to fit the experimental data. This new approach [43] has the particularity of calculating the kinetic parameters in defined work intervals of the fraction transformed curve rather than for the entire curve where the overlapping effect is present. Furthermore, these work intervals are distant from the boundary points $\xi=0$ and $\xi=1$in order to minimize the errors [44].

3.1.2 Fundamentals of the kinetic theory of two overlapping processes

In the kinetic theory for two overlapping precipitation processes [43] (in isothermal regime), the real fraction transformed is defined in function of the macroscopic parameter P(t) as:

$$\xi_r(t) = \frac{\Delta P(t)}{\Delta P(t_{end1})} \tag{4}$$

where t_{end1} is the time to complete of the first process. t_{end1} is chosen instead of t_{end2} because we thus take into account the effect of the first process during the time interval $t < t_{end1}$ when there is appreciable influence of the second process on the experimental parameter.

It is considered that variations of P(t) associated with each process are independent:

$$\Delta P(t) = \Delta P_1(t) + \Delta P_2(t) \tag{5}$$

then, the real fraction transformed is written as:

$$\xi_r(t) = \frac{\Delta P_1(t) + \Delta P_2(t)}{\Delta P_1(t_{end1}) + \Delta P_2(t_{end1})} \tag{6}$$

As the fraction transformed associated which each elementary process obeys a JMAK kinetic equation, Eq. (6) can be written:

$$\xi_r(t) = \frac{\xi_1(t) + \alpha_p \xi_2(t)}{1 + \alpha_p \xi_2(t_{end1})} \tag{7}$$

where

$$\alpha_p = \frac{\Delta P_2(t_{end2})}{\Delta P_1(t_{end1})} \tag{8}$$

The magnitude of $\xi_2(t_{end1})$ measures the degree of overlap (large or small) of the processes. Depending on the choice of P(t), the parameter α_p may be either positive or negative. When the variation of this parameter, with time, for one of the precipitation processes increases while for the other it decreases $\alpha_p<0$ but when P(t) changes in the same sense, $\alpha_p>0$.

The experimental determination of $\xi_r(t)$ would be possible if we would be able to measure the parameter P(t) from the beginning of the phase transformations (in an isothermal regime). However, the sample takes a certain time to reach the temperature of the isothermal treatment. During this small time interval, the sample is already undergoing heat treatment, so the beginning of the transformations may be prior to that of temperature stabilization corresponding to the isothermal regime, and therefore; the real initiation of the precipitation processes is unknown.

Let us consider two processes that proceed in the same sense during the isothermal treatment ($\alpha_p >0$) [33]. Thus, it is necessary to begin the study not from the origin of the data obtained from the measuring equipment but from the moment of time when the isothermal regime is reached. If we take the length of the sample as the macroscopic parameter, then $l'(0)$ will be the initial length of the sample ($l'(0)= l(t_0)$) [43]. Thus, we may compute the fraction transformed by:

$$\xi'(t') = \frac{p'(t') - p'(0)}{p'(t'_{end1}) - p'(0)} = \frac{l'(t') - l'(0)}{l'(t'_{end1}) - l'(0)} \tag{9}$$

Since ξ_r(t) cannot be obtained directly from experiment (we do not known the actual origin in time of the transformations), a relation between the fraction transformed $\xi'(t')$, (that is experimentally measurable) and ξ_r(t) is found:

$$\xi(t) = \xi'(t') = \frac{\xi_r(t)}{1-\xi_r(t_0)} - \frac{\xi_r(t_0)}{1-\xi_r(t_0)} \tag{10}$$

and a general expression correlating $\xi'(t')$ with the fractions transformed for the individual processes (obeying a JMAK expression) is obtained:

$$(1-\xi_r(t_0))(1+\alpha_l \xi_2'(t'_{end1}))\frac{d\xi'(t')}{d\ln t'} = n_1 Z(\xi_1) + \alpha_l n_2 Z(\xi_2) \tag{11}$$

where

$$Z(\xi) = \left[\frac{d(\ln\ln\frac{1}{1-\xi})}{d\xi}\right]^{-1} = -(1-\xi)\ln(1-\xi) \tag{12}$$

Our approach focusses on the behavior of the function Z(ξ) vs. ξ [33, 43]. This function is not symmetrical with reference to its maximum at ξ=0.632. It increases as ξ increase initially, but when ξ→1 it decreases very rapidly. By contrast Z(ξ) is nearly constant for values of ξ in the neighborhood of the point where this function is a maximum. In other words, we have an interval where ξ takes values (far from the boundary points ξ=0 and ξ=1) for which Z(ξ) depends weakly on ξ. If we allow time intervals where ξ_1 (t) and ξ_2(t) lie far enough from the boundary points ξ=0 and ξ=1, then we may consider that Z(ξ_1) and Z(ξ_2) are nearly constants.

In order to develop the equations for calculating the kinetic parameters for both processes, it should be kept in mind that α_1>0, (processes that progress in the same sense). Thus, considering the definition of the fraction transformed, Eq.(9), the $\xi'(t')>1$ for $t'>t'_{end1}$. For the time interval ($t'>t'_{end1}$), the second process (obeying a JMAK-type relation) develops alone, and therefore; the fraction transformed ξ' (t') should be normalized as:

$$\xi''(t') = \frac{\xi'(t') - \xi'(t'_{end2})}{1-\xi'(t'_{end2})} = p\xi'(t') - W \tag{13}$$

where at $t'=t'_{end1}$ the $\xi'(t'_{end1})=\xi''(t'_{end1})=1$ and for $t'=t'_{end2}$, $\xi''(t'_{end2})=0$.

According to reasoning followed in [33], for times $t'_{end1}<t'<t'_{end2}$, the second process develops alone in this time interval, therefore $\xi_2'(t')$ increases while $\xi''(t')$ decreases (for the case where α_1>0). Thus, it is possible to find a time interval $[t_\alpha', t_\beta']$ where the points $[Z(\xi''(t')),\xi''(t')]$ and $[Z(\xi_2'(t')),\xi_2'(t')]$ are symmetrically located about the maximum of the Z(ξ) function (stability region). This enables to consider that $Z(\xi_2'(t')) \approx Z(\xi''(t'))$ for the above time interval and Eq.(11) can be written as:

$$(1-\xi_r(t_0))(1+\alpha_l\xi_2'(t'_{end1}))\frac{d\xi'(t')}{d\ln t'}=\alpha_l n_2 Z(\xi_2')=\alpha_l n_2 Z(\xi'') \tag{14}$$

or:

$$\frac{d\xi''(t')}{d\ln t'}=\bar{N}Z(\xi''(t')) \text{ with } \bar{N}=\frac{p\alpha_l n_2}{(1-\xi_r(t_0))(1+\alpha_l\xi_2'(t'_{end1})} \tag{15}$$

In the above time interval, the experimental fraction transformed $\xi''(t')$ follows a nearly JMAK behaviour, where the $\bar{N}$ value is obtained by linear fitting of the $\ln\ln\frac{1}{1-\xi''}$ vs. $\ln t'$ for the time interval considered.

By considering the kinetics of a second process ($t'>t'_{end1}$) one may obtain $\xi_2'(t')$ (or $\xi_2(t)$) and the corresponding parameters n_2 and k_2. In this sense Eqs. (7), (10) and Eq. (15) are considered, thus it is obtained:

$$\xi_2(t)=\xi_2'(t')=p\frac{n_2}{\bar{N}}[\xi'(t')-1]+\xi_2'(t'_{end1}) \tag{16}$$

as for $t'=t'_{end2}$, the $\xi_2'(t'_{end2})=1$, and taking into account the Eq. (13) we obtain:

$$\xi_2(t)=\xi_2'(t')=\frac{n_2}{\bar{N}}[\xi''(t')-\xi''(t'_{end2})+1 \tag{17}$$

For computing the kinetics of a first process ($\xi_1'(t')$) at $0<t'<t'_{end1}$, Eqs.(16) and (10) are considered, thus the following expression results:

$$\xi_1(t)=\xi_1'(t')=\xi'(t')[1-\xi_r(t_0)][1+\alpha_l\xi'_2(t'_{end1})]+\xi_r(t_0)[1+\alpha_l\xi'_2(t'_{end1})]-\alpha_l\xi'_2(t') \tag{18}$$

We consider two situations of overlapping for the processes [33, 43].

3.1.2.1 Small overlap

A second process occurs, but it manifests itself only weakly during the interval $t_0<t<t_{end1}$($0<t'<t'_{end1}$); i.e., $\xi_2(t_{end1})$ is small so, the second process disturbs the first one. In this case [43]:

$$\xi_1(t)=\xi_1'(t')=1+\frac{n_1}{n}[\xi'(t')-1] \tag{19}$$

where it is assumed that $\xi_2(t_{end1})$- $\xi_2(t)\approx 0$.

3.1.2.2 Large overlap

In this case a second process is manifested strongly during the time when the first one is occurring ($t_0<t<t_{end1}$). As already remarked, it is necessary to pick a time interval such that $\xi_2'(t')>0.25$, i.e., $Z(\xi_2')$ is nearly constant and, consequently, $\xi_1'(t')$ and $\xi'(t')$ are below 0.9. Therefore for times far from the boundary points $\xi'=0$ and $\xi'=1$, the $Z(\xi_1')\approx Z(\xi_2')\approx Z(\xi')\approx$constant and Eq.(11) simplifies to:

$$\frac{d\xi'(t')}{d\ln t'} = nZ(\xi') \text{ with } n = \frac{n_1 + \alpha_l n_2}{(1\text{-}\xi_r(t_0))(1+\alpha_l \xi_2'(t'_{end1}))} \tag{20}$$

and according to Eqs.(15) and (20); Eq.(18) reduces to:

$$\xi_1(t) = \xi_1'(t') = 1 + \frac{pn_1}{pn - \bar{N}}[\xi'(t') - 1] + \frac{n_1 \bar{N}}{n_2(pn - \bar{N})}[\xi'_2(t'_{end1}) - \xi'_2(t')] \tag{21}$$

It is established in [43] that for a time interval $t'>t'_{end1}$ the ξ'' follows a nearly JMAK behaviour such that $\bar{N}$ and n_2 must be correlated as: $\bar{N} = -n_2$.

3.1.3 Determination of the kinetic parameters from isothermal dilatometry curves by the use of the KTOPT

In order to exemplify the use of this approach, isothermal dilatometry data corresponding to the tempering treatment of a low-alloy steel were used. The chemical composition of the selected low-alloy steel is: 0.32% C, 1.12% Mn, 0.67% Si, 0.07% Ni, 0.02% P, 0.05% S and the rest of Fe.

Samples of the studied steel, with diameter of 9 mm and 10 cm in length (L_0), were austenitized in a vacuum furnace at 900°C for 30 minutes. After this, the samples were quenched in water at ambient temperature. The quenched samples were tempered isothermally at 350°C in a dilatometer manufactured at Havana University, Cuba, with an accuracy of 10^{-3} mm in length. The isothermal dilatometry data ($\Delta L/L_0$) versus time are shown in Figure 5. In this figure the best work interval corresponding to each process (I and II) are shown by applying the KTOPT.

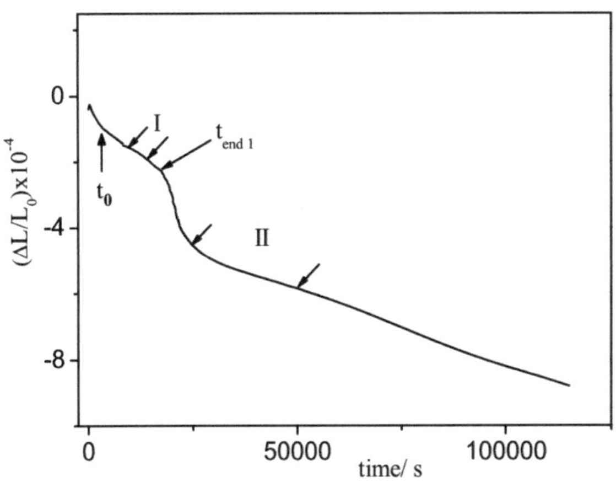

Fig. 5. Isothermal dilatometry data (experimental). shows the best work interval for both processes on

The procedure for computing the kinetic parameters of the second process ($\alpha_l>0$) begins by calculating the fraction transformed values $\xi''(t')$ through Eq.(13) from the dilatometry results. After this, the best time interval, far from the boundary points t'_{end1} and t'_{end2}, is selected by the best linear fitting values for $\ln\ln\frac{1}{1-\xi''}$ vs. lnt′ for this second process. Thus, the $\bar{N}$ value is calculated as the slope of the previous linear fitting. Then, knowing the $\bar{N}$, $\xi''(t'_{end2})$ and $\xi'(t')$ (or $\xi''(t')$) values for this best time interval, the fraction transformed values $\xi'_2(t')$ are calculated by an iterative procedure using Eq. (17) until a desired accuracy is reached: ($\bar{N}$ =-n_2). In this procedure the first n_2 value is arbitrary.

In order to evaluate the kind of processes overlap, $\xi'_2(t')$ is determined for longer times (t'_g) within of the work interval where the kinetic parameters are calculated for the first process. Thus, $\xi'_2(t'_g)$ is calculated to be approximately: 0.35, which can be considered as a small perturbation to the first process by the second. In this manner, the kinetic parameters of the first process are obtained initially, by selecting the best linear fitting of the lnln(1/1-$\xi'(t')$) versus lnt′, far from the points t′=0 and t'_{end1}. Then, the n value obtained from the slope of the above best linear fit, and Eq.(19)(small overlap) are used to generate the $\xi'_1(t')$ for this time interval using as iterative procedure, as already used in calculating the kinetic parameters for the second process. In this procedure the first n_1 value is arbitrary. The kinetic parameters corresponding to both processes by applying the KTOPT, are listed in Tables 10-a and 10-b.

Best interval in $\xi'(t')$	2.864≤ $\xi'(t')$≤3.963
Normalized interval in $\xi''(t')$	0.4506≤ $\xi''(t')$ ≤0.6544
Work interval in $\xi'_2(t')$ from the final iteration	0.5975≤ $\xi'_2(t')$ ≤0.8010
$\bar{N}$	-0.669
Correlation coefficient R for $\bar{N}$	0.998
n_2 (Avrami-exponent)	0.669
Correlation coefficient R for n_2	0.998
k_2 (s^{-1})	$4.4*10^{-5}$

Table 10.a Kinetic parameters for the second precipitation process according to the KTOPT, [33].

Best interval in $\xi'(t')$	0.5050≤ $\xi'(t')$≤0.7625
Work interval $\xi'_1(t')$ from the final iteration	0.5310≤ $\xi'_1(t')$ ≤0.7750
N	1.58
Correlation coefficient R for n	0.993
n_1 (Avrami-exponent)	1.49
Correlation coefficient R for n_1	0.992
k_1 (s^{-1})	$1.2*10^{-4}$

Table 10.b Kinetic parameters for the first precipitation process according to the KTOPT, [33].

3.1.4 Uncertainty in the Avrami-exponent

In order to estimate the error-prone Avrami-exponent (n′), the uncertainty in the fraction transformed $\delta\xi'(t')$ for both work interval is determined according to the procedure described in [44], where:

$$n' = n \pm \delta n \text{ with } \delta n = \frac{n[1+\ln(1-\xi')]}{(1-\xi')\ln(1-\xi')}\delta\xi' \tag{22}$$

Let us define the parameters: $y'=\Delta l'(t')/L_0=(l'(t')-L_0)/L_0$, $y'_1=\Delta l'(t'_{end1})/L_0$, and $y'_0=\Delta l'(0)/L_0$. The uncertainty $\delta\xi'(t')$ is obtained by the propagation of the uncertainty corresponding to the length changes measured directly from the dilatometry curve ($\delta l'=10^{-3}$mm). Then:

$$\delta\xi'(t') = \frac{2[(y'_1-y'_0)+(y'-y'_0)]}{L_0(y'_1-y'_0)^2}\delta l' \tag{23}$$

For the best work interval corresponding to the second process, the selected parameters for calculating the error-prone Avrami-exponent (n'_2) are: $y'=5.815*10^{-4}$, $y'_1=2.2*10^{-4}$ and $y'_0=0.98*10^{-4}$; then $\delta\xi'=0.8$. As $\alpha_l>0$, the uncertainty δn_2 for the boundary values of the fraction transformed $\xi'_2(t')$ in the above selected work interval is: $\delta n_2=-0.02$ for $\xi'_2=0.5976$ and $\delta n_2=0.18$ for $\xi'_2=0.801$ [33]. In this manner, the Avrami-exponent n_2 with its error bounds can be written as: $0.65<0.67<0.69$ or $n'_2=0.67\pm0.02$.

In order to calculate the Avrami-exponent corresponding to the first process, a similar procedure to the second one is applied. The same parameters y'_1 and y'_2 are selected, but the y' parameter is now $1.91*10^{-4}$ as the extreme value of the fraction transformed $\xi'(t')$ for the best time interval corresponding to the first process from the dilatometry curve.

The uncertainty $\delta\xi'(t')$ according to Eq. (23) for the first process is $\delta\xi'(t')\cong0.28$ and assuming a small overlap situation, $\delta\xi'_1=(n_1/n)\delta\xi'\cong0.2$. The uncertainty in the Avrami exponent n_1 resulted to be: $\delta n_1=-0.2$ for $\xi'_1(t')=0.531$ and $\delta n_1=0.4$ for $\xi'_1(t')=0.775$. The Avrami-exponent for this first process, n_1 with its smallest error bounds, can be written as: $1.29<1.49<1.69$ or $n'_1=1.5\pm0.2$.

3.1.5 Precipitation processes on tempering

The results obtained by the proposed approach ($n_1=1.5$) show that the first process identified in the dilatometry curve (by an initial contraction of length) corresponds to a diffusion-controlled precipitation process during which small particles grow with zero nucleation rate ($n_1=1.5$) [31, 45]. This first stage of tempering is recognized in the literature as the decomposition of martensite into transition carbide (ε or η-carbides) and a less tetragonal martensite [46]. The precipitation of the transition carbide could have occurred in very early stages of tempering or during quenching for this alloy; the Ms temperature is about 365°C [47]. This transition carbide is frequently observed to nucleate uniformly throughout the martensite matrix and some studies [48, 49] have indicated that the nucleation may be influenced by the modulated structure formed by spinodal decomposition that occurs before the first stage. It is also reported in other works [50], that the initial formation of the first transition carbide is due to a shear of the martensite structure which would involve neither carbon diffusion nor a significant initial fall in hardness. According to the literature [46, 48, 51], there is no evidence that the nucleation of the transition carbide (ε or η carbide) could be related to the dislocations in the martensite.

The second process identified in the dilatometry curve by a second contraction in length, corresponds to the disappearance of the transition carbide and the formation of stable cementite. Many articles [52, 53] report that cementite nucleates on dislocations, inter-lath and grain boundaries. In this sense, an Avrami-exponent n_2=0.67 is in agreement with a mechanism that involved precipitation on dislocations and diffusion-controlled growth (n=0.66) [45].

3.2 Non-isothermal tempering

3.2.1 Introduction

In the past, a number of methods have been proposed to describe the progress of a reaction in solid systems from non-isothermal experiments. Although non-isothermal experiments can use any arbitrary thermal history, the most usual in thermal analysis is to employ a constant heating rate, (β = dT/dt =const).

In order to study the kinetics of the phase transformations performed at constant heating rates (β), a wide range of methods has been established for deriving the kinetic parameters of the reactions obeying equations:

$$\frac{d\xi}{dt} = K(T)g(\xi) \tag{24}$$

and

$$K(T) = K_0 \exp[-\frac{E}{RT}] \tag{25}$$

where g(ξ) is a specific function of the fraction transformed, K(T) is the rate constant of the reaction mechanism and E, K_0 and R are the activation energy of the reaction mechanism, the frequency factor and the gas constant respectively [54]. Thus, without recourse to any kinetic model, values for effective activation energy can be obtained upon isochronal experiments from the temperatures T_ξ needed to attain a certain fixed value of ξ (temperatures corresponding to the same degree of transformation), as measured for different heating rates (β)[55]. The procedures that use the above temperatures T_ξ for calculating the effective activation energy are known as Kissinger-like methods [54, 55], and these rely on approximating the so-called temperature integral [55-58]. Another set of methods does not use any mathematical approximation for calculating the temperature integral, but instead require determinations of the reaction rates at a stage with the same degree of transformation (it corresponds to an equivalent stage of the reaction) for various heating rates. These procedures are known as the Friedman-like methods [59].

As the determination of g(ξ), E and K_0 (the so-called kinetic triplet) is an interlinked problem in non-isothermal experiments[54,60], a deviation in the determination of any of the three will cause a deviation in the other parameters of the triplet. Thus, it is important to start the analysis of a non-isothermal experiment by determining one element of the triplet with high accuracy. In this sense, it is usual to calculate the effective activation energy (E) by a Kissinger-like method for nucleation and growth reactions [55, 61, 62]. This is because the T_ξ constitutes a parameter that can be determined with high accuracy in some non-isothermal experiments, such as in non-isothermal dilatometry curves. For nucleation and growth reactions, in general, the effective activation energy is a function of both transformation time and temperature (E does not have to be constant even with constant nucleation and growth mechanisms). Because of this, the above lineal approximation of the Kissinger-like plot will be strictly valid only if the effective activation energy is constant during the entire transformation [62]. This condition is initially ignored in most papers. Thus, this research has been devised to explore the possibilities that combination of the different non-isothermal analysis methods has to obtain the kinetic parameters of the tempering reactions in low-alloy steels using non-isothermal dilatometric data [63].

3.2.2 Non-isothermal dilatometric analysis: Theoretical background and experimental procedure

The precipitation reactions in isothermal conditions are generally described by the JMAK-like relation [64-66]:

$$\xi = 1 - \exp -(\theta)^n \text{ with } \theta = K(T)t \tag{26}$$

where n is known as the Avrami exponent and t is the time.

In order to maintain the JMAK description under non-isothermal conditions, the formalism of Eq. (26) is accepted for an infinitesimal lapse of time [61, 67]:

$$d\theta = K_0 \exp(-\frac{E}{RT})dt \tag{27}$$

Integration of Eq.(27), resulted in:

$$\theta(T) = \frac{T^2}{\beta}\frac{R}{E}[K_0 \exp(-\frac{E}{RT})][1 - \frac{2RT}{E}] \tag{28}$$

After deriving $\xi(p) = \dfrac{p - p_0}{p_1 - p_0}$ twice with respect to T, evaluating the resulted equation at temperatures corresponding to the inflection points T_i [55], and taking into account several mathematical approximations, it is obtained the working expression [61, 67]:

$$Ln\left[\frac{\beta}{T_i^2}\right] = -\frac{E}{R * T_i} + Ln\left[\frac{RK_o}{E}\right] - \text{Re}s1 - \text{Re}s2 \tag{29}$$

with:

$$\text{Re}s1 = \frac{QRT_i^2}{n^2 E} \text{ and } \text{Re}s2 = 2 * \left\{1 - \frac{1}{n^2} + n\ln\left[\frac{T_i^2 RK(T_i)}{\beta E}\right]\right\}\frac{RT_i}{E} \tag{30}$$

here

$$Q(T_p) = 2\frac{\frac{dp_1}{dT} - \frac{dp_0}{dT}}{p_1 - p_0}\Big|_{Ti} \tag{31}$$

If both residuals are neglected in Eq. (29) (see appendix in [63]), the data points in a plot of the lnβ/T_i^2 versus $1/T_i$ (Kissiger-like plot) are approximated by a straight line, from the slope of which a value for the effective activation energy, E, is obtained. We settle that with the definition of the state variable θ (c.f. ref. 55 and 62), the adoption of a specific model of reaction is not a necessary condition to obtain the effective activation energy from the slope of a Kissinger-like plot.

An analytical solution of the JMAK rate equation for the general non-isothermal case at constant heating rate, assuming that the nucleation (N) and growth (G) rates have an Arrhenian dependence of temperature has been published [68]:

$$\xi = 1 - Exp\left\{-\left[K_0 C \frac{E}{\beta R} P\left(\frac{E}{RT}\right)\right]^n\right\} \tag{32}$$

where it is considered that the transformation rate is negligible at the initial temperature of the experiment.

In this expression, P(E/RT) is the exponential integral, and C a constant that depends on *n*, E_N and E_G(activation energies for nucleation and growth mechanisms). The constant C reduces to unity in particular situations when the nucleation is completed prior to crystal growth (site saturation situation) or in the isokinetic situation where E_N=E_G [68].

In order to obtain a suitable Kissinger-like plot (ln[β/T_p^2] versus $1/T_p$), the authors [68], without recourse to any kinetic model, obtained the relation:

$$\ln\left[\frac{\beta}{T_p^{\,2}}\right] = -\frac{E}{RT_p} + \ln[-\frac{RK_0 C g'(\xi_p)}{E}] \tag{33}$$

As a JMAK kinetic model is assumed, then:

$$g(\xi) = n(1-\xi)[-\ln(1-\xi)]^{\frac{n-1}{n}} \text{ and } g'(\xi) = \frac{dg}{d\xi} = \frac{g(\xi)}{1-\xi} + \frac{n-1}{[-\ln(1-\xi)]^{\frac{1}{n}}} \tag{34}$$

It has been demonstrated that in the peak temperature, ξ_P=0.632 and therefore $g'(\xi_{Tp})$=-1, [68]. Eq.(33) can be now applied to the non-isothermal dilatometry data at temperatures T_i, corresponding to the inflection points for different constant heating rates (the temperatures T_i belong with very good approximation to an equivalent stage of the reaction [55], ξ_{Ti}=0.632).

Then, as it can be seen, Eq.(33) coincides with Eq.(29) in the isokinetic case (E_N=E_G) , or in the site saturation situation (C=1), if both residuals in Eq.(29) are neglected.

In order to calculate the parameters K_0 and C a non linear regression analysis is performed using Eq. (33) where T_p has been changed by T_i and $g'(\xi_{Ti})=-1$ at the inflection points. Consequently, it is possible to obtain a transcendent equation to find *n* through the second derivative of Eq.(32):

$$-n(K_0C)^n[\frac{E}{\beta R}]^{n-1}[p(\frac{E}{RT_i})]^n\exp(-\frac{E}{RT_i})+(n-1)\frac{\beta R}{E}\exp(-\frac{E}{RT_i})+\frac{\beta Ep(\frac{E}{RT_i})}{RT_i^2}=0 \quad (35)$$

An alternative method to estimate the effective activation energy and other kinetic parameters directly from the non-isothermal dilatometry curves is presented in [69]. Furthermore, this procedure allows analyzing if the effective activation energy, obtained by the above Kissinger-like plot, is constant during the entire transformation. For this, the fraction untransformed (1-ξ), according to $\xi(p)=\frac{p-p_0}{p_1-p_0}$ with p=Δl/l (relative change of length), is defined as:

$$(1-\xi)=[(\frac{\Delta l(T)}{l})_T-(\frac{\Delta l(T)}{l})_{end}]/[(\frac{\Delta l(T)}{1})_0-(\frac{\Delta l(T)}{l})_{end}] \quad (36)$$

where $(\Delta l(T)/l)_0$ and $(\Delta l(T)/l)_{end}$ are the relative length increments of the start and end stages, at temperature T. The $(\Delta l(T)/l)_T$ is the relative length increment on dilatometric curve at temperature T, as it is depicted in Figure 6. According to this definition (1-ξ)=1 and ξ=0 at the start temperature of the transformation $T=T_S$.

This procedure was applied to nucleation and growth reactions (tempering reactions in steels) using an expression oversimplified for $g(\xi)=(1-\xi)^{n_0}$ [69]. This way, the analysis is based on the assumption of homogeneous reactions with a reaction rate:

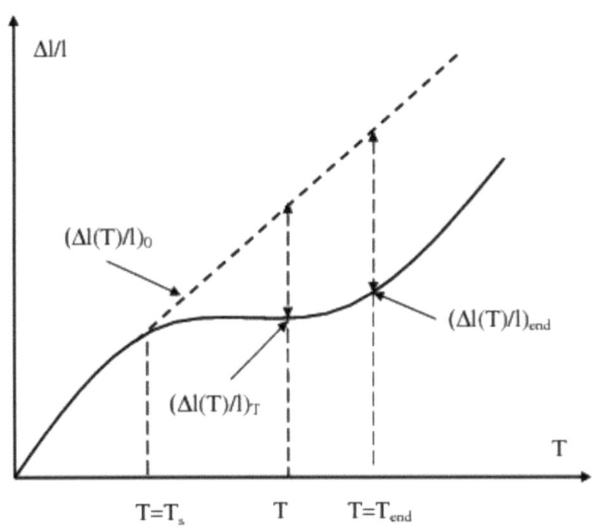

Fig. 6. Relative length change (Δl(T) / l) near the inflection point [63]

$$-\frac{d(1-\xi)}{dt} = K(1\text{-}\xi)^{n_0} \tag{37}$$

where the order of the tempering reaction under consideration is n_0.

Differentiation of Eq.(37) (with β as a constant and $K = K_o \exp(-E/RT)$), and its evaluation at the inflection point temperature (T_i) on the curve (1-ξ) versus temperature, the equation for the effective activation energy is obtained:

$$E = -\frac{n_o R T_i^2 (1-\xi)_{Ti}^{n_0-1} \left.\frac{d(1-\xi)}{dT}\right|_{Ti}}{(1-\xi)_{Ti}{}^{n_0}} \tag{38}$$

The other kinetic parameters can be calculated through the equations:

$$K_0 = -\frac{\beta}{(1-\xi)_{Ti}{}^{n_0}} \left.\frac{d(1-\xi)}{dT}\right|_{Ti} \exp(\frac{E}{RT_i}) \tag{39}$$

and:

$$(1-\xi)_T^{n_0} = -\frac{\beta}{K_0}[\frac{d(1-\xi)}{dT}]_T \exp(\frac{E}{RT}) \tag{40}$$

Although the above simplification could be questioned, this approximation may be used to investigate the variation of the effective activation energy under different experimental conditions [70]. In order to solve this difficulty it is assumed that the tempering reactions obey a JMAK kinetic model. Therefore, if the above formalism for homogeneous transformations is settled; then, a relationship between the Avrami exponent n and the kinetic order of the reaction n_0 can be obtained as a function of the fraction untransformed (1-ξ). Thus, the transformation rate for a reaction that obeys a JMAK relation is:

$$-\frac{d}{dt}(1-\xi) = K(1-\xi)n(-\ln(1-\xi))^{\frac{n-1}{n}} \tag{41}$$

From the Eqs. (37) and (41), the expression that related n, n_0 and (1-ξ) resulted to be:

$$n_0 = 1 + \frac{\ln[n(-\ln(1-\xi))^{\frac{n-1}{n}}]}{\ln(1-\xi)} \tag{42}$$

As it is shown in Figure 7, the behavior of the n_0 function depends weakly on (1-ξ) far from the boundary point (1-ξ)=1. As a consequence, n_0 can be considered as a constant during the development of the reaction for stages where the fraction transformed is greater than 0.4. Also, the n_0 values are lower the greater the Avrami exponent, and when the Avrami exponent takes the value n=1, the reaction is the first order (n_0=1). Evaluation of n_0 from the different nucleation and growth protocols, given by the n values, makes possible to use the procedure established in [69] to nucleation and growth reactions and therefore to determine if E is constant during the entire transformation. For this, the fraction untransformed values calculated from Eq. (40) (where it is used the E value determined by a Kissinger-like plot) for the entire reaction are compared with the experimental fraction untransformed data obtained from the dilatometric curve.

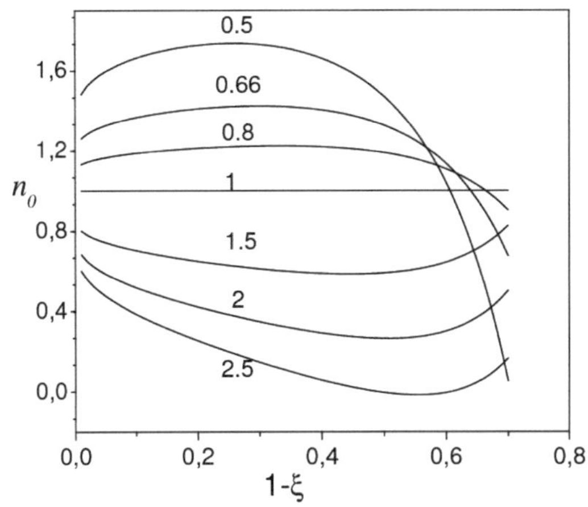

Fig. 7. Behavior of the n_0 parameter versus the fraction untransformed (1-ξ) for different Avrami exponent values [63]

The Friedman-like methods conceived to be used in processing of non-isothermal calorimetric data do not require any mathematical approximation to solve the temperature or exponential integral [54, 59]. This procedure allows obtaining the activation energy of a reaction knowing the reaction rates at a stage with the same degree of transformation for various heating rates. According to the knowledge of the present authors [63], the Friedman-like methods have not been very much used with non-isothermal dilatometry data.

In fact, if dξ/dt at temperatures corresponding to the inflection points (Ti) is known then, substituting the K(T) expression into dξ/dt equation (β= dT/dt); the following equation results after applying logarithms to both terms:

$$\ln[\frac{d\xi}{dt}]_{T_i} = \ln[\beta \frac{d\xi}{dT}]_{T_i} = -\frac{E}{RT_i} + \ln(K_0 g(\xi_i)) \tag{43}$$

As in the dilatometry record one has that $(\frac{\delta l}{l})_t = \frac{l_t - l}{l}$, where l is the initial length of the sample and l_t is the length at any instant, then:

$$\frac{dl_t}{dT} = l\left(\frac{d}{dT}\left(\frac{\delta l}{l}\right)_t\right) \tag{44}$$

Differentiating $\xi(p) = \frac{p - p_0}{p_1 - p_0}$ with respect to time (with p= $\frac{\delta l}{l}$), and taking into account Eq. (44); Eq. (43) can be written, at temperatures corresponding to the inflection points, as [63]:

$$\ln\left[\beta\frac{d}{dT}\left(\frac{\delta l}{l}\right)\Big|_{Ti}\right] = -\frac{E}{RT_i} + Const \tag{45}$$

According to Eq.(45) it is possible to calculate the effective activation energy E, (but not the frequency factor, K_0).

Commercial plain carbon steel having a carbon content of approximately 0.5 wt. % was used for the tempering analysis. The chemical composition of the (AISI 1050) steel is: 0.48-0.55% C, 0.6-0.9% Mn, 0.04% P, 0.05% S and the rest of Fe. To obtain the non-isothermal dilatometry records at constant heating rates on the tempering treatment (from ambient temperature up to 600 °C), a dilatometer Adamel-Lhomargy (model DT 1000, NY, USA) was used. The relative length changes in the specimen (with an accuracy of 10^{-4}) against temperature were obtained while the specimen is heated at a constant rate. The constant heating rates were 5, 10, 15, 20 and 30 K/min respectively.

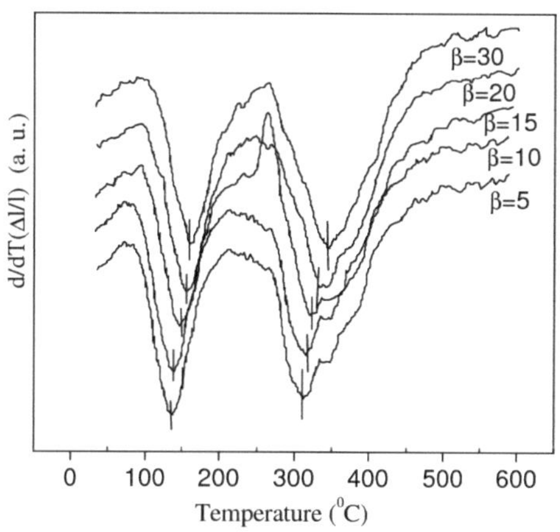

Fig. 8. The d(Δl(T)/l)/dT versus temperature at different heating rates [63].

Table11 and Figure 8 show the temperatures corresponding to the inflection points from the dilatometric curves with different heating rates for both processes. By lineal (see Table 12) and non-lineal regression analyses using Eq. (29) (neglecting the residues as it is outlined in [63]), the best E and K_0 parameters were calculated for the two identified processes on tempering in the non-isothermal dilatometry records.

β (K/min)	T_i (K) Process I	T_i(K) Process II
5	411.6	584.3
10	418.9	593.5
15	424.3	599.1
20	428.5	603.1
30	432.4	608.8

Table 11. Temperatures corresponding to the inflection points (T_i) at different heating rates (β) for processes I and II on tempering in the dilatometry records, [63].

Process	Activation Energy, E KJ/mol	Frequency Factor, K_0 min^{-1}	Confidence Interval for E	Regression Coefficient R	P value for E
I	117	$2.99*10^{14}$	104< E<130 (*) 116.5<E<117.4 (**)	0.998 (**)	$4.8*10^{-10}$ (**)
II	206	$9.53*10^{17}$	196<E<216 (*) 205.6<E<206.4 (**)	0.999(**)	$9.0*10^{-14}$ (**)

(*) Lineal regression analysis
(**) Non-lineal regression analysis (E and K_0 are the fitting parameters)
P=Distr. T(T, FD, 2): It represents the probability that a better fit to the same data can be carried out by another model. As the P values are very low, it is not very probable that another model fits the data better than the model here shown.

Table 12. Best parameters E and K_0 calculated by linear and non-lineal regression analyses respectively for the two processes on tempering using Eq.(29) where both residuals have been ignored, [63].

Although one has the possibility, with this procedure (Eqs. (29-31)), to obtain a relationship for calculating the Avrami exponent (n) of the reaction, we think that the n values are not very reliable due to the many approximations used to obtain the residues. For this reason, the use of the transcendent equation, Eq.(35), is very much reliable to calculate n, given by much less approximations in their determination. In this sense, as a first step, the K_0 and C parameters are calculated according to a procedure detailed in [63]. Table 13 shows the Avrami exponent for each considered process on tempering at different heating rates solving the transcendent equation, Eq. (35).

β (K/min)	T_i (K)	n	δn
5	411.6	1.2	0.2
10	418.9	1.1	0.2
15	424.3	1.0	0.2
20	428.5	1.1	0.2
30	432.4	1.0	0.2

First Process on Tempering

β (K/min)	**T_i (K)**	**n**	**δn**
5	584.3	0.7	0.1
10	593.5	0.7	0.1
15	599.1	0.7	0.1
20	603.1	0.7	0.1
30	608.8	0.7	0.1

Second Process on Tempering

Table 13. The Avrami exponent for the two processes on tempering, [63]

The uncertainty showed in each Avrami exponent value (δn) (Table 13) was calculated by the propagation of the uncertainty [44] in temperature (δT=0.1°C) and in the activation energy (δE=0.5 KJ/mol calculated by non-linear regression analysis of Eq.(33)) from the transcendent equation F(n, T, E)=0, [63].

After this, it is concluded that the first process on tempering considered by us corresponds to a reaction with n_0=1 according to the formalism showed in [69]. On the contrary, the second process (third stage of tempering) that in the literature is identified as the cementite precipitation, the n_0 parameter has a value close to 1.4 (n=0.66) according to Eq.(42).

Taking the above n_0 values for the first and second processes respectively, the new E values at temperatures of the inflection points can be determined by Eq.(38). As can be seen in Tables 14 and 15, these values are the same, within the error boundary, as those obtained by a Kissinger-like plot. The frequency factors for each heating rate are calculated by the use of Eq.(39).

β	Ti(K)	n_0	K_0 (min^{-1})	E(KJ/mol)	δE(KJ/mol)
5	411.6	1.0	5.4 *10^{14}	119	6
10	418.9	1.0	6.87 *10^{13}	112	6
15	424.3	1.0	3.3 *10^{14}	117	6
20	428.5	1.0	2.0 *10^{15}	124	7
30	432.4	1.0	1.1*10^{14}	113	6

Table 14. Activation energies (E) and the frequency factors (K_0) for the first process on tempering using Eq.(38) and (39), [63].

β	Ti(K)	n_0	K_0*10^{17}(min^{-1})	E(KJ/mol)	δE(KJ/mol)
5	584.3	1.4	1.3	196	11
10	593.5	1.4	1.2	196	10
15	599.1	1.4	1.3	196	10
20	603.1	1.4	1.6	197	10
30	608.8	1.4	1.3	196	10

Table 15. Activation energies (E) and the frequency factors (K_0) for the second process on tempering using Eq.(38) and (39), [63].

According to Eq. (40), the fraction untransformed values are calculated for the entire range of temperatures using the determined E, K_0 and n_0 values. Only for the above n_0 values, the calculated effective activation energy is constant for the entire reaction, Figure 9. Other nucleation and growth protocols, that generate another n_0 values, cause appreciable deviations among the experimental and calculated fraction untransformed values for the range of temperatures where the reactions are developed.

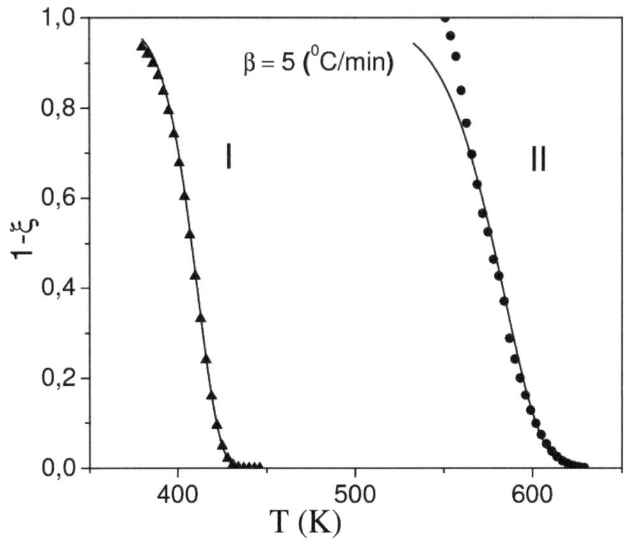

Fig. 9. Fraction untransformed versus temperature according to Eq. (40): Experimental curves for the first (I) and second (II) processes with a heating rate: β=5 °C/min. The symbols: *filled triangle* and *filled circle* are the fraction untransformed values calculated by Eq.(40) for different temperatures. Process (I): K_0=5.4x10[14] min[-1]; E=119 KJ/mol; n_o=1; d(1-ξ) /dT $|_{Ti}$ =-0.032275. Process (II): K_0=1.3 x10[17] min[-1]; E=196 KJ/mol; n_o=1.4; d(1-ξ) /dT $|_{Ti}$ =-0.0184273 [63].

The d/dT(δl/l) data used in the determination of the effective activation energy by a Friedman-like procedure is shown in [63]. By linear fitting of Eq.(45), the effective activation energy was for both processes: E=127.5 KJ/mol; δE=32.7 KJ/mol (first process) and E=202.9 KJ/mol; δE=40 KJ/mol (second process), very close to those determined by a Kissinger-like method.

3.2.3 Precipitation processes on tempering

Considering the chemical composition of steel and the results obtained by the analysis of the dilatometry records, it is assumed that the new nuclei corresponding to the first stage of tempering could have been formed during the quench. This is because in the studied steel, the Ms temperature (martensite start temperature) is closed to 300°C according to the M_s relation for 0.5 wt % C and 0.8 wt % Mn[71], and by the greater mobility of the carbon atoms through the dislocations inherited from the martensite structure[72].

In the temperature range of approximately 100°-200° C (see Figure 8) for the first processes at the non-isothermal dilatometric registers, the transition carbide (epsilon carbide) nuclei that have been formed on quenching are growing. The growth of the transition carbide produces a lost of tetragonality of the martensite matrix by the exit of the interstitial carbon in solution. The above is in agreement with the situation where existing nuclei in form of needles or plates are developed by controlled diffusion growth corresponding to the Avrami exponent close to one (n=1)[45] as it is shown in Table 13 for this first process on tempering.

As it can be appreciated in processes (I and II) on tempering, the nucleation and growth mechanisms are quite separated (site saturation situation); since, the coefficient C in Eq. (33) resulted to be C=1 by the non-linear regression analysis above discussed.

The effective activation energy found for the first process from the non-isothermal dilatometry records using the most accepted isoconversion methods was among 117~128 KJ/mol. These values are in agreement with the one reported by [67,73] for diffusion of the iron atoms along dislocations that are generated by the incoherency between the transition carbide and the matrix. Therefore, it will be the diffusion of the iron atoms and not the carbon atoms diffusion that control the reaction during this first process on tempering. In Figure 9, the curve (I) shows that the fraction untransformed values calculated by the use of Eq.(40) during the ongoing reaction are in agreement with the fraction untransformed values determined from the dilatometric record for the same interval of the transformation temperatures (experimental curve in Figure 9). Thus, the effective activation energy obtained as the slope of a Kissinger-like plot is not only valid for the temperature corresponding to the inflection point, but for the whole interval of temperatures where this reaction occur. For this reason, we assume that $E=E_G$ is constant for this first precipitation process on tempering.

For temperatures from 170° to 300°C, when the transformation (I) concludes, it is well established that the retained austenite (γ_r), with approximately 4 % in volume, transforms into cementite (θ) and bainitic ferrite (α) [74]. This process should increase the volume of the sample; however, as the amount of the retained austenite is small, the respective change in the volume of the sample is difficult to appreciate in the dilatometry curves.

In the temperature range (~300° to 350°C), process (II) (Figure 8), the transition carbide is dissolved to form cementite. This process should originate a contraction of volume that it is appreciated in the dilatometric curve. While the cementite particle grows, the transition carbide particles should disappear gradually, due to the iron atoms diffusion along dislocations to form the cementite. The decomposition processes of retained austenite and the transition carbide are generally overlapped, but as the amount of retained austenite is very small, it is assumed that the reduction of volume during the second process is due, practically, to the nucleation and growth of the cementite on dislocations near of the transition carbide. The above is supported by the fact that the Avrami exponent is close to 0.66 for this second process (third stage of tempering) (see Table 13). This value is in correspondence with the protocol where the nuclei (cementite) are formed at dislocations and the growth is controlled by diffusion of the iron atoms [45]. This last statement is argued by the value of the effective activation energy calculated for this second process, which resulted to be close to 200 KJ/mol. This intermediate value of the activation energy between 134 and 251 is in correspondence with the combination of the pipe diffusion of the iron atoms,[67, 73] and of the volume diffusion of the iron atoms in ferrite [67, 75]. This could suggest a new distribution of the iron atoms to form the cementite by dissolution of the transition carbide.

As it can be seen for this second process, according to the calculated kinetic parameters (n_0 and K_0), the effective activation energy is constant during the entire transformation as it is shown in Figure 9, curve II.

4. References

[1] T. Gladman, The Physical Metallurgy of Microalloyed Steels, The Institute of Materials, London (1997), p 28-57.
[2] H.-J. Kestenbach and E. V. Morales, Transmission Electron Microscopy of Carbonitride Precipitation in Microalloyed Steels, Acta Microscópica, (1998), 7(1): 22-33.
[3] W. B. Morrison and J.H. Woodhead, The influence of small niobium additions on the mechanical properties of commercial mild steels. J. Iron Steel Inst. (1963), 201:43-46.
[4] W. B. Morrison, The influence of small niobium additions on the properties of carbon-manganese steels. J. Iron steel Inst. (1963), 201:317-325.
[5] J. M. Gray, D. Webster, and J. H. Woodhead, J. Iron Steel Inst. (1965), 203:812-818.
[6] H.-D. Bartholot, H.-J. Engell, W. Vordemesche, and K. Kaup, Stahl Eisen, (1971), 91:204-219.
[7] T. Gladman, D. Dulieu, and I. D. McIvor, Microalloying 75, Union Carbide Corporation, New York, (1977), p 32-55.
[8] T. Gladman, B. Holmes, and I. D. McIvor, The effects of second phase particles on the mechanical properties of steel, The Iron and Steel Institute, London, (1971) p 68-78.
[9] A. Itman, K. R. Cardoso, and H.-J. Kestenbach, Mater. Sci. Technol. (1997), 13:49-55.
[10] A. T. Davenport, L. C. Brossard, and R. E. Miner, J. Met., (1975), 27:21-27.
[11] F. B. Pickering, Microalloying 75, Union Carbide Corporation, New York, (1977), p 9-30.
[12] A. J. DeArdo, 8th process technology conference, The Iron and Steel Society, Warrendale, PA, (1988) p 67-78.
[13] L Meyer, C. Strassburger, and C. Schneider, HSLA steels: metallurgy and applications, American Society for Metals, Metals Park, OH, (1986) p 29-44.
[14] E. Valencia Morales, and H.-J. Kestenbach, Rev. Metal. Madrid, (1998), 34(6):488-498.
[15] S. S. Campos, E. V. Morales, and H.-J. Kestenbach, Metall. Mater. Trans. (2001), 32A:1245-1248.
[16] S. S. Campos, J. Gallego, E. V. Morales, and H.-J. Kestenbach, HSLA Steels 2000, (ed. Liu Guoquan et al.), Metallurgical Industry Press, Beijing, (2000) p 629-634.
[17] S. Freeman, and R. W. K. Honeycombe, Met. Sci., (1977), 11:59-64.
[18] R. M. Brito, and H.-J. Kestenbach, J. Mater. Sci., (1981), 16:1257-1263.
[19] F. B. Pickering and T. Gladman, Metallurgical Developments in Carbon Steels, Iron and Steel Institute, Special Report N° 81, (1963), p 10.
[20] M. F. Ashby, Strength Methods in Crystals, A. Kelly and R. B. Nicholson, eds. Elsevier, London, (1971) p 137.
[21] A. S. Keh and S. Weissmann, Electron Microscopy and Strength of Crystals, G. Thomas and J. Washburn, eds., Interscience Publishers, New York, NY, (1963), p 231.
[22] J. M. Rosenberg and H. R. Piehler, Metall. Trans., (1971), 2: 257-259.
[23] F. R. N. Nabarro, Z. S. Basinski, and D. B. Holt, Adv. Phys., (1964), 13: 193-198.
[24] R. W. K. Honeycombe, HSLA Steels: Metallurgy and Applications, Beijin Conf., ASM, Metals Park, OH, (1986), p 243-250.
[25] A. S. Keh, Direct Observations of Imperfections in Crystals, J. B. Newkirk and J. H.Wernick, eds., J. Wiley, London, (1962), p. 213.
[26] H.-J. Kestenbach, S. S. Campos, and E. V. Morales, Mater. Sci. Technol. (2006), 22(6):615-626.
[27] A. D. Batte, and R. W. K. Honeycombe, J. Iron Steel Inst., (1973), 211:284-289.

[28] J.D. Robson, Modelling of carbide and laves phase precipitation in 9-12wt% chromium Steels, Ph.D. diss, University of Cambridge, UK, 1996.
[29] P. Wilkes, *Met. Sci. J.*, (1968), 2:8-17.
[30] J.K.L. Lai and M. Meshkat, *Metal Science*, (1978), 9:415-420.
[31] M.J. Starink, *J. Mater. Sci.*, (1997), 32:4061-4070.
[32] Eon-Sik Lee and Y.G. Kim, *Acta Metall. Mater.*, (1990), 38:1669-1676.
[33] E. V. Morales, J. A. Vega-Leiva, H.L. Lopez Salinas, and I. S. Bott, Phase Transitions, (2011), 84(2):179 -191.
[34] K. Hanawaka and T. Mimura, *Metall Trans A*, (1984), 15:1147-1153.
[35] N.J. Luiggi and A. Betancourt, *Metall. Trans. B*, (1994), 25:917-925.
[36] N.J. Luiggi and A. Betancourt, *Metall. Trans. B*, (1994), 25:927-935.
[37] N.J. Luiggi and A. Betancourt, *Metall Trans B*, (1997), 28:161-168.
[38] A.T.W. Kempen, F. Sommer and E.J. Mittemeijer, *J. Mater. Sci.*, (2002), 37:1321-1332.
[39] F. Liu, F. Sommer and E.J. Mittemeijer, *J. Mater. Sci.*, (2004), 39:1621-1634.
[40] D. Wang, Y. Liu, Z. Gao and Y. Zhang, *J. Non-Cryst. Solids*, (2008), 354:3990-3999.
[41] D. Wang, Y. Liu and Y. Zhang, *J. Mater. Sci.*, (2008), 43:4876-4885.
[42] D. Wang, Y. Liu, C. Bao, W. Tan and Z. Gao, *Appl. Phys. A*, (2009), 96:721-729.
[43] E.V. Morales, N.J. Galeano, J.V. Leiva, L.M. Castellanos, C.E. Villar and R.J. Hernández, *Acta Mater.*, (2004), 52:1083-1088.
[44] E.V. Morales, J.V. Leiva, C.E. Villar, M.J. Antiquera and R.C. Fadragas, *Scripta Mater.*, (2005), 52:217-219.
[45] J.W. Christian, The theory of transformations in metals and alloys, Part I. 2nd ed. Oxford: Pergamon Press, 1975, p 525-548.
[46] Y. Wang, S. Denis, B. Appolaire and P. Archambault, *J. Phys. IV, France*, (2004), 120:103-110.
[47] E.R. Parker, *Metall. Trans. A*, (1977), 8A:1025-1042.
[48] K.A. Taylor, G.B. Olson, M. Cohen and J.B. Vander Sande, *Metall. Trans. A*, (1989), 20A:2749-2765.
[49] S. Nagakura, Y. Hirotsu, M. Kusunoki, T. Suzuki and Y. Nakamura, *Metall. Trans. A*, (1983), 14A:1025-1031.
[50] M. Dirand and L. Afqir, *Acta Metall.*, (1983), 31:1089-1107.
[51] E. Tekin and P.M. Kelly, Precipitation from iron-base alloys. G.R. Speich and J.B. Clark Eds. Metall. Soc. Conference. Cleveland, vol. 28, 1963, p 173-229.
[52] Y. Imai, T. Ogura and A. Inoue, *Trans ISIJ.*, (1973), 13:183-191.
[53] J. Nutting, J.I.S.I., (1969), 207:872-893.
[54] M. J. Starink, *Thermochim. Acta.* (2003), 404:163-176.
[55] E.J. Mittemijer, *J. Mater. Sci.*, (1992), 27:3977-3987.
[56] H.E. Kissinger, *J. Res. Nat. Bur. Stand.*, (1956), 57:217-221.
[57] H.E. Kissinger, *Anal. Chem.* (1957), 29:1702-1706.
[58] Órfáo J.J.M. AIChE Journal, (2007), 53:2905-2915.
[59] H.L. Friedman, *J. Polym. Sci.* (1964), C6:183-195.
[60] J.D. Sewry and M.E. Brown, *Termochim. Acta,* (2002), 390:217-245.
[61] E.J. Mittemeijer, A. Van Gent and P.J. Van der Schaaf, *Metall. Trans. A,* (1986), 17A:1441-1445.
[62] F. Liu, F. Sommer, C. Bos and E. J. Mittemeijer, International Mater. Reviews, (2007), 52:193-212.

[63] J. A. V. Leiva, E. V. Morales, E. Villar-Cociña, C. A. Donis and Ivani de S. Bott, *J. Mater. Sci.*, (2010), 45:418–428.
[64] A.N. Kolmogorov, Izv. Akad. Nauk. USSR-Ser. Matemat., (1937), 1:355-360.
[65] W.A. Johnson and R. F. Mehl, *Trans. AIME.* (1939), 135:416-460.
[66] M. Avrami, *J. Chem. Phys.* (1939), 7:1103-1112.
[67] L. Cheng, C.M. Brakman, B.M. Korevaar and E. J. Mittemeijer, *Metall. Trans. A*, (1988), 19A:2415-2426.
[68] J. Farjas and P. Roura, *Acta Mater.*, (2006), 54:5573-5579.
[69] Y. Tomita, *J. Mater. Sci.*, (1989), 24:731-735.
[70] J. Sestak, A. Brown, V. Rihak and G.Berggren, (1969), *in Thermal Analysis.* Academic Press. N.Y. p 1035.
[71] W. Stevens and A.G. Haynes, *J. Iron Steel Inst.*, (1956), 183:349-357.
[72] Y. Hirotsu and S. Nagakura, *Acta Metall.*, (1972), 20:645-655.
[73] M. Cohen, *Trans. J.I.M.*, (1970), II:145-151.
[74] R.W.K. Honeycombe, (1981), *Steels Microstructure and Properties, Edward Arnold Editions*, Chapter 8, p 142.
[75] F. S. Buffington, K. Hirano and M. Cohen, *Acta Metall.*, (1961), 9:534-539.

4

Effect of Niobium on HAZ Toughness of HSLA Steels

E. El-Kashif[1] and T. Koseki[2]
[1]*Department of Mechanical Design and Production Engineering, Cairo University, Giza,*
[2]*Department of Materials Engineering, The University of Tokyo, Hongo-Bunkyo-ku,*
[1]*Egypt*
[2]*Japan*

1. Introduction

The microstructure of HSLA steels depends on the steel composition and thermo-mechanical processing route. With the recent trend towards lower carbon contents,

niobium's effect on transformation behavior has been noted with the emergence of acicular or bainitic steels. Under certain conditions, such as utilizing low interstitial contents and high austenizing temperatures, small Nb additions increase hardenability by depressing Ar3 transformation temperature. Many disagreements exist, however, on the effect of Nb on the HAZ toughness of HSLA steels. These disagreements include the cause of HAZ embrittlement in steels containing Nb. Some controversy exists in the literature concerning the influence of Nb on HAZ properties under certain conditions.

Numerous investigations (Bhadeshia et al., 1985; Nakasugi et al.,1981; Ohtani et al., 1983; Sham, 1985; Yang & Bhadeshia, 1989,1991) have discussed the effect of Nb addition on the properties and microstructure of HAZ in low Carbon - microalloyed steels. Nb is reported to be beneficial as it expands the non-recrystallization temperature range which is useful in plate rolling, furthermore, it increases the hardenability which lead to retardation of GB ferrite and enhancement of intragranular ferrite formation in low heat input HAZ (Yang et al., 1999). Other works reported that Nb is detrimental as it increases the hardenability, enahances widmanstätten ferrite and upper bainite and enhances the MA formation on the interlath and reheated region. Its effect on the hardness increase due to precipitation of fine Nb(C, N) was also reported (Hattingh & Pienaar, 1998).

The effect on HAZ microstructure varies with composition, welding heat input and etc., two researchers (Tanaka & Kosazu, 1977) reported that HAZ of SAW 0.06 C-1.5 Mn-Cu-Ni steel contains more MA than 0.1 C-1.5 Mn –Cu-Ni steel in the presence of MA. Other researchers (Sakui et al., 1977) found that the increase in Nb linearly increases MA in 0.1 C-1.5 Mn HAZ.

Many researchers have investigated the effect of Nb on the HAZ toughness. Some researchers (Fujibayashi & Endo, 2002; Yamamoto et al., 1984) found that small addition of Nb decreases toughness however other researchers (Yasuhara et al., 1999; Kawano et al., 2004) found that no significant effect of Nb addition in case of low C steels. In the first research (Yasuhara et al., 1999) it was reported that small addition of Nb (0.02%Nb) increases toughness in Low C steel. In the other research (Kawano et al., 2004) it was observed that Nb content more than 0.02 % decreases toughness especially for high heat input welds. On 1999, a research (Kusabiraki et al., 1999) confirmed the former results; in low c steels (0.03 C), small addition of Nb (0.02%) improves toughness, while in medium C steels (0.07%), addition of Nb decreases toughness monotonously.

Two researchers (Hattingh & Pienaar, 1998) studied the weld HAZ embrittlement of Nb containing C-Mn steels and they found that: (i) the C content determine the toughness properties in general and tends to be particularly detrimental to HAZ toughness at higher C levels (0.19%C) in combination with Nb, (ii) Nb doesn't have a significant effect on HAZ toughness at low C levels (0.06%C) during high heat input thermal cycling, (iii) good toughness properties can be obtained at intermediate C levels of 0.12% with intermediate to high Nb additions at lower heat input of 1.5 and 3 KJ/mm, (iv) High C levels (0.19%C) combined with a low heat input result in the formation of untempered brittle martensite and lower bainite with poor toughness properties. A research (Fujibayashi & Endo, 2002) reported that small addition of Nb (0.01%) significantly degrades HAZ toughness due to promoting MA while another research (Sugimoto et al., 2000) reported that toughness is mainly controlled by C equivalent and Nb has little effect in CG-HAZ. Other research (Furuya et al., 2000) reported that high C-Nb free steels shows better crack tip opening displacement (CTOD) values than low C- Nb bearing steels.

There is conflicting interest among researchers about how Nb affects MA formation in CG-HAZ and how MA affects HAZ toughness. (Ohya et al., 1996) investigated the microstructures relevant to brittle fracture initiation at the HAZ of weldment in low C steel. They concluded that the brittle fracture initiation sites are mostly associated with intersections of bainitic ferrite areas with different orientations. They added that MA is not likely to play an essential role in the brittle fracture initiation. However, they didn't investigate the effect of steel composition on MA content as they used only one steel composition. On the contrary, (Li et al., 2001) emphasized that the presence of MA phase is the dominant factor in determining the toughness of intercritically reheated coarse-grained HAZ. In addition, they found correlations between the toughness and area fraction of MA phase. The presence of MA is widely reported as the main factor in determining the HAZ toughness. Due to the conflict mentioned above among researchers, the present study was carried out in order to reveal the effect of MA formation on the HAZ toughness of the low carbon - microalloyed steels.

2. Experimental work

The investigation involved eleven different steels. Plain C-Mn steel was chosen as the base metal. To find out the effect of Nb addition to different C equivalent, three groups of steels have the same Mn content with three levels of C (0.05, 0.1 and 0.15 mass%) were prepared. Each group has three steels, One without Nb addition and for the other two, Nb was added by 0.015 and 0.03 mass% respectively. Each group of the three groups has different C equivalent which is 0.3, 0.35 and 0.4 respectively. The carbon equivalent is expressed as: Ceq. = C+Si/24+Mn/6+Ni/40+Cr/4+V/14 (wt%). For comparison, two more steels with the same Nb content (0.015) and different Mn content were prepared. The chemical compositions of the eleven steels used are given in Table 1. The steels were designated as KT1~ KT11.

The simulated HAZ experiments were carried out for all Steels. The simulation route was chosen to obtain the most adverse HAZ properties. This is obtained by a single thermal cycle with a peak temperature of 1400°C representing the high heat input coarse-grained HAZ. The thermal cycle is shown in **Fig. 1**; the cooling rate between 1400°C and 800°C is 20 K/s followed by cooling rate of 2k/s below 800°C. After the thermal cycle, the Charpy specimens were prepared in the standard form of 10X10X55 mm. Notches are located in the uniform microstructure region at the center of the specimens. Charpy test was carried out at 0, -20 and –40 °C and impact transition curves were constructed.

The optical metallography specimens were prepared and then etched with 2% nital and Le Pera solution (Le Pera, 1980). Quantitative analysis of MA was conducted for all steels. Both macro and micro-hardness were measured on optical specimens using 10 kg and 10 g load respectively. The driving force of NbC precipitation was calculated using Thermo-Calc version P. The fracture surface of the Charpy specimens was observed using SEM.

Steels	C	Si	Mn	P	S	Al	Nb	N	Cequ
KT1	0.05	0.2	1.5	0.01	0.005	0.03	0	0.003	0.3
KT2	0.05	0.2	1.5	0.01	0.005	0.03	0.015	0.003	0.3
KT3	0.05	0.2	1.5	0.01	0.005	0.02	0.03	0.003	0.3
KT4	0.1	0.2	1.5	0.01	0.005	0.02	0	0.003	0.35
KT5	0.1	0.2	1.5	0.01	0.005	0.02	0.015	0.003	0.35
KT6	0.1	0.2	1.5	0.01	0.005	0.02	0.03	0.003	0.35
KT7	0.15	0.2	1.5	0.01	0.005	0.03	0	0.003	0.4
KT8	0.15	0.2	1.5	0.01	0.005	0.02	0.015	0.003	0.4
KT9	0.15	0.2	1.5	0.01	0.005	0.02	0.03	0.003	0.4
KT10	0.05	0.2	2.1	0.01	0.005	0.02	0.015	0.003	0.4
KT11	0.15	0.2	0.9	0.01	0.005	0.02	0.015	0.003	0.3

Table 1. The chemical compositions (mass %) of the steels used.

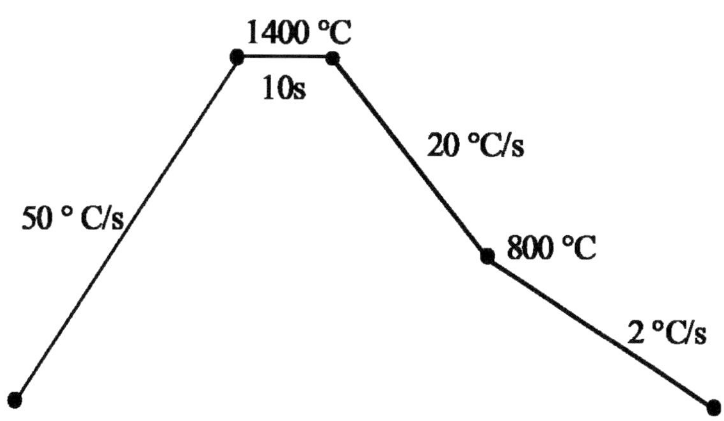

Fig. 1. Thermal cycle of HAZ simulation used.

3. Results and discussion

3.1 Optical metallography

Figures 2a ~2c show the optical microstructure of steels KT4 ~ KT6. All of them have the same carbon content 0.1% and different Nb content 0.05, 0.1 and 0.15 respectively. The typical microstructure of the low-alloy steel HAZ is characterized by a very coarse prior austenite grain size with grain boundary ferrite layers and acicular ferrite. The microstructures in Fig.2 vary according to the chemical composition. Steel KT4 (free Nb) consists of large areas of grain boundary ferrite (equiaxed and extended) and widmanstätten ferrite. As the Nb added, the large areas of allotriomorphic ferrite decreases and the widmanstätten ferrite are replaced by intragranular acicular ferrite as shown for Steel KT5 (medium Nb). Acicular ferrite is recognized in most weld microstructures as intragranulary nucleated bainite. Steel KT6 (high Nb) shows a similar microstructure to KT5 however the increase in Nb content (0.03%) leads to thinner allotriomorphic layers and increases the bainitic products of transformation as well; the intragranular acicular ferrite are replaced by lath structure. The results from optical metallography indicate that the Nb addition retard $\gamma \rightarrow \alpha$ transformation kinetics and as a result the allotriomorphic layers decrease and the transformation products tend to be bainitic.

Figures 3a ~ 3c show the optical microstructure of steels KT2, KT5 and KT8. All of them have the same Nb content (0.015%) and different carbon content 0.05,0.1,0.15% respectively. The effect of the increase in Carbon content shows similar effect as the increase in Nb content for steels KT4~KT6 which have medium carbon content. As the carbon content increases the microstructure varies from widmanstatten ferrite to acicular ferrite and finally bainitic lath structure and the grain boundary ferrite layers decreases. The only difference is that in Steel KT8 (high C - medium Nb), the bainitic structure is more dominant and the grain boundary layer is thinner than Steel KT6 (medium C- high Nb).

The dilatometric results, shown in Fig. 4 are in consistent with the microstructure results. Figure 4 shows that the grain boundary ferrite start temperature decreases with Nb and C addition, which verify the delaying effect of Nb on ferrite nucleation. The presence of grain boundary ferrite at the austenite grain boundaries induces the transition from a lath structure, bainite, to acicular ferrite. According to the overall microstructure obtained for the eleven steels, the different steels can be categorized to three categories, namely large grain boundary ferrite with widmnastatten ferrite plus areas of pearlite, which is denoted as category I; thin grain boundary ferrite plus intragranular acicular ferrite which is called category II; and finally the bainitic lath structure which is called category III. It is important to note that the grain boundary ferrite in the last category is very thin and not continuous which means that not all the grains are covered by boundary ferrite. Category I represents the base steel KT1, while category II represents free Nb- medium and high Carbon steels plus medium Nb- low Carbon steel. The last category represents Nb added - medium and high Carbon equivalent steels plus low Carbon - high Nb steels. A summary of these microstructures is shown in Fig. 5.

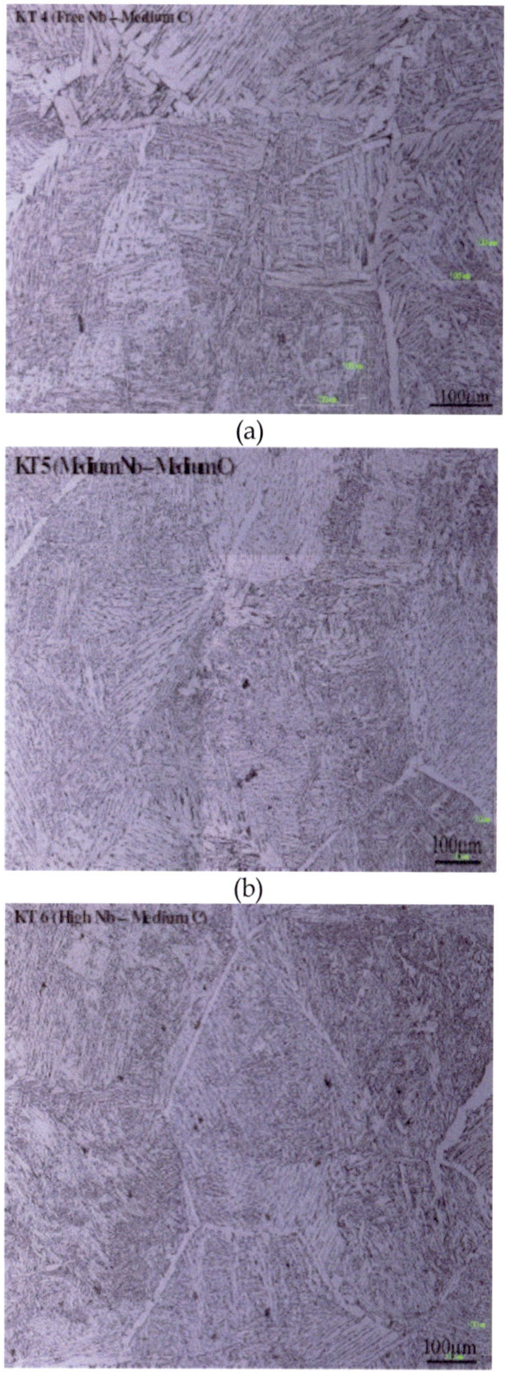

Fig. 2. (a) Optical photographs of steel KT4 (free Nb- medium C) after simulated cycle (b) Optical photographs of steel KT5 (medium Nb- medium C) after simulated cycle (c) Optical photographs of steel KT6 (high Nb- medium C) after simulated cycle.

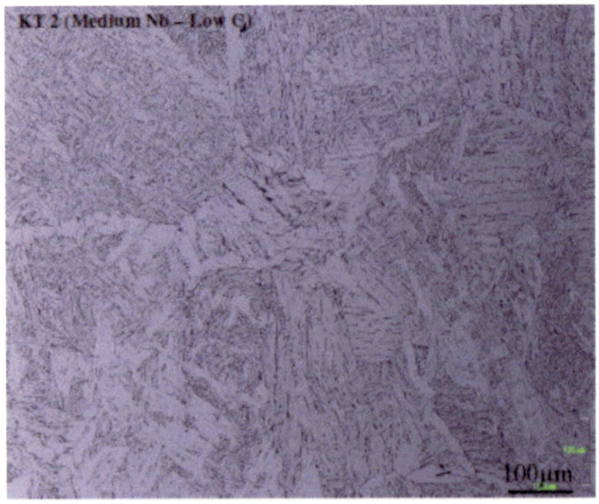

(a)

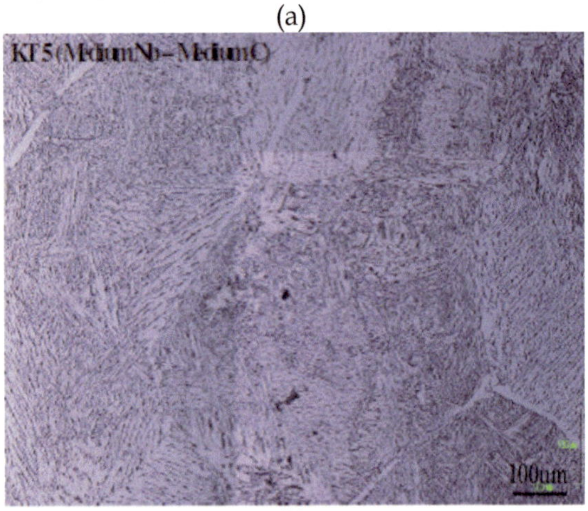

(b)

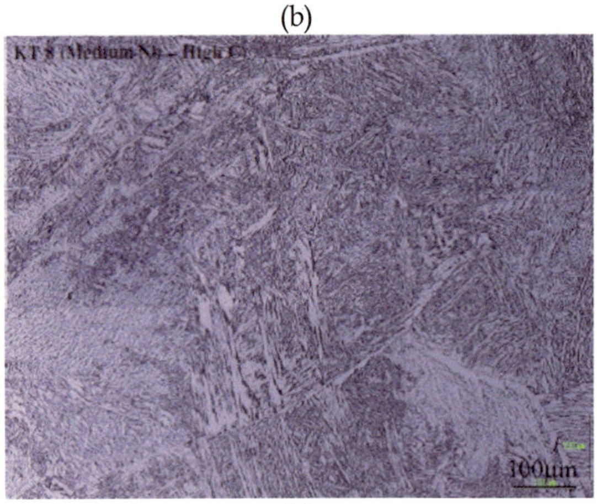

(c)

Fig. 3. (a) Optical photographs of steel KT2 (medium Nb- low C) after simulated cycle.
(b) Optical photographs of steel KT5 (medium Nb- med. C) after simulated cycle.
(c) Optical photographs of steel KT8 (medium Nb- high C) after simulated cycle.

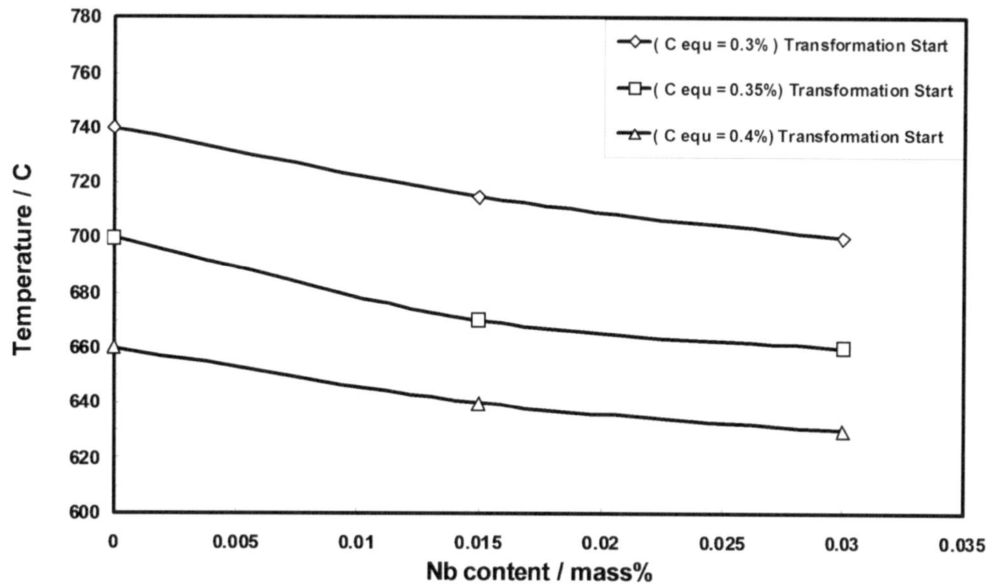

Fig. 4. Effect of C and Nb content on the transformation starting temperatures for grain boundary ferrite.

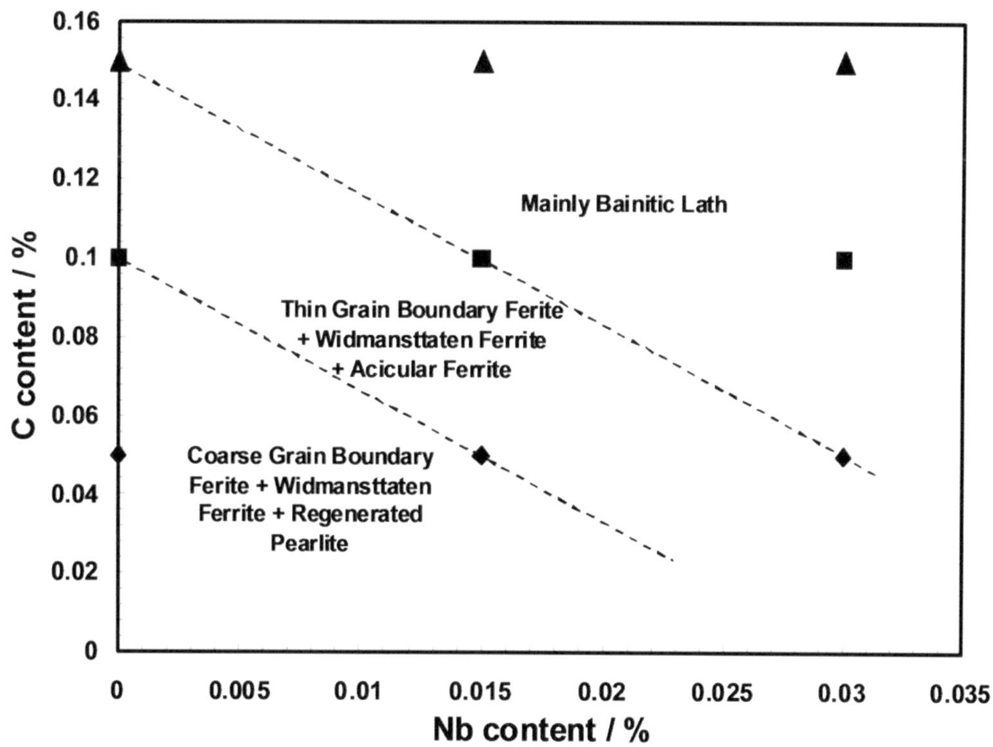

Fig. 5. Summary for the microstructures obtained under different C and Nb content for Steels KT1 ~ KT9.

Figure 6 shows Le Pera etching micrographs for Steels KT1, KT3, KT4, KT6, KT7 and KT9. It appears that the area fraction of MA markedly increases with increasing Nb content for both low carbon and medium carbon steels, whereas high carbon steels didn't show an increase of MA phase with Nb content which may be due to the consumption of the Nb in carbides precipitation.

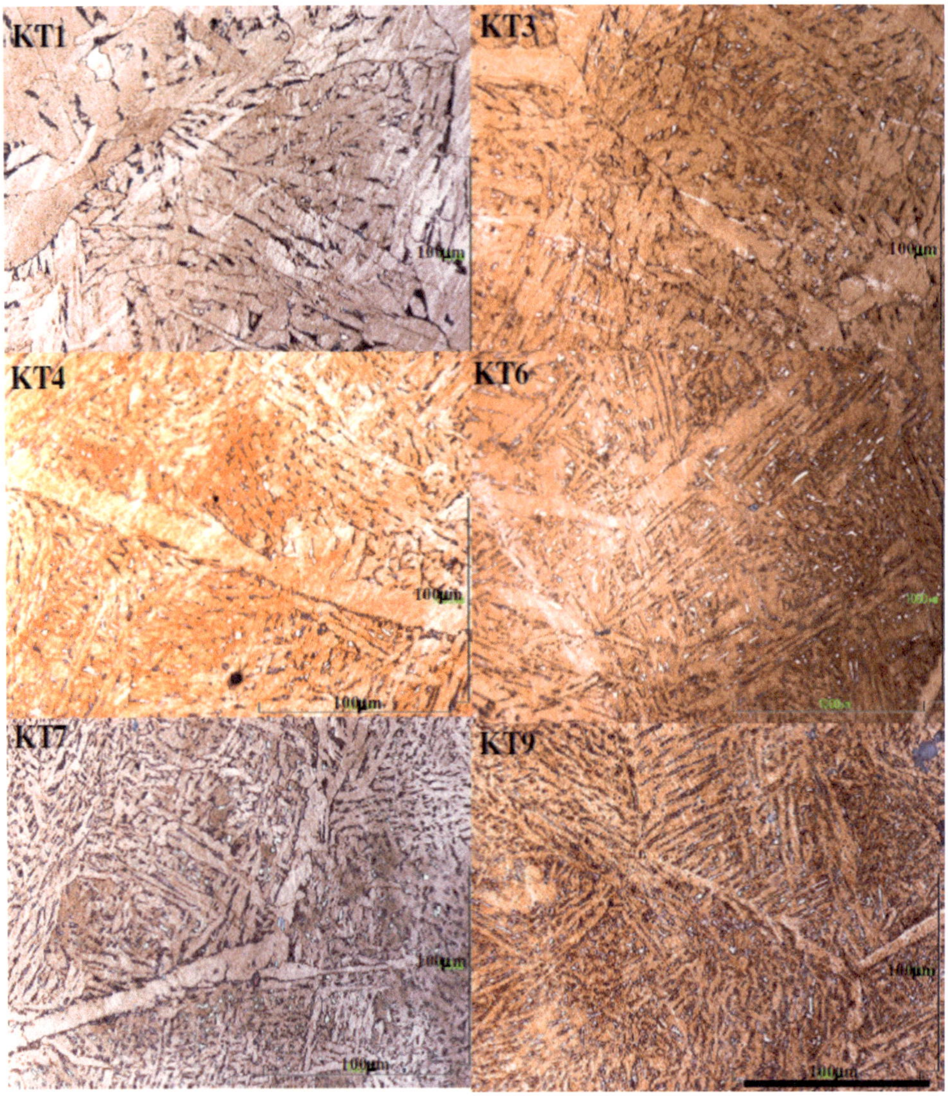

Fig. 6. Optical micrographs of Steels KT1, KT3, KT4, KT6, KT7 and KT9 showing MA after Le Pera etching.

Two researchers (Hattingh & Pienaar, 1998) observed relatively large undissolved Nb rich precipitates in 0.12%C-0.06%Nb and 0.19%C-0.03Nb steels, which could have an effect for reducing the MA area fraction for high Carbon steels in the following study. Calculation of the driving force of NbC precipitation For Fe-C-1.5%Mn-0.03%Nb system using Thermo-

Calc. showed that the driving force increases with the increase in C content and shows maximum value at 0.015 C content which is in good agreement with (Hattingh & Pienaar, 1998) work and can explain the reducing effect of MA area fraction for high carbon Steels.

Figure 7 shows La Pera etching micrographs for Steels KT2, KT11, KT8 and KT10. Steels KT2 and KT11 have the same carbon equivalent 0.3 % but the Mn content is 1.5 % for the former and 0.9% for the latter. While, Steels KT8 and KT10 have the same carbon equivalent 0.4 % but the Mn content is 1.5 % for the former and 2.1 % for the latter. High Mn content Steel (KT10) shows the highest fraction of MA, however the low Mn content Steel (KT11) doesn't show MA at all.

This implies that Mn content has more detrimental effect on the formation of MA than Nb. It was observed that the distribution of MA for all steels is random and MA concentrates at the interface between grain boundary ferrite and the interior of the grain which consist with the microsegregation mechanism for MA formation proposed by (Furuya et al., 2000). The only steel shows homogeneous distribution of MA is KT10, which has the maximum area fraction as well.

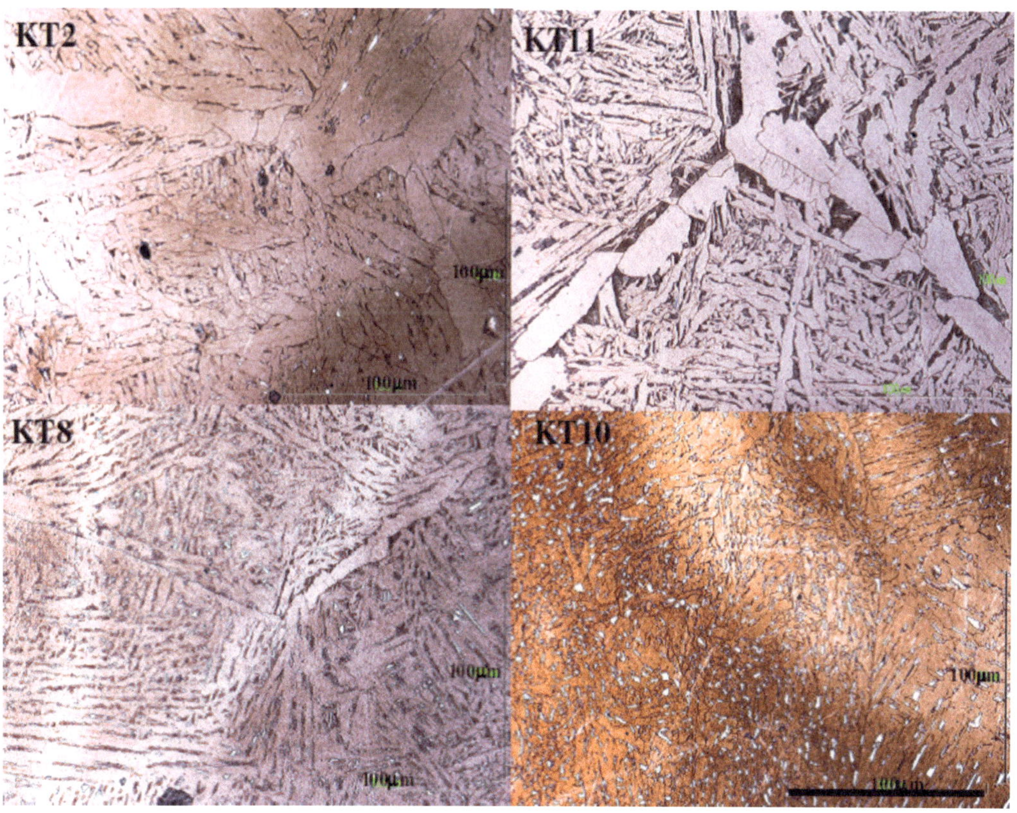

Fig. 7. Optical micrographs of Steels KT2, KT11, KT8 and KT10 showing MA after Le Pera etching.

Figure 8 shows the variation of grain boundary ferrite area fraction with Nb content for Steels KT1 ~ KT9. It can be shown that the increase in Nb content for low C steels shows a large decrease in grain boundary ferrite, whereas a slight decrease is resulted for medium and high carbon steels, which may be attributed to the segregation of Nb to the grain boundaries. (Yasuhara et al., 1999) suggest that because the Nb atoms have a large misfit within the iron lattice, austenite grain boundaries are favorable sites for the location of niobium atoms and as a result the nucleation and growth of polygonal ferrite will be retarded. The slight decrease in polygonal ferrite for medium and high carbon steels may be attributed to the precipitation of niobium carbides, which consume most of the Nb. The change of Mn and C content within the same carbon equivalent doesn't show a great effect on the allotriomorphic ferrite area fraction.

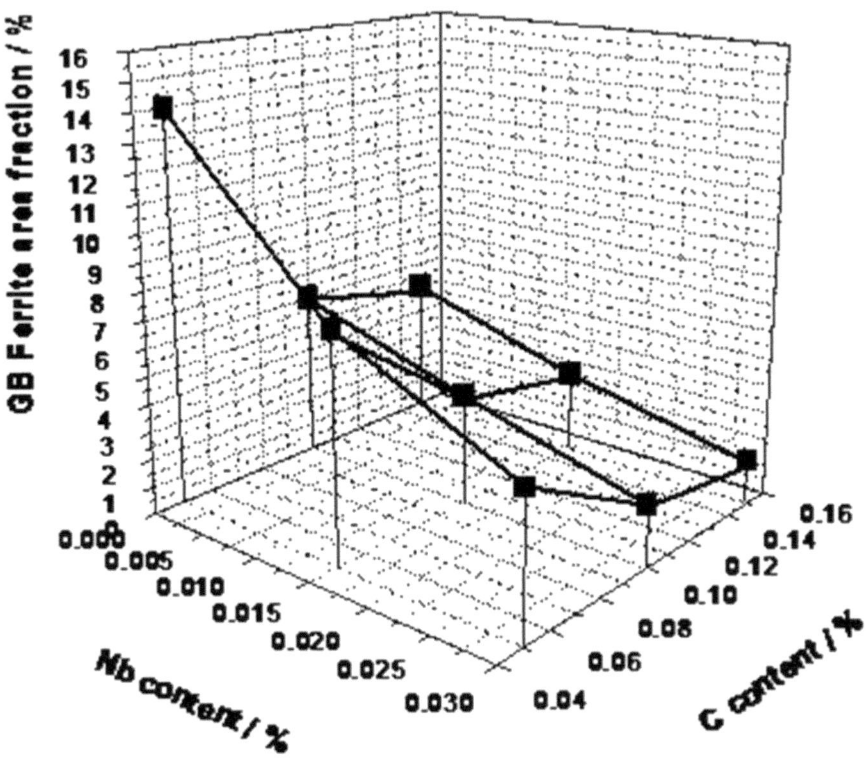

Fig. 8. Variation of grain boundary ferrite area fraction with Nb content for Steels KT1 ~ KT9.

Figure 9 shows the variation of MA area fraction with Nb content for steels KT1 ~ KT9, the increase in Nb content shows an increase in MA area fraction for low and medium carbon steels but in case of high carbon steels the increase in Nb content doesn't show an increase in MA area fraction due to the precipitation of niobium carbides as mentioned before.

Figure 10 shows the variation of MA area fraction with the change of C and Mn content within the same carbon equivalent. Steels KT2, KT11, KT8 and KT10 have the same Nb content (0.015%). The increase in Mn content shows sharp increase in MA area fraction at the same carbon equivalent. Lowering the Mn content to 0.9%, the MA was not observed at all and this was confirmed by more than 15 micrographs.

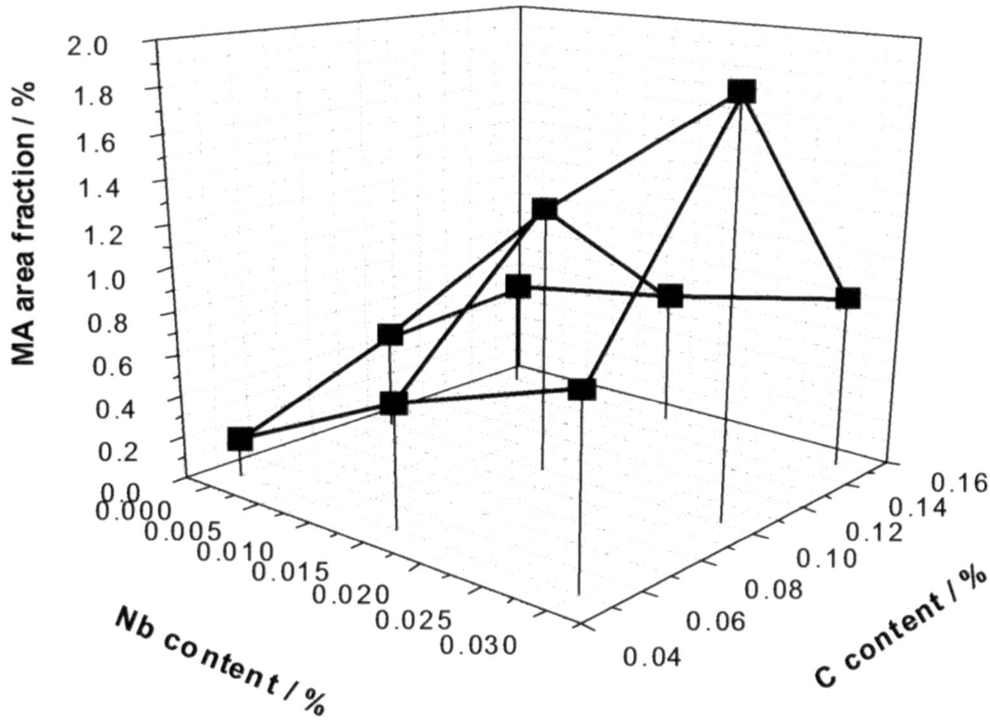

Fig. 9. Variation of MA area fraction with Nb content for steels KT1 ~ Kt9.

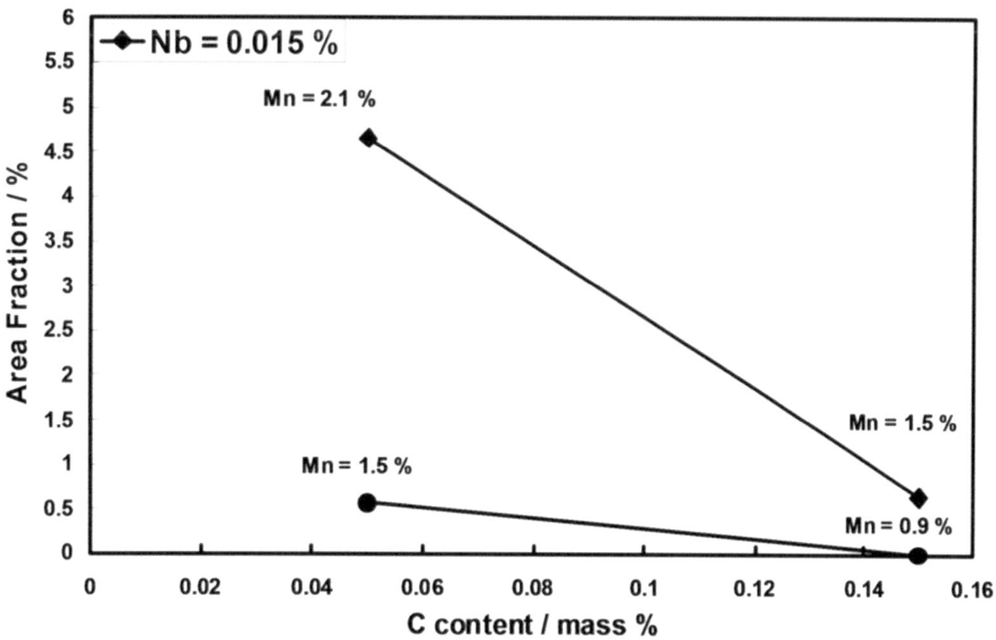

Fig. 10. Variation of MA area fraction with the change of C and Mn content.

3.2 Hardness and charpy impact testing

The variation of hardness with Nb content for Steels KT1 ~ KT9 is shown in Fig. 11. Hardness of the simulated HAZ increases with increasing C and Nb content. The change of hardness is corresponding to the change of microstructure, which was observed before. Figure 12 shows the fracture appearance as a function of test temperature. Steel KT4 (medium C-free Nb) shows the lowest transition temperature followed by KT2 (low C-medium Nb) then KT11 (high C- medium Nb). All the three steels have almost the same microstructure, which is grain boundary ferrite and acicular ferrite. Acicular ferrite has been known to provide an optimal combination of high strength and good toughness due to its refined and interwoven structure. They have a similar hardness with similar grain size as well. Steels KT5 and KT6 shows almost the same transition temperature as they have a bainitic microstructure with similar hardness and grain size. Steels KT8 and KT10 show the worst transition temperatures as they have the highest C equivalent (0.4%) but different Mn and C content. Steel KT10 has the highest MA area fraction but shows a similar transition temperature to KT8, which has low MA fraction, which imply that the microstructure features control the transition temperature and not MA fraction.

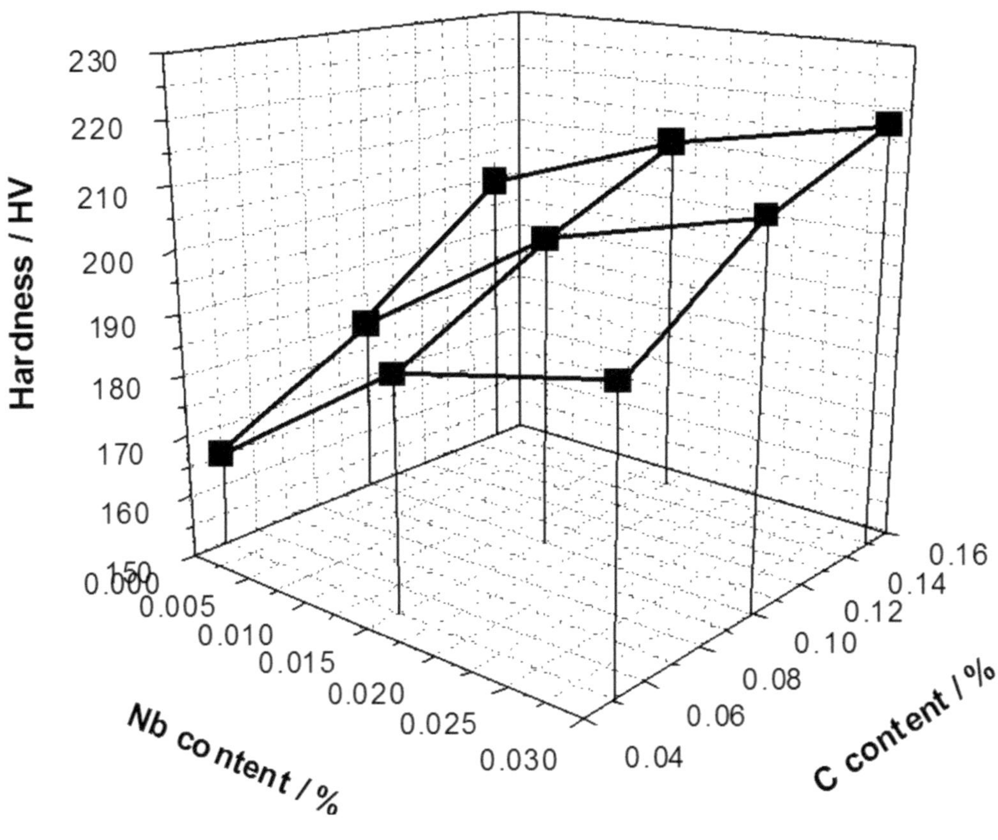

Fig. 11. Variation of hardness with Nb content for Steels KT1 ~ KT9.

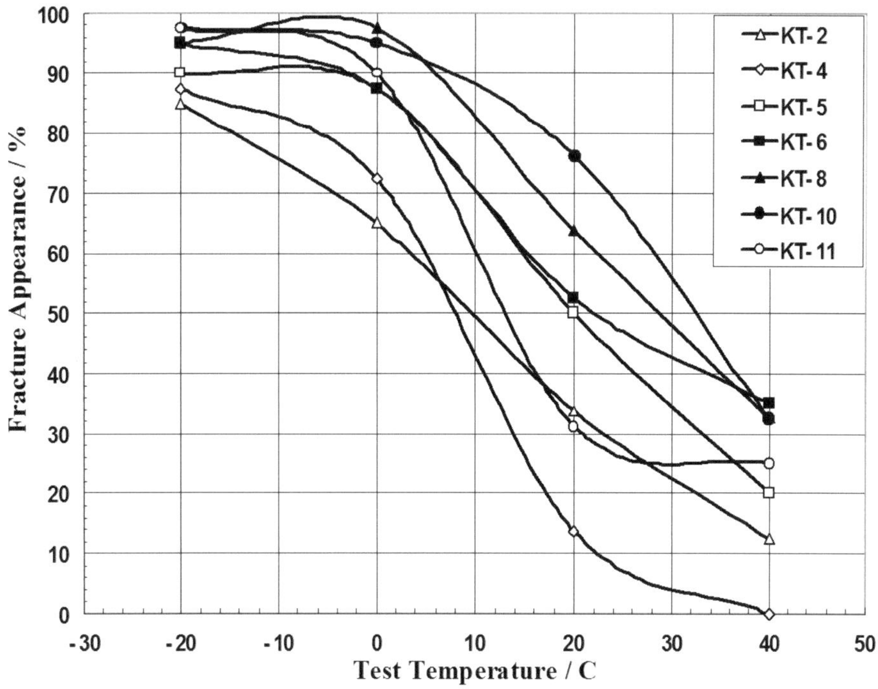

Fig. 12. Fracture appearance as a function of test temperature for some steels.

3.3 Fracture characteristics

Fracture characteristics were observed for all steels based on the initiation site of fracture. Three categories of initiation sites were observed based on the microstructure.

Category I, which is the base metal (KT1), the fracture initiates at the regenerated pearlite near grain boundary as shown in Fig. 13. The circle indicates the brittle fracture initiation site.

Category II, was obtained for welds whose microstructure contains relatively large areas of grain boundary ferrite and interior acicular ferrite. To this category belong steels KT2, KT4, KT7 and KT11. In these steels fracture initiates at the grain boundary ferrite. The fracture surface of Steel KT4 is shown in Fig. 14 as an example for this category.

Category III, was obtained at Nb level of 0.015 and 0.03 with 1.5% Mn content, to this category belong Steels KT3, KT5, KT6, KT8, KT9 and KT10. The fracture initiates at the intersection of bainitic packets with different orientation and propagate through the bainitic packets. An example of the fracture surface of this category is shown in Fig. 15.

The transition from one category to another was affected by the C equivalent and Nb weld metal levels which affects the microstructure products. Measurements of the fracture unit for samples impact fractured at -40°C shows that all steels have the same fracture unit (around 90 μm) and there is no effect of Nb on the fracture unit. The existence of pearlite colonies, which is well known with its bad effect on toughness acts as initiation site for the brittle fracture for base steel as pearlite colonies don't exist in Nb containing steels.

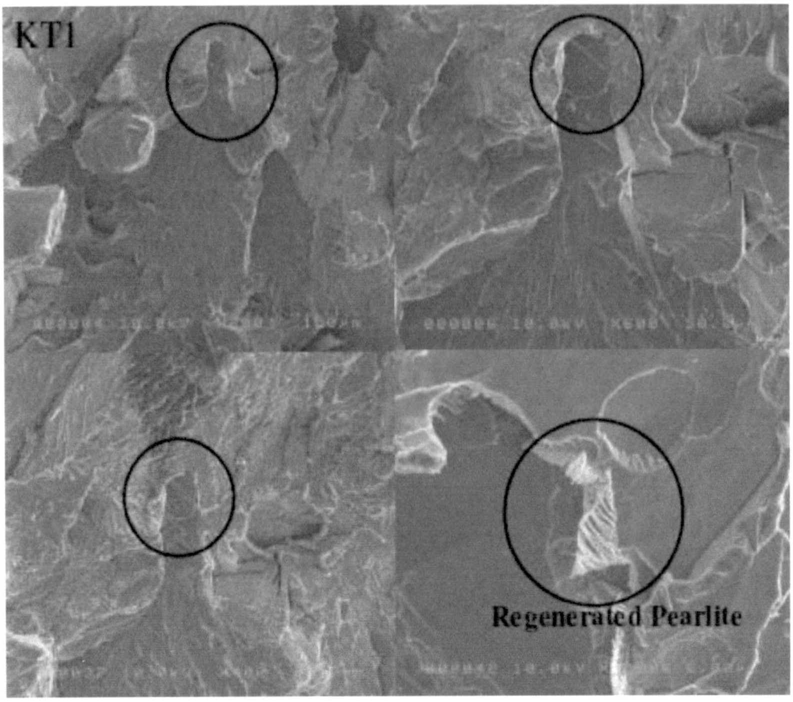

Fig. 13. Fractograph for steel KT1 (base steel).

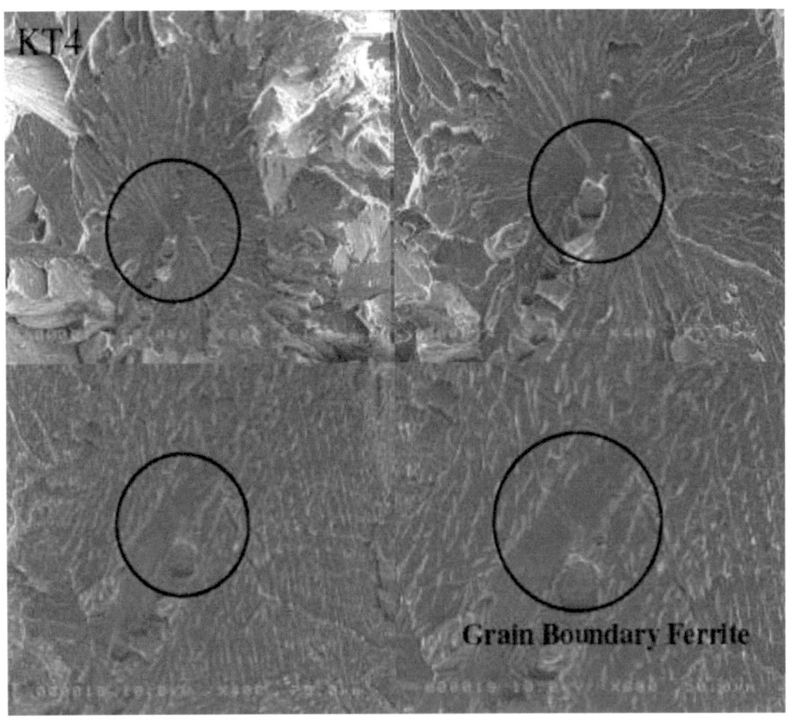

Fig. 14. Fractograph for steel KT4 (medium C-free Nb).

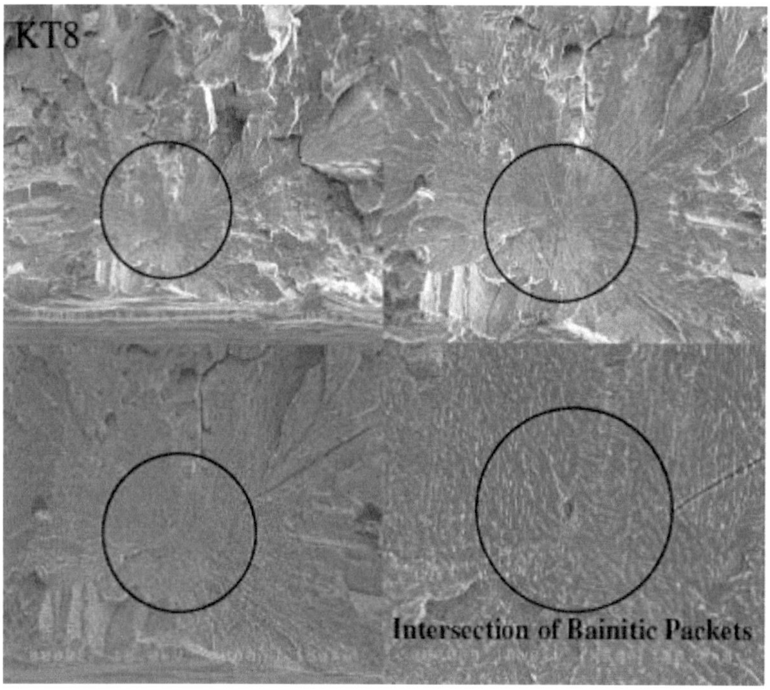

Fig. 15. Fractograph for steel KT8 (high C - Medium Nb).

To correlate between the fracture characteristics and the microstructure obtained for all steels. All the metallurgical aspects have been studied, MA area fraction, grain boundary ferrite area fraction and micro-hardness differences among different phases. It was found that the initiation point of fracture depends on the grain boundary area fraction as shown in Fig. 16.

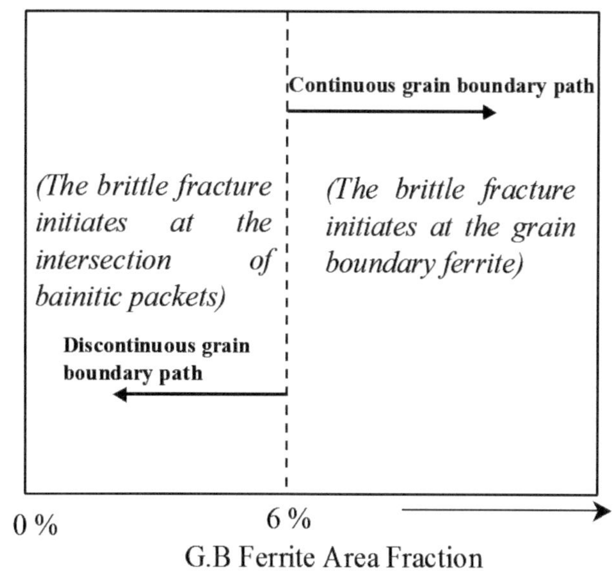

Fig. 16. Relationship between initiation sites and G.B ferrite area fraction.

There are two fracture characteristics, which is determined according to the grain boundary ferrite area fraction. The first one in which the fracture initiates at the grain boundary ferrite and this happened for steels with grain boundary ferrite fraction than 6% and the grain boundary ferrite path is continuous. However if the grain boundary ferrite is less than 6% and discontionus, the fracture initiates at the intersection of bainitic packets. To make sure with this experimental observation, the thickness of grain boundary ferrite layer is calculated for 6% area fraction for 300 μm grain size, it was found that the layer is about 0.03 μm. This thickness is with good agreement with the experimental observation and imply that if the area fraction is less than 6%, it will be difficult to cover all the grains which will result in discontinous grain boundary ferrite path and the initiation point will be changed. Steels free Nb - medium carbon, free Nb - high carbon and low carbon - medium Nb steels (KT4, KT7 and KT2 respectively) shows grain boundary area larger than 6% and this is because the relatively high transformation temperature. The large thickness of grain boundary ferrite with about 100 HV lower microhardness value than the interior of the grain create a stress concentration at the grain boundary ferrite which leads to the brittle fracture initiation at such sites. Although MA was not observed at the initiation sites under SEM, which may be due to debonding during fracture, the concentration of MA observed after Le Pera etching around the grain boundary ferrite for these steels was confirmed by observation of more than 15 positions.

For steels with grain boundary ferrite area fraction less than 6%, the grain boundary path is not continous and the matrix is mainly bainite, the MA has the same alignment as the bainitic laths and the stress is concentrated at the intersection of bainitic packets, which leads to fracture initiation at these intersections. A summary of the microstructure features and corresponding initiation sites are shown in Table 2.

(Hill & Levine, 1977) emphasized that the relative amount of boundary ferrite and the availability of continuous boundary path determine the initiation of the cracking in their study made on the effects of Nb and V on weld metal structure and impact properties.

(Ohya et al., 1996) have studied the microstructures relevant to brittle fracture initiation at the HAZ of low alloy steel (0.1% C, 1.5 % Mn, 0.5 % Cu, 0.55 % Ni, 0.012 % Nb); they found that the brittle fracture initiation sites are mostly associated with intersections of bainitic ferrite with different orientations. They added that MA constituent is not likely to play an essential role in the brittle fracture initiation.

Some researchers (Ji-ming et al., 2010) have studied the microstructure and mechanical properties of new ultrahigh strength pipeline steel (X120) with high yield strength and high impact toughness. In their work, mechanical properties and microstructure of the steel were investigated. The steel exhibited outstanding mechanical properties with yield strength levels of up to 951 MPa and tensile strength levels up to 1023 MPa. The sharp notch toughness with absorbed energy values of 227 J/cm2 at -30 °C and shear area of up to 95% in drop weight tear test (DWTT) at temperature of -20 °C were achieved. It was found that microstructure of the steel comprises a majority of low-carbon lath bainite with different sublaths and subsublaths, meanwhile there is a high density of dislocation between laths and the dispersed film-like martensite austenite (M-A) constituents. They didn't investigate the phenomenon of M-A formation as they used only one chemical composition which is as following: C 0.05, Mn 1.9, Nb 0.048, Ti 0.015, Mo 0.03, Cr 0.22, B 0.0013, Fe balance. They also didn't investigate the effect of welding on the mechanical properties and microstructure which was investigated in the present work.

Other researchers (Liang-yun et al., 2011) have studied the correlation between microstructures and mechanical properties of a Nb-Ti micro-alloyed pipeline steel. The results revealed that with decreasing the finish rolling temperature and the cooling stop temperature, the matrix microstructure was changed from quasi-polygonal ferrite to acicular ferrite, as a result of improvement of both strength and low temperature toughness. Their steel was only 1.1% Mn and as a result no M-A was observed in it which agrees with the present work.

Darcis et al., 2008 have investigated the crack tip opening angle (CTOA) in five pipeline steels and they concluded that the low carbon (0.05%), fine grained ferrite-pearlite pipeline steel has the highest resistance to crack growth. For steels with the same carbon content, the lower Mn contents steels show higher resistance to crack growth which agrees with the present work that the high Mn content has a great effect on MA formation and leads to poor toughness. In weld simulated HAZ, it is difficult to obtain ferrite-pearlite microstructure but Widmnasttaten structure or bainitic structure is usually obtained and it is well known that both Widmnsttaten and bainitic structure have low resistance to crack growth.

As mentioned before, the relative amount of grain boundary ferrite, which is 6 % for these steels determine the fracture behavior, and as a result the toughness. It is important to stress that this relative amount varies according to the chemical composition, thermal cycle and grain size. Therefore this relative amount can't be used for other series of steels with different chemical composition, thermal cycle and grain size.

In the present study, a good correlation between fracture properties and microstructure was established and the confliction among researchers about the brittle fracture initiation sites and its relation to microstructure in Nb added HSLA could be clarified.

Steel	**C%**	**Cequ**	**Nb%**	**Grain size, μm**	**Microstructure**	**G.B.F %**	**MA %**	**Initiation Point**
Low C- Free Nb	0.05	0.3	0	348	W.F + G.B.F+ small areas of Pealite	14	0.2	Regenerated pearlite near G.B
Low C+ Nb	0.05	0.3	0.03	207	Lath Structure	5	1	Intersection of bainitic packets
Med. C - Free Nb	0.1	0.35	0	336	G.B.F + W.F + acicular Ferrite	6	0.5	At Grain boundary
Med. C + Nb	0.1	0.35	0.03	248	Lath Structure	2	2	Intersection of bainitic packets
High C - Free Nb	0.15	0.4	0	342	G.B.F + W.F + acicular Ferrite	6	0.5	At Grain boundary
High C + Nb	0.15	0.4	0.03	230	Lath Structure	1.8	1	Intersection of bainitic packets
Low C- High Mn	0.05	0.4	0.015	236	Lath Structure	2.5	5	Intersection of bainitic packets
High C- LowMn	0.15	0.3	0.015	330	G.B.F + W.F + acicular Ferrite	6	0	At Grain boundary

Table 2. The initiation sites observed for different steels after impact fractured at -40°C.

3.4 The formation mechanism of MA

The grain size for all steels was measured and found to be mainly dependent on the C equivalent and there is no major change on the grain size for the 0.4% carbon equivalent steels. For the low carbon equivalent steels (0.3%), the decrease of Mn content from 1.5 to 0.9 % results in an increase of grain size by about 20%. Many researchers reported a refining effect for the Nb on the grain size but for this series of steels, the Nb content is the same and as a result no effect of the Nb on the grain size was observed. The change of grain size with C and Mn content at the same carbon equivalent is shown in Fig.17.

Some researchers (Furuya et al., 2000) proposed a mechanism for the formation of MA, this mechanism is illustrated schematically in Fig.18. It can be summarized in that the grain boundary ferrite causes an enrichment of alloying elements and C at the interface between the grain boundary ferrite and the bainite and also near the interface region. This enrichment is mainly due to the low solubility of alloying elements in ferrite compared to that in austenite. This micro-segregation enhances the formation of MA or M* as they named it. To confirm this mechanism, EPMA technique has been carried out for all steels to explain the formation mechanism of MA. The difference in chemical composition of these steels was selected to investigate the effect of change in Mn and C content at the same carbon equivalent on MA formation mechanism. A line analysis method with 10 μm step was used; the starting point was the center of 10X10X1 mm samples which were subjected to the simulated thermal cycle mentioned before.

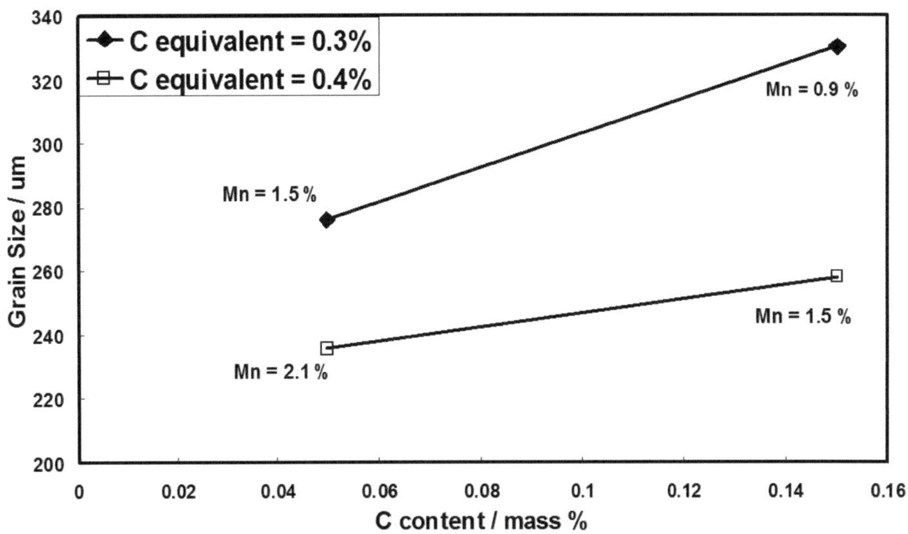

Fig. 17. Effect of C content on the grain size.

The EPMA results for Steel KT4 (2.1% Mn - 0.05% C) are shown in Figs. 19 and 20. A considerable degree of segregation was obtained for C and Mn. The degree of Mn segregation is much higher than that of C which may be attributed to the larger content of Mn. To confirm the furuya's mechanism, steel KT11 (0.9% Mn – 0.15% C)) was also examined using EPMA, the degree of Mn segregation was much lower than that of steel KT10 (2.1% Mn – 0.05% C) however, the degree of C segregation is higher than that of Steel KT10 (2.1% Mn – 0.05% C). This behavior is mainly due to the higher C content of Steel

KT11 and lower Mn content. From microstructure observation, Steel KT11 shows no MA at all and at the same time, it has a considerable degree of microsegregation for C but low degree for micro segregation for Mn and this is in good agreement of furuya's mechanism which requires both enrichement of C and microsegregation of alloying elements as conditions for MA formation. In steel KT11, only the enrichement of C was confirmed by EPMA but the segregation of alloying elements was not obtained in a considerable degree which results in no MA formation. As for Nb, no microsegregation was observed for both steels KT11 and KT10, which may be due to the low content of Nb (0.015%).

Finally, the mechanism proposed by furuya and coworkers was confirmed by good experimental evidence which is the EPMA results of this series of steels.

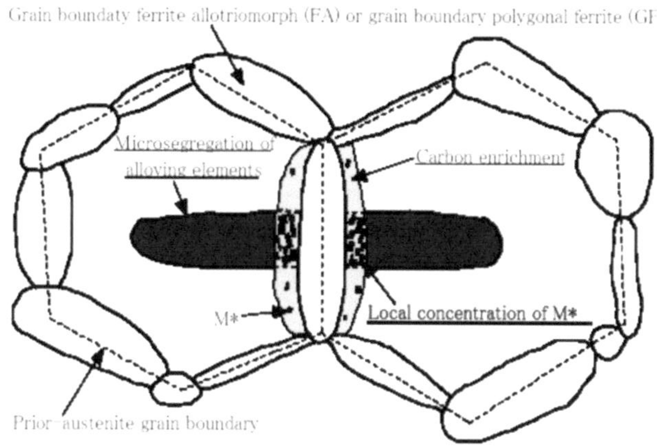

Fig. 18. MA formation mechanism proposed by furuya et al.

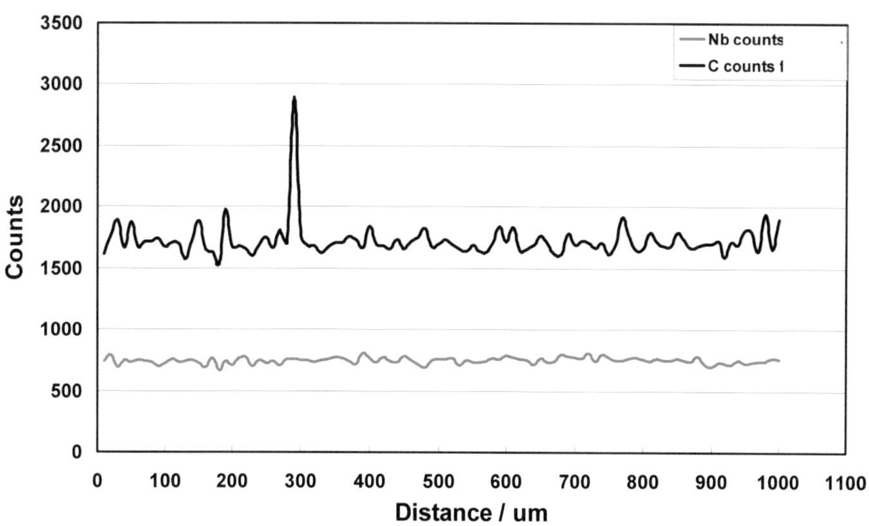

Fig. 19. The EPMA results for Nb and C for Steel KT10.

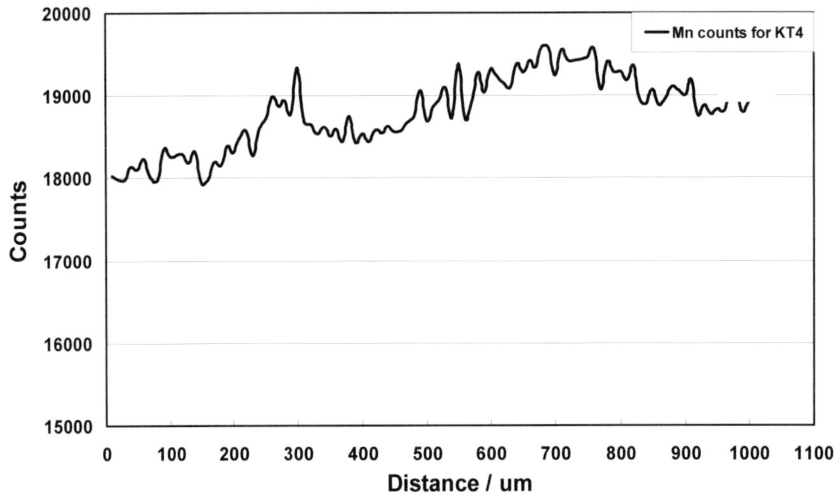

Fig. 20. The EPMA results for Mn for Steel KT10.

4. Summary

The effect of Nb, C and Mn on HAZ Toughness of HSLA Steels was investigated, and the following conclusions can be derived:

1. The microstructures of HAZ varies according to the Carbon equivalent and Nb content and it was found to be consisted of three categories; Category I (base steel), consists of grain boundary ferrite, Widmanstatten ferrite, MA constituents and small areas of pearlite; Category II, consists of grain boundary ferrite, Acicular ferrite, Widmanstatten ferrite and MA constituents; Category III, consists of very thin grain boundary ferrite, lath structure and MA constituents.
2. The CVN transition temperatures reflect the microstructure dependence; category II shows relatively good toughness compared with category III microstructure. The brittle fracture was observed to be initiated at the pearlite colonies, at the grain boundary ferrite and at the intersection of the bainitic packets for the three categories respectively.
3. Mn is the most dominant element affects the formation of MA at the same Carbon equivalent; lowering the Mn content to 0.9% at the same carbon equivalent results in nul MA while increasing it to 2.1% at the same Carbon equivalent results in large increase in MA area fraction. The area fraction of MA increases by Nb addition for low and medium Carbon steels but for high Carbon steels, MA area fraction shows a slight increase due to carbides precipitation.
4. A relative amount of grain boundary ferrite, which is 6 % for these steels determine the initiation site, and as a result the fracture behavior and toughness. A good correlation between fracture properties and microstructure was established based on this criterion.
5. The micro-segregation of C and Mn at the interface between the grain boundary ferrite and the bainitic matrix could explain the formation mechanism of (MA). Enrichment of C alone near the grain boundary didn't result in MA formations, which imply that both C enrichment and alloying elements micro-segregation should be achieved simultaneously to form MA.

5. References

Bhadeshia, H. , Sevensson, L. & Gretoft, B. (1985). A model for the development of microstructure in low-alloy steel (Fe-Mn-Si-C) weld deposits. *Acta Metall.*, Vol. 33, pp. 1271-1283, ISSN 0001-6160

Darcis, Ph. P. ,McCowan, C.N., Windhoff, H. ,McColskey, J.D. & Siewert, T. A. (2008). Crack tip opening angle optical measurement methods in five pipeline steels. *Engineering Fracture Mechanics,* Vol. 75, pp. 2453 - 2468, ISSN 0013-7944

Fujibayashi, S. & Endo, T. (2002). Creep Behavior at the Inter-critical HAZ of a 1.25Cr-0.5Mo Steel. *ISIJ Int.* ,Vol. 42, No. 11, pp. 1309-1317, ISSN 0915-1559

Furuya, H., Uemori, R., Tomita, Y., Aihara, S. & Hagiwara, Y. (2000). Ductile crack propagation characteristics and mechanism of structural steel under high strain rate. *Technical report by Steel Research Laboratories, Tetsu-to-hagané, Nippon Steel Corporation*, Vol. 86, No.6, pp. 409-416, ISSN 0021-1575

Hattingh, R. & Pienaar, G. (1998). Weld HAZ embrittlement of Nb containing C–Mn steels. *International Journal of Pressure Vessels and Piping,* Vol. 75, pp. 661-677, ISSN 0308-0161

Ji-ming, Z., Wei-hua, S. & Sun Hao, S. (2010). Mechanical Properties and Microstructure of X120 Grade High Strength Pipeline Steel. *Journal of Iron and Steel Research International,* Vol. 17. No. 10, pp. 63-67, ISSN 1006-706X

Kawano, H., Shibata, M., Okano, S., Kobayashi, Y. & Okazaki, Y. (2004). TMCP Steel Plate with Excellent HAZ Toughness for High-rise Buildings. *R&D Kobe Steel Engineering Reports, Tokyo,* Vol. 54,pp. 110-113, ISSN 0373-8868

Kusabiraki, K. , Saji, S. & Tsutsumi, T. (1999). Effects of cold rolling and annealing on the structure of γ'' precipitates in a Ni-18Cr-16Fe-5Nb-3Mo alloy. *Metallurgical and Materials Transactions A,* Vol. 30 , 1923-1931, ISSN 1073-5623

Le Pera, F. (1980). Metallographic methods for revealing the multiphase microstructure of steels. *Journal of Metals,* Vol. 32, pp. 38-39, ISSN 1047-4838

Levine, E. & Hill, D. (1977). Structure- Property relationships in low c weld metal. *Metallurgical Transactions A,* Vol. 8A, pp.1453-1463, ISSN 1073-5623

Li, Y., Crowther, D., Green, M., Mitchell, P. & Baker, T. (2001). The Effect of Vanadium and Niobium on the Properties and Microstructure of the Inter-critically Reheated Coarse Grained Heat Affected Zone in Low Carbon Micro-alloyed Steels. *ISIJ Int.,* Vo. 41, No. 46, ISSN 0915-1559

Liang-yun, L., Chun-lin, Q., De-wen, Z. & Xiu-hua,G. (2011). Microstructural Evolution and Mechanical Properties of Nb-Ti Microalloyed Pipeline Steel. *Journal of Iron and Steel Research International,* Vol. 18, No. 2, pp. 57-63, ISSN 1006-706X

Nakasugi, H., Matsuda, H. & Tamehiro, H. (1981). Steels for Line Pipe and Pipeline Fittings. *Proceedings of an international conference and held at Grosvenor House, The Metals Society* , ISBN 9780904357455, London, October, 1981

Ohtani, H. , Hashimoto, T. & Kyogoku, T. (1983). HSLA. Steel Technology and Applications. *ASM, Philadelphia, PA,* p. 843, ISSN 0097-3912

Ohya, K., Kim, J. , Yokohama, K. & Nagumo, M. (1996). Microstructures relevant to brittle fracture initiation at the heat-affected zone of weldment of a low carbon steel. *Metallurgical and Materials Transactions A,* 27A , pp. 2574-2582, ISSN 1073-5623

Sakui, S., Sakai, T. and Takeishi, K. (1977). Hot deformation of austenite in a plain carbon steel. *Trans. Of Iron Steel Institute Japan,* Vol. 17, pp. 718-725, ISSN 0021-1583

Sham, N. (1985). A comparative study of the HAZ properties of B-containing low alloy steels. *Journal of Metals*, Vol. 12, No. 21, ISSN 1047-4838

Sugimoto, K., Iida, T., Sakaguchi, J. & Kashima, T. (2000). Retained Austenite Characteristics and Tensile Properties in a TRIP Type Bainitic Sheet Steel. *ISIJ Int.*, Vol. 40, No. 9, pp. 902-908, ISSN 0915-1559

Tanaka, J. & Kosazu, I. (1977). Acicular ferrite HSLA for line pipe. *Journal of Metal Science and Heat Treatment*, Vol. 19, No.7, pp. 559-572, ISSN 0026-0673

Yamamoto, S., Ouchi, C. & Osuka, T. (1982). The effect of microalloying elements on the recovery and recrystallization in deformed austenite, Proc. Thermo-mechanical Processing of Micro-alloyed Austenite. Ed. by A. J. DeArdo et al., AIME, Warredale, PA, pp.613-619, ISBN 9780895203984

Yang, J. & Bhadeshia, H. (1989). Orientation relationships between adjacent plates of acicular ferrite in steel weld deposits. *Material Science and Technology*, Vol. 5, pp.93-97, ISSN 0267-0836

Yang, J. & Bhadeshia, H. (1991). Acicular ferrite transformation in alloy-steel weld metals. *J. Mater. Sci.*, Vol. 26, pp. 839-845, ISSN 0022-2461

Yang, J., Hyang, C. & Chou, C. (1999). Microstructures of Heat-Affected Zone in Niobium Containing Steels. *Materials Transactions, JIM*, Vol. 40, No.3, ISSN 0916-1821

Yasuhara, H., Okuda, K., Tosaka, A. & Furukimi, O. (1999). Carbon-manganese wrought steel with inoculated acicular ferrite microstructure. *CAMP-ISIJ*, Vol.12, ISSN 0915-1559

Part 3

Behaviour of Steels: Influence of the Enviroments

5

Environmentally Assisted Cracking Behavior of Low Alloy Steels in Simulated BWR Coolant Conditions

J. Y. Huang, J. J. Yeh, J. S. Huang and R. C. Kuo
Institute of Nuclear Energy Research (INER), Chiaan Village, Lungtan, Taiwan

1. Introduction

SA533, low carbon and low alloy steel, is most commonly used for nuclear reactor pressure vessels (RPV) whose integrity governs the safety of nuclear power plants. Fatigue is one of the main degradation mechanisms affecting the pressure vessel integrity of pressurized water reactors (PWRs) and boiling water reactors (BWRs) (Shah, 1993 ; Huang, 2001, 1999, 2007 ; Seifert, 2001, 2003). Fatigue life can generally be divided into two distinct phases—initiation and propagation. The factors affecting the initiation and propagation lives may not be the same. Recent test data indicate the initiation life is significantly decreased when the applied strain range, temperature, DO level in water, sulfur content in steel and strain rate are simultaneously satisfied(Chopra, 1997). The fatigue crack growth rates of RPV materials in simulated light water reactor (LWR) coolant environments are influenced by sulfur content, sulfide morphology (Huang, 2003, 2004 ; Van Der Sluys, 1985 ; Combrade, 1987) and orientation, water chemistry, loading frequency separately and synergistically (Van Der Sluys, 1985, 1987 ; Amzallag, 1983 ; O'Donnell, 1995 ; Eason, 1993 ; Roth, 2003 ; Chopra, 1998 ; Atkinson, 1986 ; Shoji, 1986). The sulfur content has been reported to enhance the corrosion fatigue crack growth rates of low alloy steels (Tice, 1986). Hydrogen water chemistry (HWC) has proved to be a powerful method for mitigating environmentally-assisted cracking (EAC) of stainless steel and nickel-base alloy components. Ritter (Ritter & Seifert, 2007) demonstrated HWC resulted in a significant drop in low-frequency corrosion fatigue crack growth rates by at least one order of magnitude with respect to normal water chemistry (NWC) conditions for pressure vessel steels with sulfur content lower than 0.02 wt %. It is of practical and academic interest to study the effect of HWC on the mitigation of corrosion fatigue initiation and propagation for low alloy steels. The slip-oxidation mechanism has been widely accepted to account for the crack propagation of the carbon and low alloy steels in LWR water systems (Ford, 1987). This mechanism relates crack advance to the oxidation rate that occurs at the crack tip. In order to better predict the crack growth rates of low alloy steels, the interaction between the oxide films and sulfur ion dissolved from steels in the oxygenated or HWC water environments should be clarified.

In this study, the low cycle fatigue and corrosion fatigue crack growth tests were performed on A533B3 low alloy steels with different sulfur contents under simulated BWR coolant conditions. Corrosion fatigue tests were conducted in different water chemistry including air saturated, deoxygenated by nitrogen and hydrogen. The fracture features of fatigued specimens were studied with optical stereography and scanning electron microscopy (SEM).

2. Experimental procedures

2.1 Materials

A533B3 steel plates with three sulfur content levels ranging from 0.008 wt % (weight percent) to 0.027 wt % were manufactured according to the specifications of ASTM A533. The materials were rolled from a thickness of 150 mm to 30 mm and solution treated at 900°C for 1.5 hours, then quenched and tempered at 670°C for 1.5 hours. Their chemical compositions and mechanical properties are given in Tables 1-2, respectively.

Designation	Composition (wt%)									
	C	Si	Mn	P	S	Ni	Mo	A1	N	Fe
Y1	0.19	0.22	1.22	0.015	0.008	0.60	0.49	0.035	0.005	Bal.
Y2	0.19	0.22	1.27	0.015	0.016	0.60	0.49	0.035	0.005	Bal.
Y3	0.21	0.23	1.25	0.015	0.027	0.60	0.49	0.035	0.005	Bal.

Solution treated at 900°C for 1.5 hour, then quenched and tempered at 670°C for 1 hour.

Table 1. Chemical compositions of A533B3 steels

Designation	Temperature (℃)	Ultimate tensile strength (MPa)	Yield strength (MPa)	Total elongation (%)	Uniform elongation (%)
Y1	25	711	625	30.3	10.5
	300	692	510	31.1	13.2
Y2	25	686	600	30.6	10.3
	300	680	516	30.1	12.1
Y3	25	722	630	28.6	10.1
	300	722	550	31.2	13.2

Table 2. Mechanical properties of A533B3 steels

2.2 Metallographic examination

To reveal the MnS morphology and carbide/nitride precipitate distribution of the low alloy steels, the as-received specimens were polished following a standard metallographic practice, then etched in a 5 vol. % Nital solution (5 vol. % nitric acid and 95 vol. % ethanol) for about 20 seconds and examined with optical microscopy.

2.3 Low-cycle fatigue test

According to the ASTM E 606 specifications, round bar fatigue specimens with a gage length of 16 mm and diameter of 8 mm were machined, as shown in Fig. 1. Before fatigue testing, all specimens were well polished as per the recommendations of ASTM E 606. The strain was measured by an LVDT extensometer which was calibrated by a strain gage.

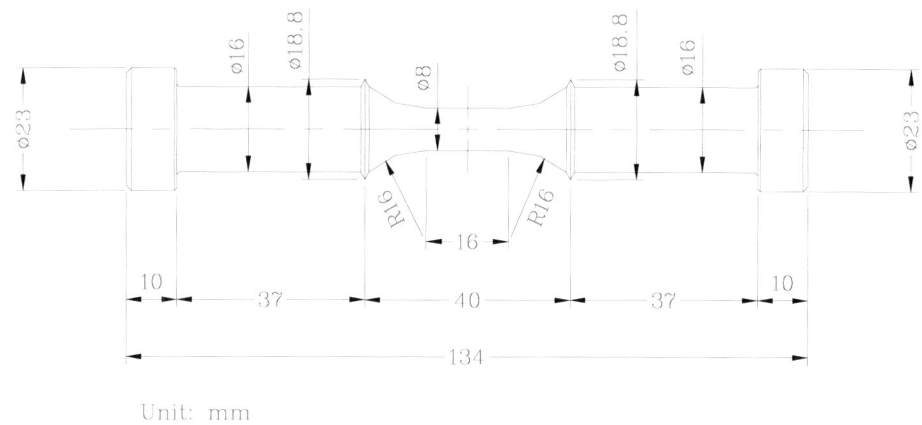

Fig. 1. Round bar specimen for low-cycle fatigue test.

The low-cycle fatigue tests were conducted on a 100 kN close-loop servo-electric machine under strain control and at a strain rate of 4×10^{-3}/sec. All fatigue tests were performed with a fully reversed axial strain (i.e., strain ratio R = -1) and a triangular waveform in air and water environments, respectively. The strain amplitude was controlled at ±0.1 %, ±0.2 %, ±0.4 %, or ±0.7 %, respectively. The steel specimens were loaded in the rolling direction. Fatigue tests were stopped when specimens broke or the fatigue cycles reached 5×10^{5} cycles.

2.4 Fatigue crack growth rate test

According to the ASTM E 647 specifications, compact-tension type (CT) specimens with a thickness of 12.5 mm and a width of 50 mm were machined. Before fatigue testing, all specimens were lightly polished with emery paper to #600. The specimens were pre-cracked by cyclic loading with decreasing ΔK (stress intensity factor range), at a load ratio (R, Pmin/Pmax) of 0.1, till a precracked length of 3 mm and ΔK of 10 MPa $\sqrt{m}$ were reached. In order to have valid test results, the specimen was designed to be predominantly elastic for the applied ΔK values less than 50 MPa $\sqrt{m}$ as per the size requirements of the ASTM E647. The corrosion fatigue tests were conducted on a closed-loop, servo-electric machine with a water circulation loop under constant load amplitude control with a sinusoidal wave form and at the frequencies of 0.02 and 0.001 Hz, respectively. The constant load amplitude was set at an R ratio ($Load_{min}/Load_{max}$) of 0.2 by an inner load cell control, which deducted the friction force between the pulling rod and the sealing material. The external load cell measurements including the friction force were also monitored for a comparison with the ones taken by the inner load cell. The electrochemical corrosion potential (ECP) was measured with Ag/AgCl/0.1 M KCl reference electrode. The conditions of the water environment are summarized in Table 3. The crack length was measured by an alternative current potential drop (ACPD) technique. The final fatigue crack length measurement was further calibrated against the average value of five measurements taken along the crack front on the fracture surface by a microscope at a magnification of 20 according to ASTM E

647. After corrosion fatigue tests, the oxide layers on the fracture surfaces were descaled with an electrolyte of 2 g hexamethylene tetramine in 1000 cm^3 of 1N HCl (Chopra, 1998) and further investigated with scanning electron microscopy.

Test parameters	Air-saturated (with filtered air)	Deoxygenated (with nitrogen)	HWC
Pressure, MPa	10	10	10
Temperature, ℃	300	300	300
Conductivity(inlet), μScm^{-1}	0.8	0.08	0.065
Conductivity(outlet), μScm^{-1}	1.2	0.16	0.072
Hydrogen(inlet),	N.A.	N.A.	48 ppb
Hydrogen(outlet),	N.A.	N.A.	39 ppb
Oxygen(inlet),	7.4 ppm	1~10 ppb	1~10 ppb
Oxygen(outlet),	6.7 ppm	1~10 ppb	1~10 ppb
ECP(SHE)	0.2 volt	-0.55volt	-0.6 volt
pH(inlet)	5.95	6.88	6.58
pH(outlet)	6.17	6.74	6.76
Autoclave exchange rate	1 time/h	1 time/h	1 time/h

Table 3. Test conditions of high-temperature water environments

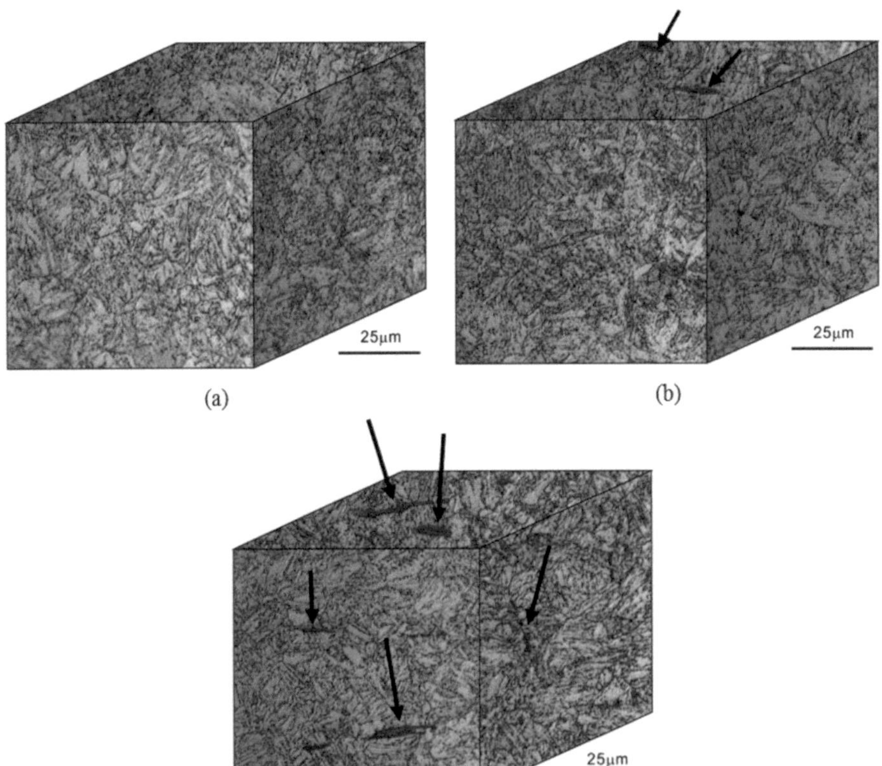

Fig. 2. Metallographs of A533B3 steels with different sulfur contents, (a) 0.008 wt%, (b) 0.016 wt%, (c) 0.027 wt%. (The arrows indicate the sulfides.)

3. Results and discussion

3.1 Metallographic examination

The metallographs of A533B3 steels with different sulfur contents are shown in Fig. 2. The sulfide laths were identified by energy dispersive spectroscopy (EDS) and observed to be oriented in the rolling direction for A533B3 steels. Little or no sulfides were observed for the steel with the lowest sulfur content. Tempered martensite was prevalent in the steels.

3.2 Low cycle fatigue life

The low-cycle fatigue lives of the steel specimen with 0.016 wt% sulfur under different environments are shown in Fig. 3. The fatigue life was significantly affected by the test temperature, strain amplitude, and oxygen level in the water environment. Regarding the environment effect, the steel specimens showed the shortest fatigue life in the air saturated water environment and the longest life in air at 300°C. The fatigue life with the lower oxygen level in the water environment showed the longer fatigue life. At the strain amplitude of ±0.2 %, the steel specimens tested in air had a much longer fatigue life at 300°C than at room temperature. This could be accounted for by the occurrence of dynamic strain aging and the effect of grain size reduction at 300°C (Huang, 2003, 2004). Lee and Kang (Lee, 1995 ; Kang, 1992) demonstrated that dynamic strain aging would improve the low-cycle fatigue resistance and degrade the fracture toughness at 300°C for the pressure vessel steel SA508 class 3 in air. It is consistent with the results of this study that a higher fatigue life was observed at 300°C than at room temperature in air.

The temperature effect on the fatigue life at a strain amplitude of ±0.4 % in the air saturated water environment and air is presented in Fig. 4. The longest fatigue life occurred at 245°C in the both environments. A similar result was reported by Chopra and Shack (Chopra & Shack, 1998) that the steel A333-Gr 6 exhibited the best fatigue resistance at 250°C in air. A large difference in fatigue life between 250°C and 300°C was also noted. The reason that fatigue life was enhanced at 245°C in the air-saturated water environment could be related to the negative strain rate sensitivity in the temperature range from 150°C to 300°C, as shown in Fig. 5. Nakao et al. (Nakao, 1995) also had a similar inference from a study of A333-6 steel. The effect of temperature on fatigue life in the water environment with 200 ppb oxygen is shown in Fig. 6. At the strain amplitude of ±0.2 % or ±0.4 %, the effect of temperature on fatigue life was not significant in the range from 150°C to 300°C. It could be accounted for by a remarkable reduction in fatigue-corrosion interaction due to the formation of dense oxide film on the specimen thereby extending the steel fatigue life. Consequently the fatigue life showed no much change from 150°C to 300°C under the present water environment. A similar dependence of fatigue life on temperature was also predicted from a statistical model reported by Chopra and Shack (Chopra & Shack, 1998), but they further indicated that the fatigue life would decrease significantly with increasing temperature in the range from 150°C to 300°C when the applied strain rate was reduced to 4×10^{-5} s^{-1}. Fig. 7 illustrates the relationship of the fatigue life against strain amplitude for A533B3 steels with different sulfur contents tested in 300°C water environments with saturated air and hydrogen, respectively. The low-cycle fatigue life of the steel specimen in the hydrogenated water increased significantly relative to that obtained in the air-saturated water environment. Under the strain rate of $4x10^{-3}s^{-1}$, the fatigue life of the specimen was not varied with its sulfur content in both water environments. This result is consistent with the report of Chopra and Shack (Chopra & Shack, 1998). The data further indicated that the fatigue life of low alloy steel with the sulfur content 0.003 wt% was ten times longer than that with the sulfur content 0.010 wt% when the strain rate was decreased to $4x10^{-6}s^{-1}$. On the contrary, for the steel with the sulfur content higher than 0.012 wt%, the effect of sulfur content on the low cycle fatigue life was insignificant. Their limited data suggested that environmental effects on fatigue life may become saturated for the specimens with sulfur contents higher than 0.012 wt% when the strain rate was controlled at $4x10^{-3}s^{-1}$.

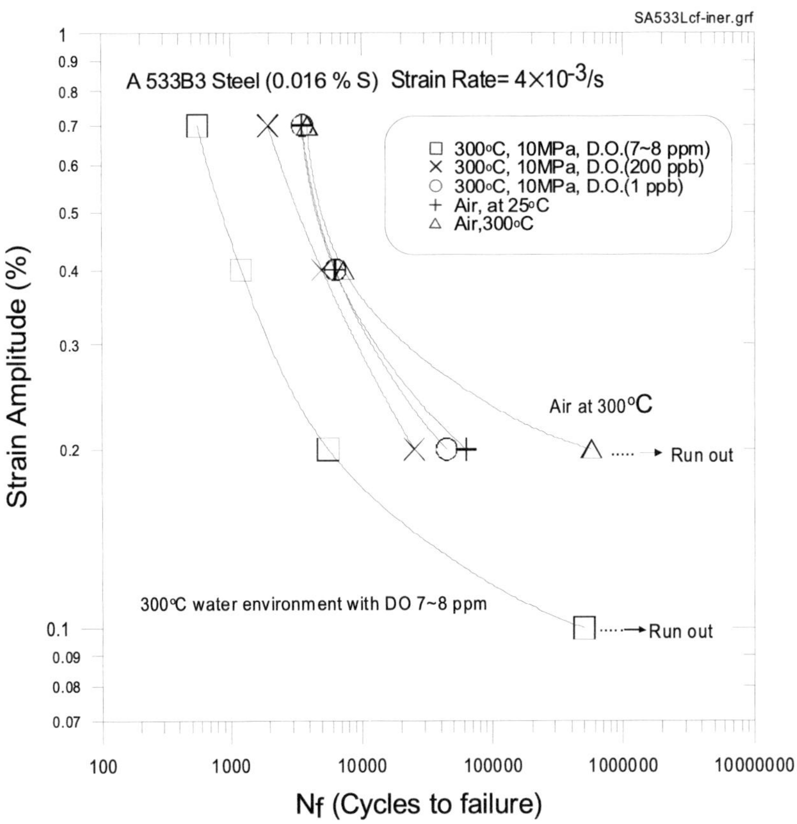

Fig. 3. Relationships between the low-cycle fatigue life and strain amplitude for A533B3 steel with 0.016 wt% sulfur tested in different environments.

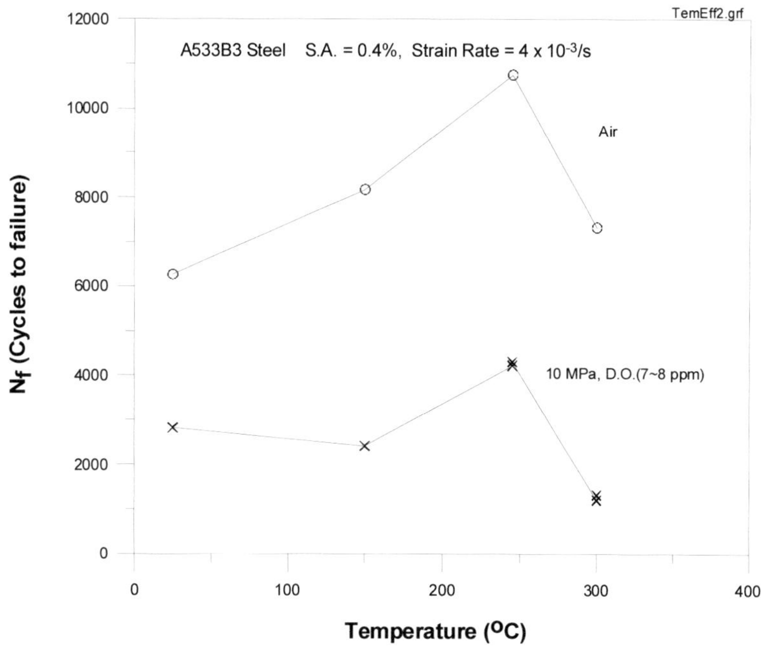

Fig. 4. Effect of temperature on fatigue life of A533B3 steel in the air-saturated water environment and in air.

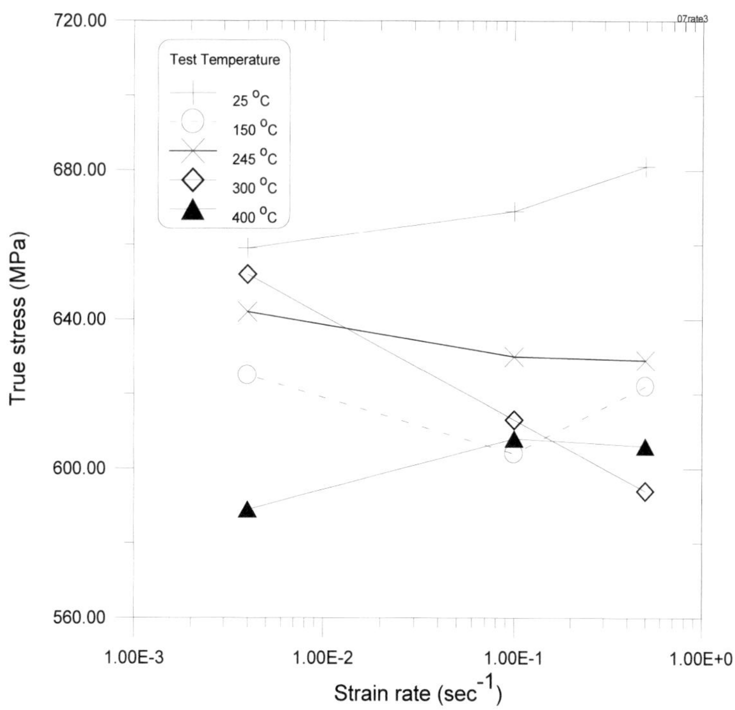

Fig. 5. Effect of strain rate on the true stress of A533B3 steel at a true strain of 2% tested at different temperatures in air.

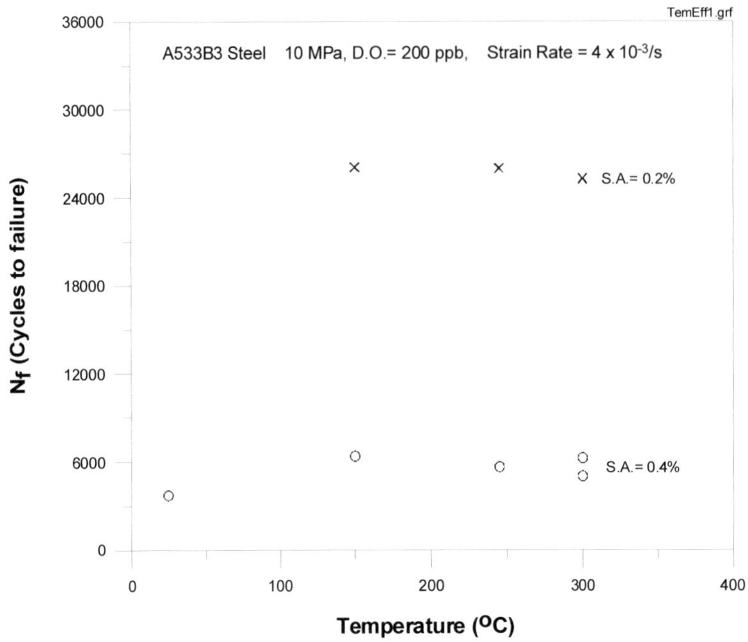

Fig. 6. Effect of temperature on fatigue life of A533B3 steel in the water environment with an oxygen content 200 ppb.

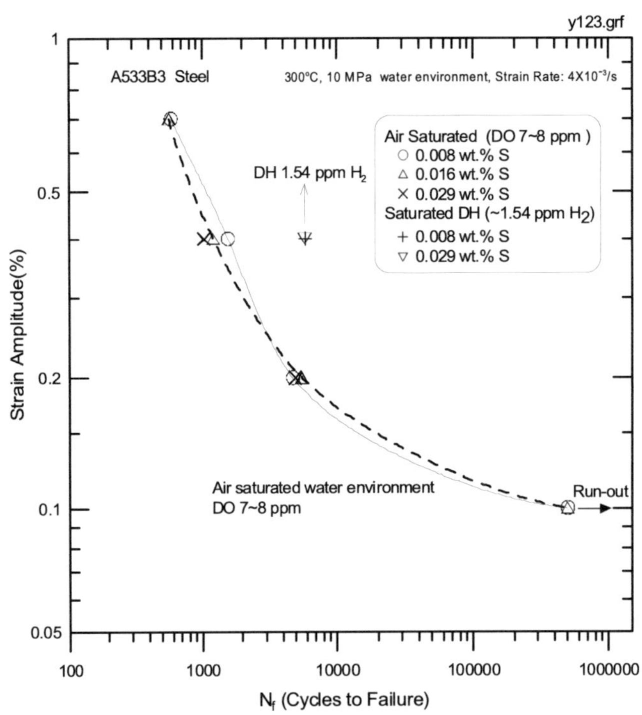

Fig. 7. Relationships between the low-cycle fatigue life and strain amplitude for A533B3 steels with different sulfur contents tested in 300°C water environments .

3.3 Effects of sulfur content on fatigue crack growth rate of low alloy steels

The fatigue crack growth rates of A533B3 steels with three sulfur content levels in water environments at 300℃ are shown in Fig. 8. In the air-saturated water environment, Fig. 8(a), there was no significant difference in the fatigue crack growth rates of the steels with different sulfur contents at a loading frequency of 0.02 Hz. But in the deoxygenated water environment, the lower corrosion fatigue crack growth rate was observed for the steels with lower sulfur contents. For the steel with 0.016 wt% sulfur, the corrosion fatigue crack growth rate increased significantly when the stress intensity factor range was larger than 38MPa $\sqrt{m}$, as shown in Fig. 8(b). Fig. 9 presents the results of A533B3 steel tested in an air-saturated water environment at a loading frequency of 0.001 Hz. The data outside the bounds of the ASME XI wet curves are not conservative. It was noted that the fatigue crack growth rates were almost the same for the steels of medium(0.016 wt% sulfur) and high sulfur content (0.027 wt% sulfur), but that those with low sulfur content showed the lowest crack growth rate. From the above results, it can be concluded that the high sulfur in low alloy steels or the high dissolved oxygen in the water coolant or their synergistic effects would degrade the corrosion fatigue resistance of low alloy steels. Therefore, it is essential to secure the integrity of pressure vessel by reducing the steel sulfur content during the steel manufacturing and by decreasing oxygen content in the reactor coolant, for instance, by HWC.

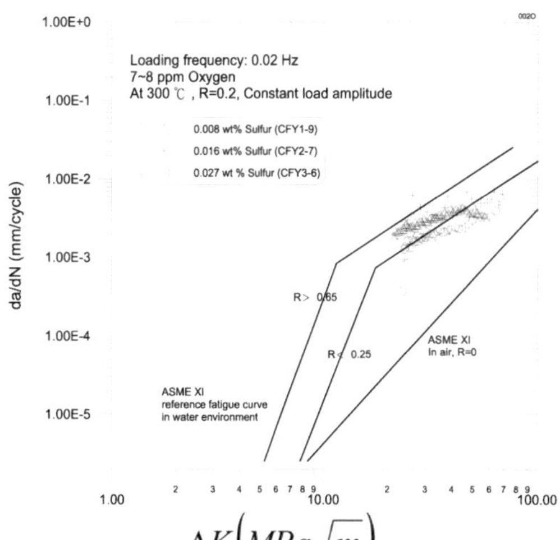

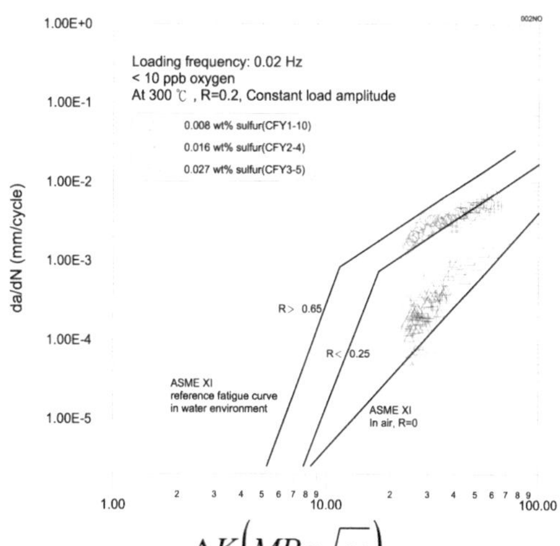

Fig. 8. Sulfur content effects on the corrosion fatigue crack growth rates of A533B3 steels in the water environments of different oxygen contents at a loading frequency of 0.02 Hz: (a) 7~8 ppm oxygen and (b) <10 ppb oxygen.

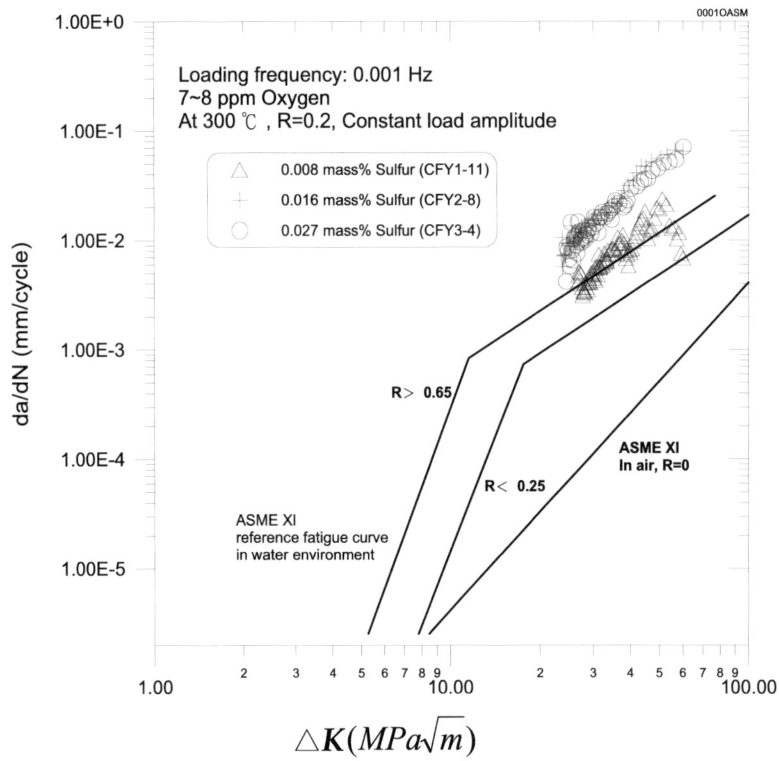

Fig. 9. Sulfur content effects on the corrosion fatigue crack growth rates of A533B3 steels in an oxygen saturated water environment at a loading frequency of 0.001 Hz.

3.4 Water chemistry effects on fatigue crack growth rate

Fig. 10 makes a comparison of fatigue crack growth rates for A533B3 steels with three sulfur levels tested in the deoxygenated and air-saturated water environments. It is clear that an increase in oxygen level accelerates crack growth in the steel specimens with sulfur contents up to 0.016 wt %. For the steel with 0.027 wt% sulfur, the opposite is true. Their corrosion fatigue crack growth rates are faster in the water environment with near zero dissolved oxygen concentration than with 7~8 ppm dissolved oxygen concentration, as shown in Fig. 10(c). Figure 11 shows a comparison of fatigue crack growth rates for the high sulfur steels tested in the high temperature water environments deoxygenated by nitrogen and hydrogen, respectively. The fatigue crack growth rate is in good agreement with each other. Both curves show a surged crack growth rate phenomenon, similar to Fig. 10(b). In Fig. 10(b), for the steel specimens with 0.016 wt% sulfur content, the fatigue crack growth rate in the deoxygenated water environment surged to the same levels as those tested in the air saturated water environment when the concentration of sulfate ion reached a critical concentration in the crack tip. A surge of the fatigue crack growth rate occurred, when the applied $\triangle K$ reached a value of ~35 MPa$\sqrt{m}$. Correspondently, a pseudo boundary was identified on the fracture surface of the specimen, as shown in Fig. 12(a). The boundary was further examined by SEM at higher magnifications to consist of a band of microcracks, which is a unique feature not observed in other regions, as shown in Figs. 12(b) and (c).

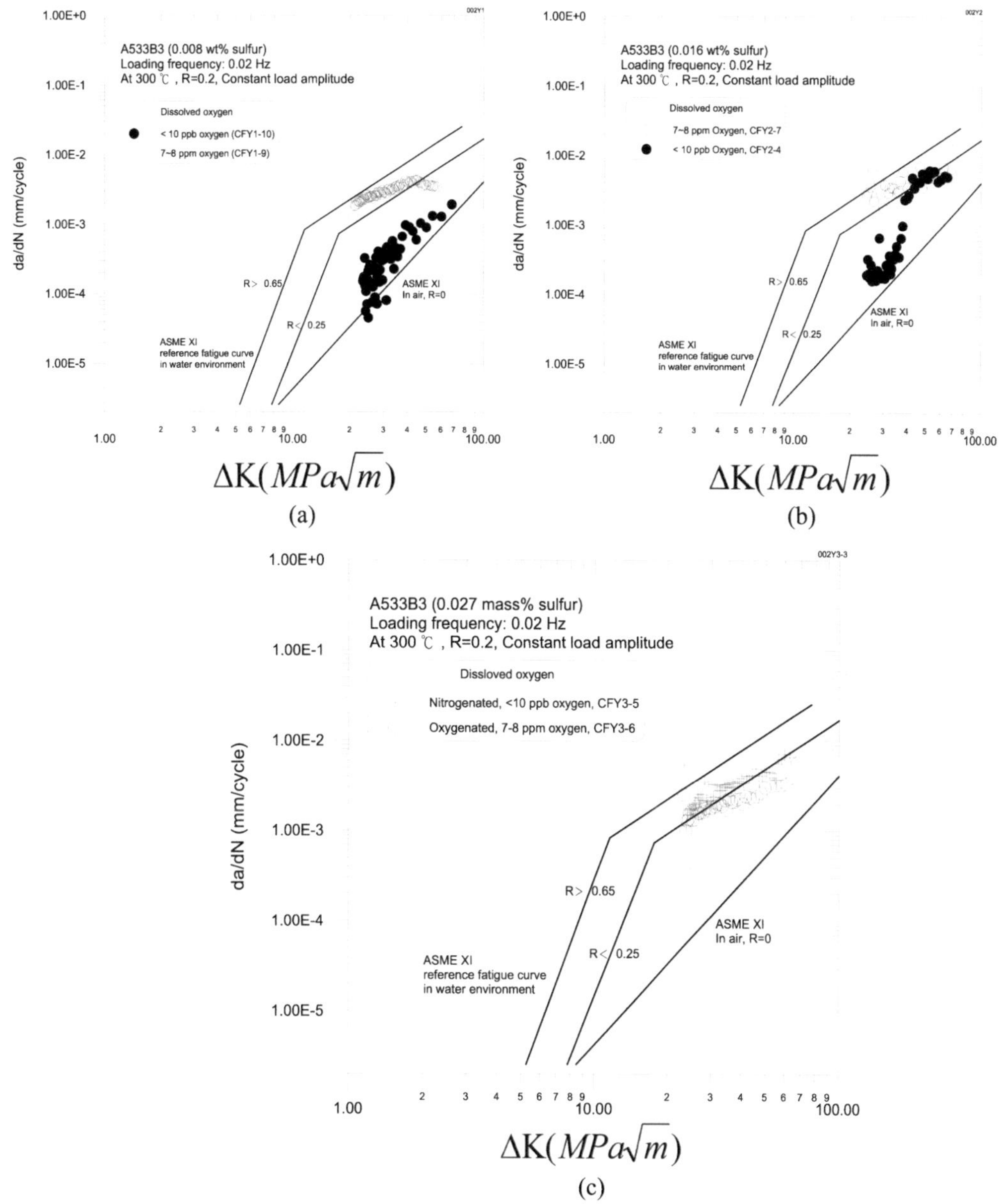

Fig. 10. A comparison of fatigue crack growth rates for A533B3 steels tested in high temperature deoxygenated and air-saturated water environments, (a) steel sulfur content 0.008 wt%, (b) steel sulfur content 0.016 wt%, and (c) steel sulfur content 0.027 wt%.

There were some inclusions imbedded in the microcracks. The inclusions containing sulfides were identified by EDS. The surged fatigue crack growth rate and a pseudo boundary were also observed with the 0.027 wt % sulfur specimens in the deoxygenated water environment, as indicated in Fig. 11. The probabilities of finding a surged fatigue crack growth rate were strongly related to the specimen sulfur levels. The greater the sulfur level, the greater the probability to find a surged growth rate. It implies the sulfides dissolved in the coolant environment around the crack tip may speed up the crack growth rate when sulfur ion reached a critical quantity. A previous study (Yeh, 2007) showed that the dense oxide film, Fe_3O_4, was formed when low alloy steel specimens were tested in the deoxygenated water environments at 300℃. The corrosion products in a 10 MPa water environment with saturated air at the test temperature 300℃ were identified to be a porous mixture of Hematite (α-Fe_2O_3) and Maghemite (γ-Fe_2O_3). The porous oxide layer allows the coolant water access to the fresh metal beneath it and provides less protection than the dense one does. As a result, there was a relatively larger proportion of fresh metal with the porous oxide layers on the specimens than those with dense ones exposed to the coolant, as illustrated in Fig. 13. A hypothesis is proposed to elucidate the observation under the assumptions that sulfate ions have higher affinity to fresh metal than to the oxide layer and that the quantities of sulfate dissolved from specimens in the air saturated and deoxygenated water environments are comparable. The sulfate ion in the coolant will transport to the fresh surface around the crack tip and the fresh metal beneath the porous oxide layer. In relative terms, there was less fresh metal surface of the specimen exposed to the deoxygenated water coolant than to the air-saturated one. The concentration of sulfate ion in the coolant in front of the crack tip of the specimens with dense oxide layer would be relatively higher than that with porous oxide layer. The attack of sulfate ions on the crack tip of the specimens tested in deoxygenated water environment would be more aggressive than those tested in air saturated water environment when sulfate ions reached a critical concentration. The sulfur dissolved in the high temperature water environment from the high-sulfur steels was sufficient to acidify the crack tip chemistry. Therefore, the deoxygenated water environment showed little or no beneficial effect for the high sulfur steels. To mitigate the environmentally assisted cracking of low alloy steels, the factor of sulfate ion in the coolant should be taken into account.

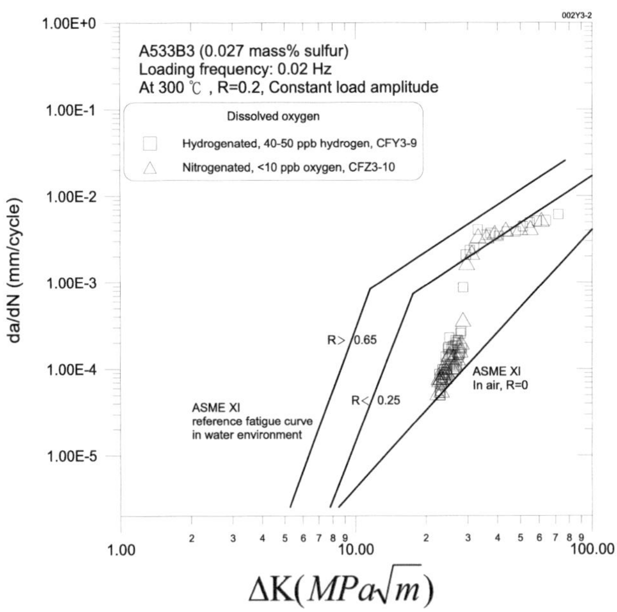

Fig. 11. A comparison of fatigue crack growth rates for A533B3 steels tested in high temperature water environments deoxygenated by nitrogen and hydrogen, respectively.

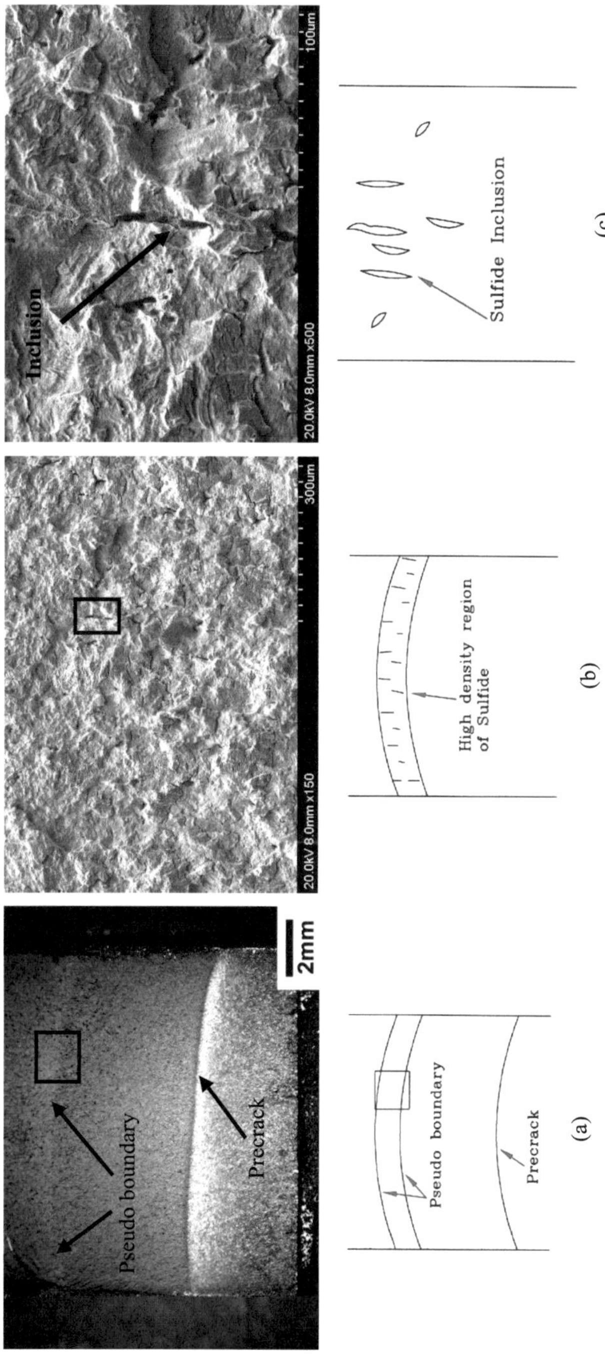

Fig. 12. A pseudo boundary observed on the fracture surface of the steel specimen with sulfur content 0.016 wt % tested at a loading frequency of 0.02 Hz in deoxygenated water environment, (a) fractographic feature by optical stereography, (b) a band of microcracks boxed up in (a) examined by SEM, (c) microcracks boxed up in (b) at a higher magnification.

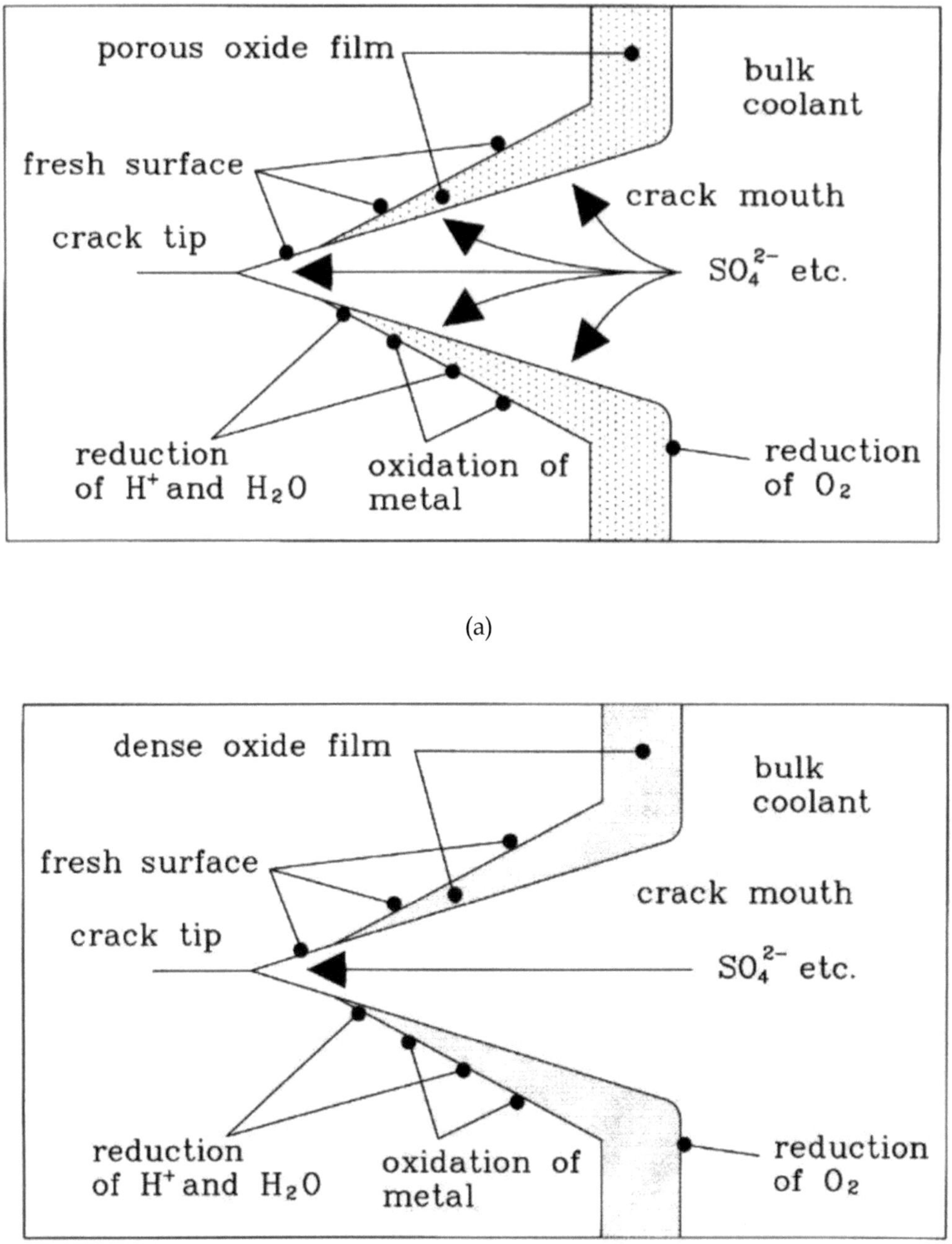

Fig. 13. A schematic illustration of (a) porous oxide film formed in an oxygenated water environment and (b) dense oxide film formed in a deoxygenated water environment.

3.5 Frequency effects on fatigue crack growth rate

Fig. 14 illustrates the frequency effects on the fatigue crack growth rate of A533B3 steels in the air saturated water environment. The lower the frequency, the higher the fatigue crack growth rate was observed for all the three steel specimens with different sulfur levels.

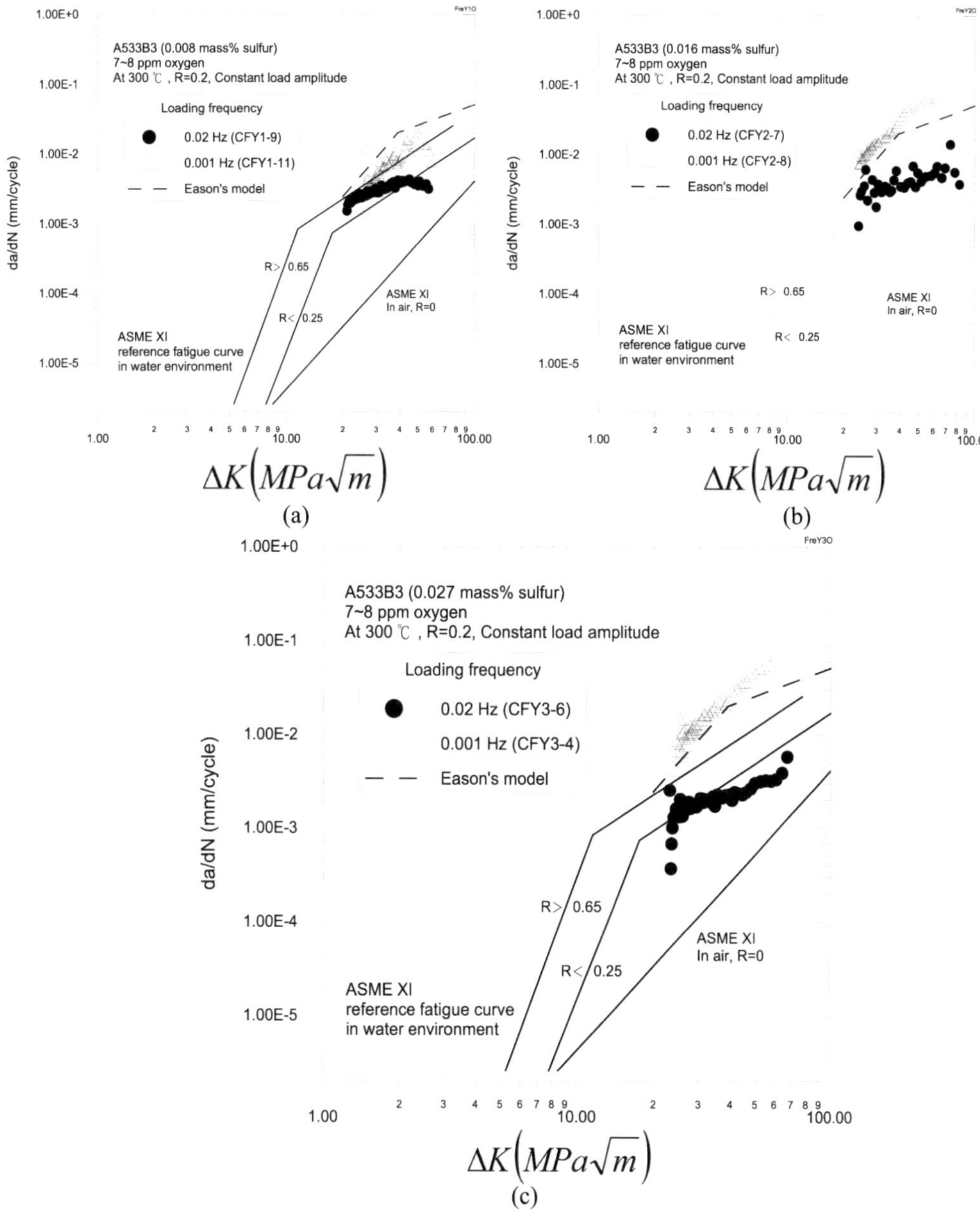

Fig. 14. Frequency effects on corrosion fatigue crack growth rates for A533B3 steels with different sulfur levels tested in 300℃ air-saturated water environment: (a) 0.008 wt% S, (b) 0.016 wt% S, and (c) 0.027 wt % S.

It means that the corrosion fatigue resistance of A533B3 steels was significantly affected by the loading frequency in the air-saturated water environment. In the GE-model (Ford, 1987) the corrosion fatigue (CF) crack growth through anodic dissolution is controlled by the crack-tip strain rate and the sulfur anion activity/pH in the crack tip electrolyte that govern the oxide film rupture and the dissolution/repassivation behavior after the film rupture event. From Fig. 13, the data points not bounded by the Eason's model (Eason, 1998) were attributed to the factors of the low frequency and higher sulfur content as well as high dissolved oxygen in the water coolant. These results are different from the observation that the initiation life is significantly decreased when the applied strain range, temperature, DO level in water, sulfur content in steel and strain rate are simultaneously satisfied. One individual factor will enhance the crack growth rates when it is higher than their threshold. Therefore, it is inferred that the benefit factors for the crack initiation are also beneficial for the crack growth of low alloy steels.

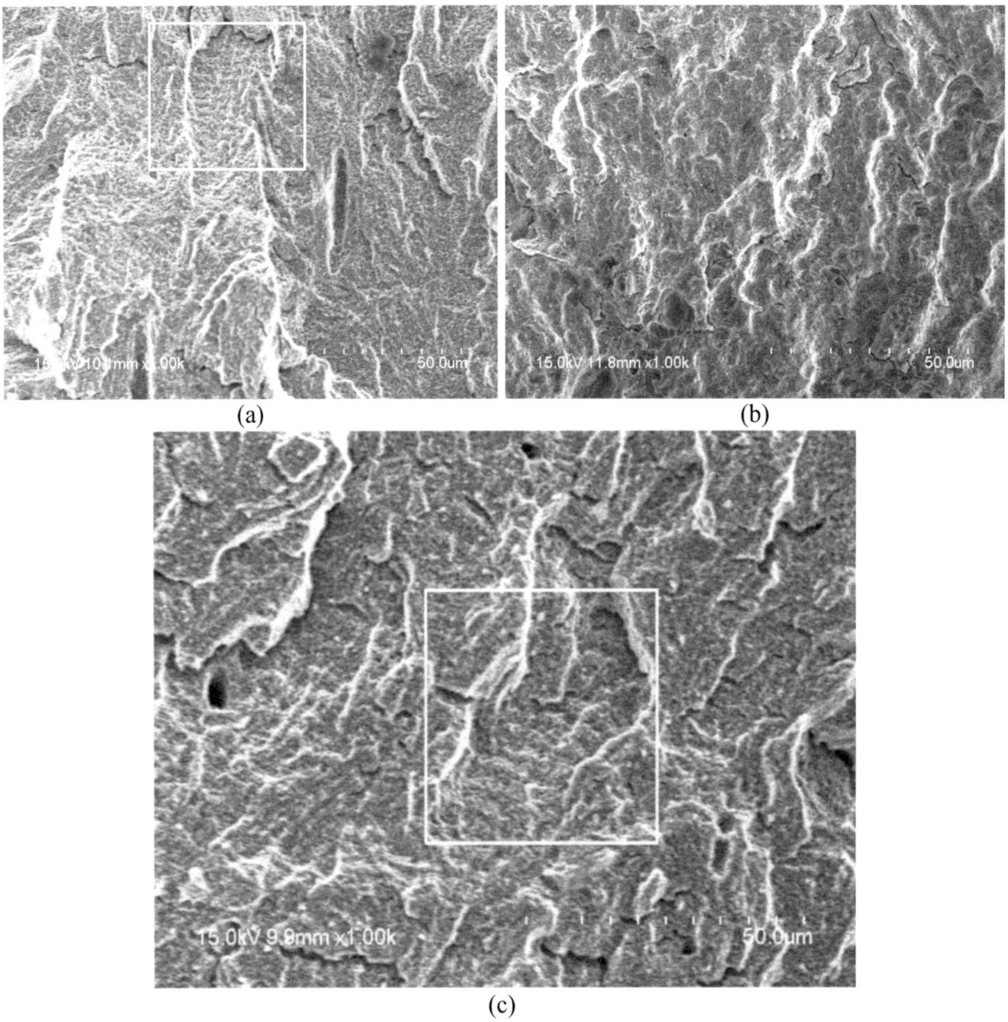

Fig. 15. SEM fractographs for A533B3 steel specimens with sulfur content 0.016 wt % tested in high-temperature water environments: (a) 0.02 Hz, air-saturated (b) 0.001 Hz, air-saturated, (c) 0.001 Hz, deoxygenated.

Distinct fracture features were revealed by SEM for the fatigue specimens tested in the high-temperature water environments at different frequencies. The fracture features were apparently related to the oxidation behavior. For the tests at a loading frequency of 0.02 Hz in the air-saturated water environment, the specimen revealed fatigue striation pattern, as shown in Fig. 15(a). The striations could have been corroded due to a long stay at a loading frequency of 0.001 Hz in an air-saturated water environment, illustrated in Fig. 15(b). By contrast, the striations were manifested in the deoxygenated water environment at a loading frequency of 0.001 Hz, as shown in Fig. 15(c).

4. Conclusions

1. The low cycle corrosion fatigue results show that a significant temperature effect on the fatigue life was observed for the specimen tested in the oxygenated water environment at the temperature range from 150℃ to 300℃, but little or no dependence of fatigue life on temperature was noted in the deaerated water environment with an oxygen content 200 ppb.
2. Under the strain rate of $4x10^{-3}s^{-1}$, the fatigue life of the specimen was not varied with its sulfur content in the air saturated and hydrogenated water environments.
3. Corrosion fatigue crack growth rates of A533B3 steels were significantly affected by the steel sulfur content, loading frequency and oxygen content in the high-temperature water environment. The data points outside the bound of Eason's model could be attributed to the low frequency, higher sulfur content and high dissolved oxygen water coolant.
4. The sulfur dissolved in the water environment from the higher sulfur steels was sufficiently concentrated to acidify the crack tip chemistry even in the HWC water. Therefore, nitrogenated or HWC water did not show any beneficiary effect on the high-sulfur steels.

5. Acknowledgments

The authors would like to acknowledge the financial support from Taiwan Power Company and the Institute of Nuclear Energy Research, Taiwan.

6. References

Shah, V. N. & Macdonald, P. E. (1993). *Aging and Life Extension of Major Light Water Reactor Components*, Amsterdam, Elsevier Science Publishers.

Huang, J. Y., Li, R. Z., Chien, K. F., Kuo, R. C., Liaw, P. K., Yang, B. & Huang, J. G. (2001), Fatigue behavior of SA533-B1 steels, in: *Fatigue and Fracture Mechanics, 32nd vol., ASTM STP 1406,* R. Chona (Ed.), Philadelphia, PA, pp. 105-121.

Huang, J. Y. ; Chen, C. Y. ; Chien, K. F. ; Kuo, R. C.; Liaw, P. K. & Huang, J. G., in: *Proc. Julia Weertman Symp., TMS Fall Meeting,* Oct. 31- Nov. 4, 1999, pp. 373-384.

Seifert, H. P. ; Ritter, S. & Heldt, J. (2001), Strain induced Corrosion Cracking of Low-alloy Reactor Pressure Vessel Steels under BWR Conditions, in: F. P. Ford, G. S. Was and J. L. Nelson (Eds.), *Proceedings of the 10th International Conference on Environmental Degradation of Materials in Nuclear Power System-water Reactors.*

Seifert, H. P.; Ritter, S. & Hickling, J. (2003), Environmentally-assisted Cracking of Low-alloy RPV and Piping Steels under LWR Conditions, in: G. S. Was, L. Nelson, and P. King (Eds.), *Proceedings of the 11th International Conference on Environmental Degradation of Materials in Nuclear Power System-water Reactors*, pp. 73-89.

Huang, J. Y.; Young, M. C.; Jeng, S. L.; Yeh, J. J.; Huang, J. S. & Kuo, R. C. (2007), Corrosion Fatigue Behavior of an Alloy 52-A508 Weldment under Simulated BWR Coolant Conditions, in: P. King, T. Allen, J. Busby (Eds.), *13th International Conference on Environmental Degradation of Materials in Nuclear Power System*, Whistler, Canada, August 19-23.

Chopra, O. K. & Shack, W. J. (1997), Evaluation of Effects of LWR Coolant Environments on Fatigue Life of Carbon and Low–alloy Steels, in : *Effects of the Environment on the Initiation of Crack Growth*, W. A. V. D Sluys, R. S. Piascik & R. Zawierucha, (Eds), ASTM STP 1298, pp. 247-266.

Huang, J. Y.; Hwang, J. R.; Yeh, J. J.; Kuo, R. C. & Chen, C. Y. (2003), Low Cycle Fatigue Resistance of SA533 Steels, *Mater. Sci. and Tech.* 19, pp. 1575-1584.

Huang, J. Y.; Hwang, J. R.; Yeh, J. J.; Chen, C. Y.; Kuo, R. C. & Huang, J. G. (2004), Dynamic Strain Aging and Grain Size Reduction Effects on the Fatigue Resistance of SA533B3 Steels, *J. Nucl. Mater.* 324, pp. 140-151.

Van Der Sluys, W. A. & Emanuels, R. H. (1985), The Effect of Sulfur Content on the Crack Growth Rate of Pressure Vessel Steels in LWR Environments, in: J. T. A. Roberts (Eds.), *Proceedings of the 2nd International Symposium on Environmental Degradation of Materials in Nuclear Power Systems-water Reactors*, Monterrey, California, September, pp. 100-107.

Amzallag, C.; Bernard, J. L. & Slama, G. (1983), Effect of Loading and Metallurgical Parameters on the Fatigue Crack Growth Rates of Pressure Vessel Steels in Pressurized Water Reactor Environment, in: J. T. A. Roberts (Eds.), *Proceedings of the International Symposium on Environmental Degradation of Materials in Nuclear Power Systems-water Reactors*, Myrtle Beach, South Carolina, August, pp. 727-745.

Combrade, P.; Foucault, M. & Slama, G. (1987), Effect of Sulfur on the Fatigue Crack Growth Rates of Pressure Vessel Steel Exposed to PWR Coolant: Preliminary Model for Prediction of the Transitions between High and Low Crack Growth Rates, in: G. J. Theus. And J. R. Weeks (Eds.), *Proceedings of the Third International Conference on Environmental Degradation of Materials in Nuclear Power System-water Reactors*, pp. 269-276.

Van Der Sluys, W. A. & Cullen, W. H. (1987), Fatigue Crack Growth of Pressure Vessel Materials in Light-water-reactor Environments, in: R. Rungta, J. D. Gilman, and W. H. Bamford (Eds.) *Performance and Evaluation of Light Water Reactor Pressure Vessels*, PVP-119, San Diego, CA., ASME, p. 63-71.

O'Donnell, T. P. & O'Donnell, W. J. (1995), Cyclic Rate-dependent Fatigue Life in Reactor Water, in: S. Yukawa (Eds.), *Performance and Evaluation of Light Water Reactor Pressure Vessels*, PVP-306, Honolulu Hawaii, ASME, July, pp. 59-69.

Eason, E. D. & Nelson, E. E. (1993), Analysis of Fatigue Crack Growth Rate Data for A508 and A533 Steels in LWR Environments, *EPRI TR-102793, Project 2006-20 Final Report, August.*

Roth, A.; Hänninen, H.; Brümmer, G.; Wachter, Ilg, U.; Widern, M. & Hoffmann, H. (2003), Investigation of Dynamic Strain Aging Effects of Low Alloy Steels and Their Possible Relevance for Environmentally Assisted Cracking in Oxygenated High-temperature Water, in: G. S. Was, L. Nelson, and P. King (Eds.), *Proceedings of the 11th International Conference on Environmental Degradation of Materials in Nuclear Power System-water Reactors*, pp. 317-329.

Chopra, O. K. & Shack, W. J. (1998), *Effects of LWR Coolant Environments on Fatigue Design Curves of Carbon and Low-alloy Steels*, NUREG/CR-6583, ANL-97/18, US Nuclear Regulation Commission.

Atkinson, J. D. & Forrest, J. E. (1986), The Role of MnS Inclusions in the Development of Environmentally Assisted Cracking of Nuclear Reactor Pressure Vessel Steels, in: W. H. Cullen (Eds.), *Proceedings of the Second International Atomic Energy Agency Specialist's Meeting on Subcritical Crack Growth*, Sendai Japan, NUREG/CP-0067, US Nuclear Regulation Commission, Vol. 2, pp. 153-178.

Shoji, T.; Komai, K.; Abe, S. & Nakajima, H. (1986), Mechanistic Understanding of Environmentally Assisted Cracking of RPV Steels in LWR Primary Coolants, in: W. H. Cullen (Eds.), *Proceedings of the Second International Atomic Energy Agency Specialist's Meeting on Subcritical Crack Growth*, Sendai Japan, NUREG/CP-0067, US Nuclear Regulation Commission, Vol. 2, pp. 99-117.

Tice, D. R. (1986), Influence of Mechanical and Environmental Variables on Crack Growth in PWR Pressure Vessel Steels, *Int. J. Pres. Ves. and Piping*, 24, pp. 139-173.

Ritter, S. & Seifert, H. P. (2007), Evaluation of the Mitigation Effect of Hydrogen Water Chemistry in BWRs on the Low-frequency Corrosion Fatigue Crack Growth in Low-alloy Steels, *J. Nucl. Mater.*, 360, pp. 170-176.

Ford, F. P.; Taylor, D. E. & Andresen, P. L. (1987), *Corrosion-assisted Cracking of Stainless and Low-alloy Steels in LWR Environments*, EPRI NP-5064M.

Standard Recommended Practice for Constant-amplitude Low-cycle Fatigue Testing (1997), *ASTM E606-80*, pp. 629-641.

Lee, B. H. & Kim, I. S. (1995), *J. Nucl. Mater.*, pp. 226 216.

Kang, S. S. & Kim, I. S. (1992), *Nucl. Technol.* 97, pp. 336-343.

Chopra, O. K. & Shack, W. J. (1998), *Effects of LWR Coolant Environments on Fatigue Design Curves of Carbon and Low-alloy Steels*, Report NUREG/CR-6583, ANL-97/18, pp. 47-52.

Nakao, G.; Kanasaki, H.; Higuchi, M.; Iida, K. & Asada, Y. (1995), Environmental Effects, Modeling Studies, and Design Considerations, American Society of Mechanical Engineers, in: *Fatigue and Crack Growth,* S. Yukawa (ed.),: New York, PVP Vol. 306, pp.123-128.

Yeh, J. J.; Huang, J. Y. & Kuo, R. C. (2007), Temperature Effects on Low-cycle Fatigue Behavior of SA533B Steel in Simulated Reactor Coolant Environments, *Mater. Chem. and Phys.*, 104, pp. 125-132.

Eason, E. D.; Nelson, E. E. & Gilman, J. D. (1998), Modeling of Fatigue Crack Growth Rate for Ferritic Steels in Light Water Reactor Environments, *Nucl. Eng. and Des.*, 184, pp. 89-111.

6

Influence of Dissolved Hydrogen on Stress Corrosion Cracking Susceptibility of Nickel Based Weld Alloy 182

Luciana Iglésias Lourenço Lima, Mônica Maria de Abreu Mendonça Schvartzman, Marco Antônio Dutra Quinan, Célia de Araújo Figueiredo and Wagner Reis da Costa Campos
Nuclear Technology Development Centre- CDTN/CNEN
Brazil

1. Introduction

Nuclear power stations are electrical energy stations that use nuclear fission reaction as a source of heat to produce energy. Most of the world's nuclear steam supply systems for generating electricity are based on water cooled and moderated systems of which the most commom design are the Ligth Water Reactors (LWR) varieties: Pressurized Water Reactor (PWR) and the Boiling Water Reactor (BWR). In the construction of these reactors nickel-based alloy 600 and its associated alloys 82 and 182 weld metals were initially selected because of their ability to withstand a variety of severe operating conditions involving corrosive environment, high temperatures, high stresses, and combinations thereof (Gomez-Briceño & Serrano, 2005, Peng, et al., 2007).

Alloy 82 and 182 weld metals are widely used to join austenitic stainless steel to low alloy steel components of PWRs. However, after many years of plant operation, those materials showed susceptibility to stress corrosion cracking (SCC). Since 1994, cracks and leaks have been discovered in about 300 welds of nickel-based alloys 82 and 182 at different PWR plant primary coolant system locations (Scott, 2004, Scott & Meunier, 2007).

Despite many studies have been performed on the SCC behavior of nickel-based alloys the mechanisms are still not well understood. It was observed that the variables that influence the SCC susceptibility of alloy 182 weld metal are quite similar to those associated for nickel-based alloy 600 such as cold work, alloy metallurgical factors, applied or residual stresses, and environmental factors including primary water hydrogen partial pressure, water temperature and chemistry, (Rebak & Szklarska-Smialowska, 1996, Rebak & Hua, 2004).

Actions to mitigate the SCC have been undertaken comprising changes in the chemical environment (optimizing dissolved hydrogen levels, zinc addition and noble metal chemical addition). The use of water chemistry optimization is attractive because many or all components can benefit. In view of the successful application of water chemistry on SCC mitigation in BWR reactors and its system wide benefit, mitigation by adjusting the primary water chemistry in PWR reactors has being considered (Andresen, et al., 2005).

Hydrogen gas is added to the primary circuit coolant at concentrations usually ranging between 25 and 50 cm^3 $H_2.kg^{-1}$ H_2O at standard temperature and pressure (STP). This concentration range is sufficient to inhibit radiolysis of water and thereby minimize corrosion of structural materials (Garbett, et al., 2000, Lima et al., 2011). Besides this, dissolved hydrogen (DH) might affect the SCC behavior of nickel-based alloys. Many experimental studies (Morton et al., 1999, Totsuka, et al., 2000, Attanasio & Morton, 2003) have indicated that the crack growth rate varies with the DH concentration at a given temperature.

In this chapter it is proposed to present the initial results of an evaluation of the dissolved hydrogen influence on the SCC susceptibility of the nickel-based weld alloy 182 using slow strain rate tensile tests (SSRT) conducted at Nuclear Technology Development Centre - CDTN/CNEN.

1.1 Dissimilar metal Weld in pressurized water reactors

A weld between different metals is known as dissimilar metal weld (DMW) and is used in western LWR plants. In Westinghouse and French nuclear power plants design the most important application of this type of weld is between low alloy steel nozzles to austenitic stainless steel pipelines. Typical locations are J-groove welds of nickel alloy 600 vessel head penetrations, pressurizer penetrations and instrument nozzles. In addition, they are used in butt welds in RPV and steam generator inlet and outlet nozzles, pressurizer surge line, safety and relief valve nozzles (Figure 1) (Banford & Hall, 2003, King, 2005, Seifert, et al., 2008).

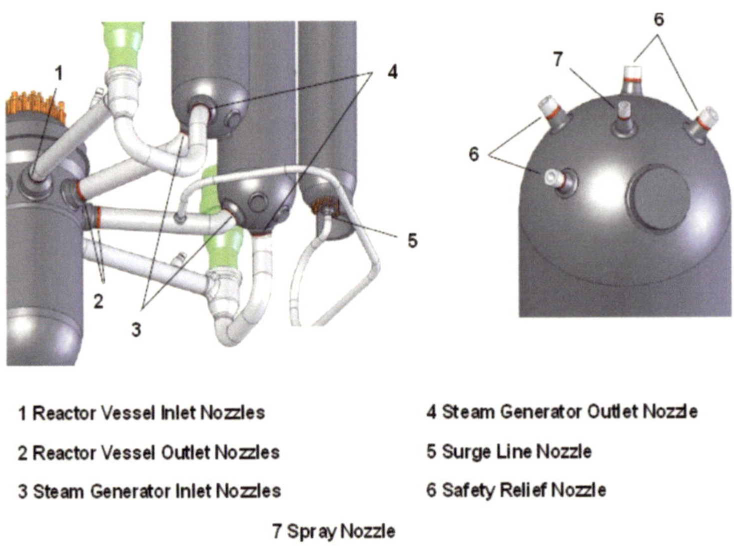

Fig. 1. Alloy 82 and 182 butt welds locations in PWR Westinghouse design plants (King, 2005).

The type and characteristics of the DMW depend on a range of factors; including the specific reactor design, the welding procedure and the weld material (International Atomic Energy Agency [IAEA], 2003, Jang et al., 2008). The welding of two different materials may occur directly, after applying a buttering layer in one of them or by using a transition piece, also called safe end, between the two dissimilar materials (Figure 2). Nickel-based alloy 82 or 182 is generally used as weld metal material in the buttering layer and in the J-groove weld (IAEA, 2003, Miteva & Taylor, 2006).

Fig. 2. Dissimilar metal weld variations of the Westinghouse plant design, (a) weld with a safe end and (b) weld without a safe end (Miteva & Taylor, 2006).

The use of safe end avoids making of DMW on site, during the component installation in the nuclear power plant. The transition piece is welded to low alloy steel nozzles in the manufacturer´s shop under controlled conditions, so the subsequent weld to connect the component is a conventional similar weld that can be made on site. The use of the buttering layer is an alternative way to accommodate the differences in the composition and properties of the two base metals, such as the melting point, coefficient of thermal expansion and to make difficult the migration of undesirable alloying elements from base metal to weld metal. Buttering can also be used to alleviate the requirements of postweld heat-treatment (PWHT) (King, 2005, Davis, 2006).

In the Westinghouse plant design the dissimilar metal welds are made in three steps. In the first stage a buttering layer is applied to the low alloy steel with a final thickness varying between 5 and 8 mm. The layer is applied by gas tungsten arc process (GTAW) with nickel-based alloy 82 and 182 as weld metal. The second stage consists of the stress relief heat treatment and the machining of the buttered surface to obtain the weld edges. The temperature and the duration of the heat treatment depend on the thickness of the low alloy steel component and the number of buttering layers. Usually the stress relief heat treatment is undertaken at temperatures from 580°C. In the third stage the buttered low alloy steel is welded to austenitic stainless steel by the GTAW process with nickel-based alloy 82 for the root pass. The joint is completed with nickel-based alloy 182, using the shielded metal arc process (SMAW) (Fallatah et al., 2002, Miteva & Taylor, 2006).

The use of pre-heating is recommended for materials with a high level of carbon equivalent and components of high thickness. Pre-heating encourages a decrease in the cooling speed, which reduces the likelihood of martensite forming in the heat affected zone (HAZ) of the low alloy steel and reduces the occurrence of hot cracking in the materials involved. Heat treatment after welding may or may not take place (Schaefer, 1979, Kou, 2003, Miteva & Taylor, 2006).

1.2 Stress corrosion cracking in pressurized water reactors

Environmentally assisted cracking (EAC) in the form of stress corrosion cracking (SCC) is one of the most critical kinds of damage experienced in nuclear power plants. It is a potentially critical issue concerning safety of plant operation and plant life extension (American Society of Metals [ASM], 2006). SCC phenomenon consists of a degradation process resulting from a combined and synergistic interaction of aggressive environment, tensile stresses and susceptible material, as well as the time for the phenomenon to occur. This phenomenon has been reported in dissimilar metal welds in various nuclear power plants all over the world. Various studies carried out since the 1980s have demonstrated that nickel-based alloys 82 and 182 are susceptible to the SCC phenomenon (Andresen et al., 2002, Fukumura & Totsuka, 2010). In the end of the year 2000, three PWRs experienced cracking concerned to dissimilar metal welds between the main austenitic stainless steel primary circuit piping and the outlet pressure vessel nozzles, believed to be stress corrosion cracking, of major primary circuit welds made from nickel-based alloy 82 or 182. (Pathania et al., 2002 Amzallag et al., 2002, Banford & Hall, 2005, Alexandreanu et al., 2007).

In the case of susceptible material the main factors that influence the SCC susceptibility of nickel-based alloy 600 and its associated alloys 82 and 182 weld metals are chemical composition, microstructure and heat treatment of the material. The levels of carbon and chromium are important variables when evaluating the chemical composition. Data from laboratories have been demonstrated that the increase of chromium content correlates with decreasing in the SCC susceptibility (White, 2004, White, 2005). The grain size and the presence and location of inclusions and precipitates are also relevant variables in the evaluation of materials susceptibility to this phenomenon. The presence of continuous or semicontinuous chromium carbides at the grain boundaries and the absence of these carbides in the grain matrix tend to increase the resistance of nickel-based alloys to SCC susceptibility. Various models have been proposed to explain this effect, but the reasons for this beneficial effect are not completely understood. (Rebak, et al., 1993, Aguillar et al., 2003, Tsai et al., 2005, Andresen & Hickling, 2007).

Two principal sources of tensile stress are able to cause the SCC phenomenon - the tensile stresses resulting from the operating conditions (pressure, temperature and mechanical load) and the residual stresses resulting from the original fabrication process, such as welding. The tensile stresses that exist during an operation are taken into account when planning nuclear power stations and must comply with the specific standards and codes (American Society of Mechanical Engineering [ASME], 2004). However, high residual stresses may be created during the manufacturing and welding processes. The residual stresses that arise from welding may be higher than the operating stresses and tend to be dominant driving force behind the crack growth. It is the combination of operating condition stresses and residual stresses that lead the occurrence of the SCC phenomenon (Sedricks, 1990, ASM, 1992, Speidel & Magdowski, 2000, Gorman, et al., 2009).

Among all of the factors that affect SCC susceptibility, the effect of environmental conditions is particularly important. The concentrations of oxygen and hydrogen, the corrosion potential, temperature and the pH balance of the solution play an important role in this process. The SCC phenomenon is a thermally activated process and can be represented by the Arrhenius law. Places which the operating temperatures are high develop cracks more rapidly than regions where the operating temperatures are lower. It has also been observed that crack growth rates of nickel-based alloys and weld metals in simulated PWR primary water environments generally increase with increasing temperature (Nishikawa et al., 2004, Lu et al., 2008, Schvartzman et al., 2009).

1.3 Stress corrosion cracking in pressurized water reactors mitigation process

The mitigation of primary water stress corrosion cracking is defined as a adoption of remedial measures to reduce the frequency and/or to delay the stress corrosion crack initiation and/or propagation. The term mitigation does not mean completely elimination or prevention of the phenomenon, imply only in the control of the variables that affect the SCC susceptibility (Scott; Meunier, 2007).

The metodologies developed to prevent and to mitigate primary water SCC of the nickel-based alloy 600 and its associated alloys 82 and 182 weld metals include:

Materials Replacement: replacement in the new nuclear power plants of nickel-based alloy 600 and alloys 82 and 182 components for materials more resistant to SCC, nickel-based alloy 690 and its associated alloys 52 and 152 weld materials, has been considered. This remedial measure sometimes can be unavoidable in nuclear power plants currently in operation due to the high cost and difficulties in relation to the operation of this modification (security issues due to the high radiation level of the components) (King, 2005).

Surface Treatment: consist in a process that reverse the unfavorable residual stress field, leaving a compressive stresses on the surface. The presence of the compressive stress inhibits initiation and propagation of SCC. Among this method can be cited: weld overlay, induction heating stress improvement and mechanical stress improvement process, (Giannuzzi, et al., 2004).

Water Chemistry Changes: mitigation can be obtained by implementing changes to the operating environment that reduces the material's susceptibility to SCC. Among this method can be cited: addition of zinc, reduction of operating temperature and control of the dissolved hydrogen concentration (Andresen, Hickling, 2007).

In recent years, it became evident that the dissolved hydrogen concentration added to the PWR primary circuit coolant can influence the nickel-based alloy stress corrosion cracking susceptibility by changing the thermodynamic equilibrium of nickel oxide (NiO) phase formation (Morton et al., 2001, Takiguchi et al., 2004). Given the fact that the corrosion potential of nickel-based alloys in deareted water is controlled by the H_2/H_2O reaction, which represents two related reactions: oxidation ($2H^+ + 2e^- \leftrightarrow H_2$) and reduction ($H_2 + 2OH^- \leftrightarrow 2H_2O + 2e$), and the line that representing this reaction is parallel to the metal/metal oxide (Ni/NiO) phase boundary (Figure 3) variations in the dissolved hydrogen concentrations can shift the corrosion potential making that this potential reach values close to the Ni/NiO phase transition (Attanasio & Morton, 2003, Andresen et al., 2008).

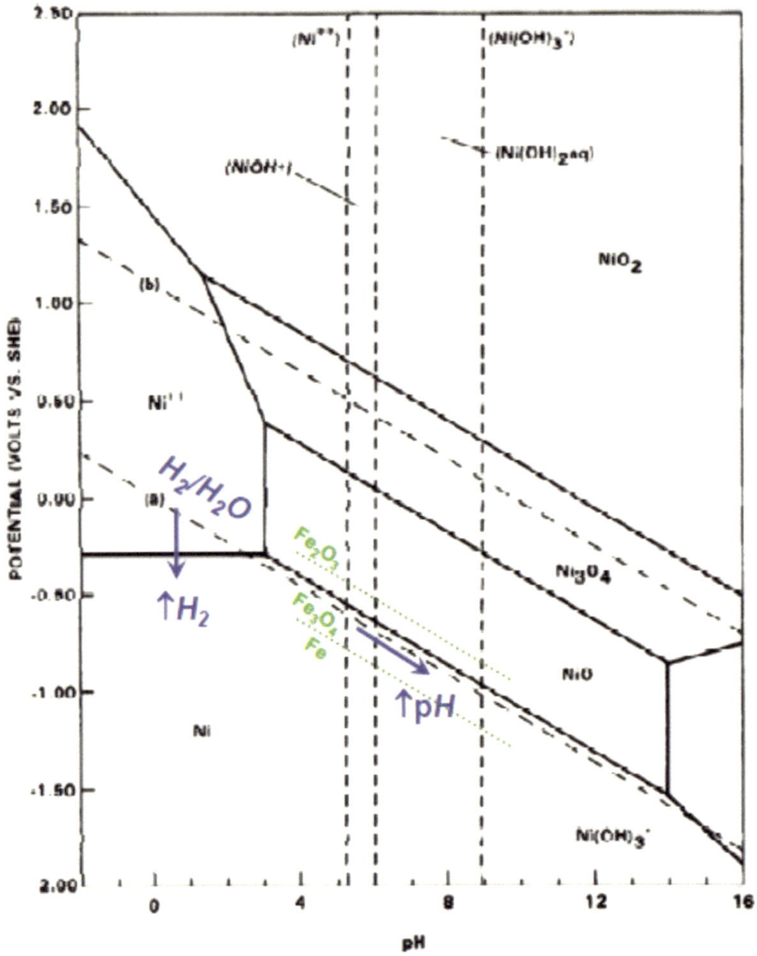

Fig. 3. System Ni – H_2O Pourbaix Diagram at 300°C (Andresen, et al., 2008).

Testing conducted by Attanasio and Morton (Attanasio & Morton, 2003) at various aqueous DH levels and temperatures has shown that a maximum in crack growth rate (CGR) occurs for nickel-based alloys in proximity to the Ni/NiO phase transition. Because the Ni/NiO boundary change with temperature the hydrogen level required to transit Ni/NiO boundary is not fixed. For nickel alloys at 325°C the peak in crack growth rate occurs at a hydrogen content of about 10 cm^3 H_2 (STP).kg^{-1} H_2O. As hydrogen concentration is increased from low values the growth rate begins to increase and then decrease, as shown schematically in Figure 4 (Andresen et al., 2005, Andresen et al., 2008).

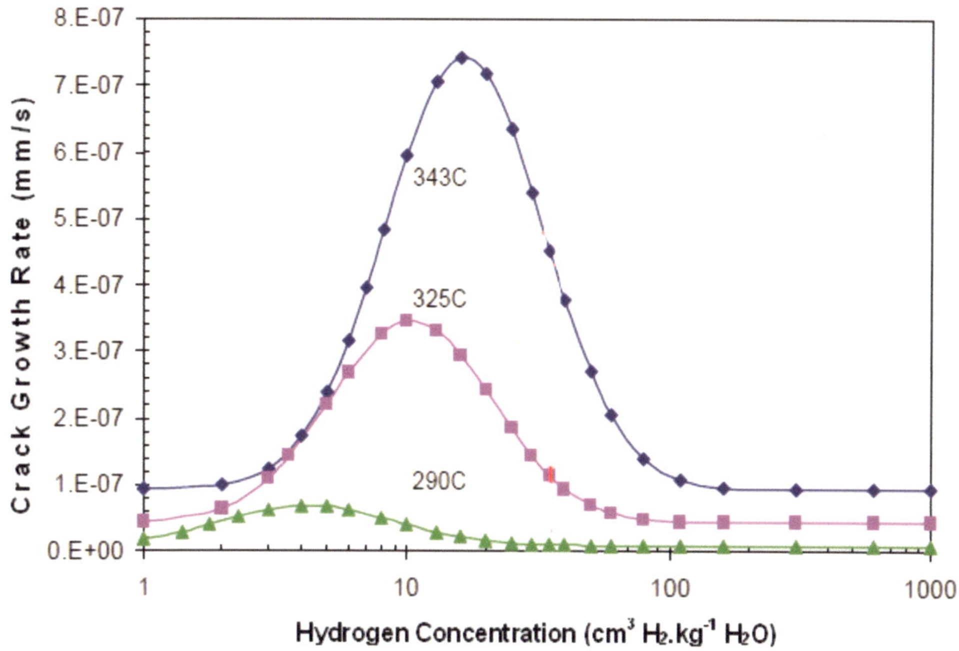

Fig. 4. Predicted effect of hydrogen concentration on the nickel-based alloys crack growth rate (Andresen et al., 2008).

Although many laboratory studies have been performed, the mechanistic basis for the effect of the primary water dissolved hydrogen on the susceptibility of nickel based alloys to SCC is still not yet completely understood.

2. Effect of dissolved hydrogen on stress corrosion cracking of nickel weld alloy 182

2.1 Experimental

The alloy 182 material was retrieved from a J groove weld joint that was made by joining two thick plates (ASTM A-508 class 3 - 130 x 300 x 36 mm and AISI 316L - 130 x 300 x 31 mm) with alloy 182, thus forming a dissimilar metal weld. The two base metals were used as received, AISI 316L in the rolled condition and ASTM A-508 class 3 in the forged condition. Five passes of buttering were applied on the ASTM A-508 class 3 plate by Gas Tungsten Arc Welding (GTAW) with Alloy 82 wire (AWS A5.14 ENiCr-3). The thickness of the buttering layer was about 8 mm. A chamfer was machined on the buttered side and the plate was subjected to a post weld heat treatment at 600°C for 2 hours to relieve the residual stresses. The final weld joint was produced by three root passes by GTAW with alloy 82 filler and thirty-seven weld passes by Shielded Metal Arc Welding (SMAW) with an alloy 182 shielded electrode (A5.11 ENiCrFe-3). The chemical composition of both base metals and filler wires is shown in Table 1. Figure 5 shows the microstructure of the alloy 182 weld joint.

	C	Mn	Si	P	S	Cr	Ni	Nb	Ti	Mo
316L	0.023	1.46	0.48	0.02	0.003	16.7	9.8	0.02	0.03	2.10
A508	0.21	1.34	0.23	0.005	0.003	0.09	0.68	0.002	0.001	0.51
182	0.05	6.16	0.34	0.01	0.009	14.3	70.3	2.07	0.05	0.24
82	0.04	3.4	0.14	0.01	0.005	18.9	73	2.47	0.25	0.16

Table 1. Chemical composition of the base and filler metals (wt%). (Fe – Bal.)

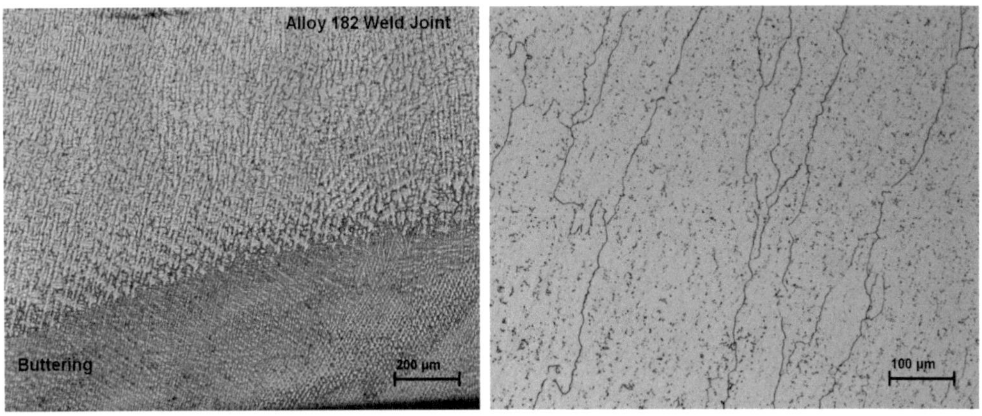

Fig. 5. Microstructure of Alloy 182 weld joint.

The finished weld joint (from now on referred to as Alloy 182 weld) was not heat-treated. It was submitted to nondestructive tests such as dye penetrant and radiographic tests, and no weld defects were revealed. Figure 6 shows a schematic of the weld design. The key welding parameters are summarized in Table 2.

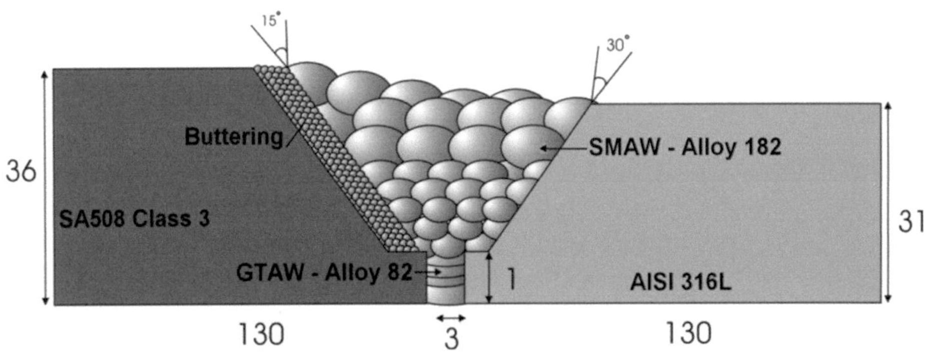

Fig. 6. Schematic of the weld design (dimensions in mm).

Weld Pass	Process	Filler Metal	Electrode Size (mm)	Current (A)	Voltage (V)	Travel Speed (mm/s)
Buttering	GTAW	82	2.5	90 – 130	17.5 – 18	1.8 – 3.0
1 - 3	GTAW	82	2.5	126 – 168	20 – 22	1 – 1.2
4 - 37	SMAW	182	4	- 135	22 – 26	1 – 3.5

Table 2. Welding parameters.

The influence of dissolved hydrogen on SCC susceptibility of the specimens was assessed by means of slow strain rate tensile (SSRT) test. The SSRT tests were carried out in accordance with ASTM G 129-95 standard (ASTM [American Society for Testing and Materials], 1995). The applied strain rate was 3×10^{-7} s^{-1}, which is an adequate strain rate to promote SCC of Alloy 182 weld in PWR primary water environments (Totsuka et al., 2003).

Specimens for the stress corrosion tests were taken from the Alloy 182 weld in the longitudinal direction using electro discharge machining (EDM). They were machined into tensile specimens with 25 mm gauge length and 4 mm gauge diameter (Figure 5), in conformity with ASTM G49-2000 and ASTM E8-2000 standards (ASTM, 2000a, ASTM, 2000b). Prior to the test, the specimens gauge length surface was polished with # 2000 silicon carbide (SiC) paper, degreased with acetone in an ultrasonic cleaner, washed with distilled water and finally dried in air. All the tests were performed at an open circuit potential (E_{OCP}) and the specimens were exposed to the environment for at least 24 hours (h) before applying load to stabilize the surface oxide layer. Three testing were performed for each condition studied.

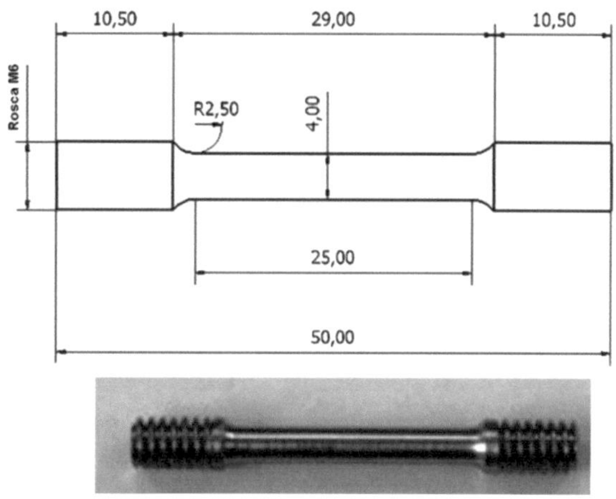

Fig. 7. Slow Strain Rate Test specimen (dimensions in mm).

The specimens were tested in simulated beginning-of-cycle PWR primary coolant environment (1000 ppm B as boric acid and 2 ppm Li as lithium hydroxide) at temperature of 325ºC and pressure of 12.5 MPa. Deionized water and analytical grade reagents were used. The test solution pH and conductivity at room temperature were 6.5 and 21 $\mu S.cm^{-1}$, respectively. The dissolved oxygen (DO) concentration was less than 5 ppb oxygen and it was obtained by bubbling pure nitrogen gas (N_2) in the work tank. After the DO content was < 5 ppb, the desired DH level were adjusted by bubbling hydrogen gas (H_2). Tests were carried out at the levels of dissolved hydrogen of 2, 10, 25 and 50 cm^3 H_2 (STP).kg^{-1} H_2O.

The SCC tests were conducted in 1.5 L type AISI 321 stainless steel autoclave, with a high temperature water circulation system. The facility was designed for SCC testing in simulated PWR or BWR environments. It is equipped with a servohydraulic loading system controlled by displacement or load. The displacement is measured by a linear variable differential transformer (LVDT) and the load is measured by a load cell. The autoclave is heated externally by an electric oven controlled continually by a proportional-integral-differential (PID) system. During the execution of the tests, on-line measurements of load, displacement, temperature, pressure, conductivity and oxygen concentration are taken. A software application developed in the LabVIEW environment is responsible for acquiring the data and for their graphic representation. Figure 8 shows a photo of the installation.

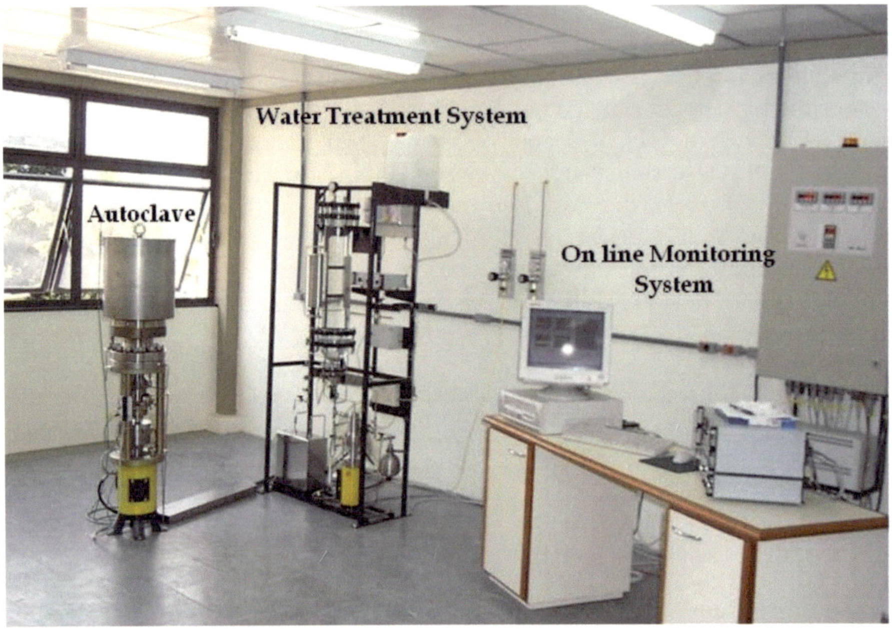

Fig. 8. Photo of the installation of SCC tests.

The susceptibility to SCC was evaluated by means of the ductility parameters obtained from the stress-strain curves and the fracture surface analyses. The fracture surfaces of the samples were examined using scanning electron microscope (SEM). All results were compared to baseline tests conducted in inert medium (N_2 – nitrogen gas) at the same strain rate. The ratio of time to failure at test condition to baseline ($t_{fsolution}/t_{fN2}$) and the ratio of elongation at test condition to baseline ($\varepsilon_{solution}/\varepsilon_{N2}$) were used as parameters to measure the degree of SCC susceptibility. In general, ratios near 100 indicate higher ductility and less susceptibility to SCC (ASTM, 1995, Nace [National Association of Corrosion Engineers], 2004).

The SCC susceptibility was also estimated using SSRT crack growth rate, which is defined as:

$$\text{SSRT crack growth rate} = \text{Crack length} \times \text{SCC fracture ratio} \div \text{fracture time} \quad (1)$$

The influence of dissolved hydrogen on Alloy 182 weld stress corrosion cracking behavior was also studied using electrochemical technique. For this purpose, open circuit potential (E_{OCP}) versus time measurements were performed in 325°C simulated PWR chemistry at hydrogen levels of 2, 10, 25 and 50 cm^3 (STP) H_2.kg^{-1} H_2O. Specimens were cut from the Alloy 182 weld in the longitudinal direction using electro discharge machining (EDM). They were machined into cylindrical specimens (5 mm diameter and 10 mm length) were mechanically ground to a surface finishing equivalent to # 600 SiC paper.

The open circuit potential (E_{OCP}) was measured after about 500 h of exposure. Measurements were conducted in a conventional three electrode cell. The electrochemical composition consisted of the working electrode, an yttrium stabilized zirconia (YSZ) electrode filled with a mixture of Ni/NiO powder and a platinum counter electrode (Figure 9). A conversion factor of - 0.800 mV taken from Bosch´s work (Bosch et al., 2003) for the temperature an pH conditions of the test was used to convert the measured values to hydrogen electrode potential values (V_{SHE}). The measurements were performed using a potentiostat system.

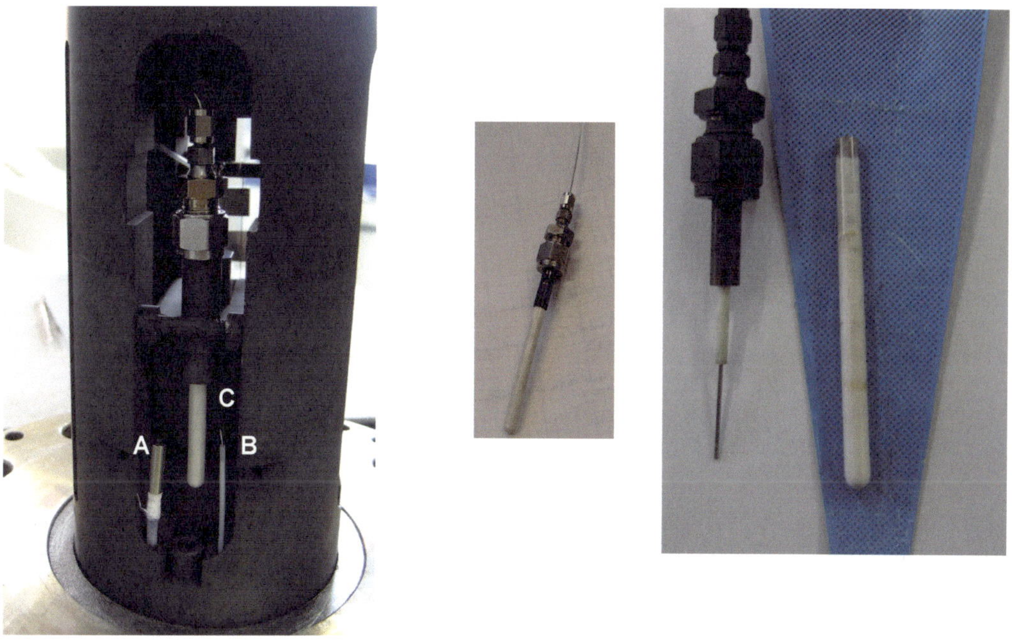

(a) (b)

Fig. 9. (a) Electrodes disposition in autoclave, A: work electrode, B: platinum counter electrode, C: yttrium stabilized zirconia (YSZ) reference electrode and (b) detail of yttrium stabilized zirconia (YSZ) reference electrode.

2.2 Results and discussion

The open circuit potential measurements results of Alloy 182 weld in 325°C simulated PWR chemistry at hydrogen levels of 2, 10, 25 and 50 cm^3 (STP) H_2.kg^{-1} H_2O are given in Figure 10. It is important to note that at 10 cm^3 (STP) H_2.kg^{-1} H_2O the potential took a longer time to stabilize, the E_{OCP} was considered stabilized when no significant variation was observed.

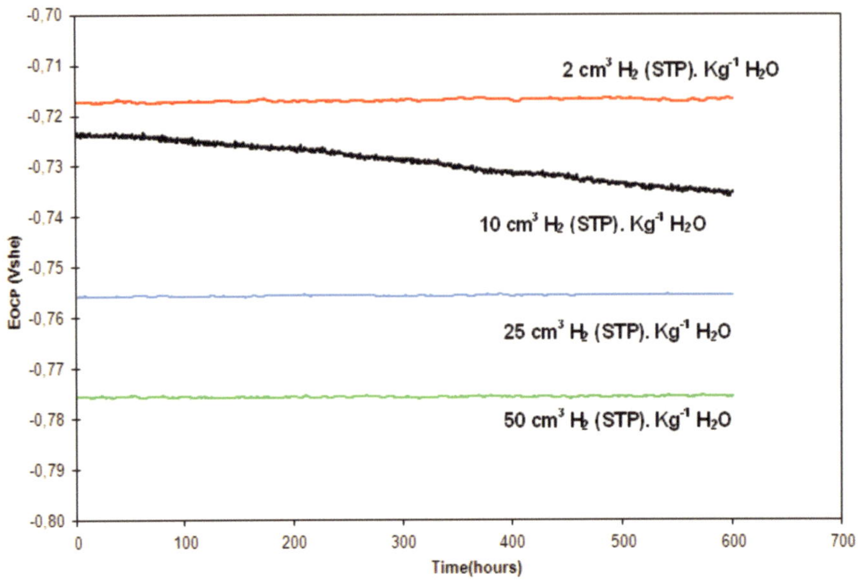

Fig. 10. OCP measurements of Alloy 182 weld in 325°C simulated PWR chemistry at hydrogen levels of 2, 10, 25 and 50 cm^3 (STP) $H_2.kg^{-1}$ H_2O.

Figure 11 (a) shows the E_{OCP} x pH diagram of nickel at 300°C in pure water obtained from EPRI´s report (Andresen & Hickling, 2007) and Figure 11 (b) shows the relationship between the corresponding E_{OCP} values obtained in this work for Alloy 182 weld in PWR environment as a function of the test solution pH (pH = 7) and the dissolved hydrogen level.

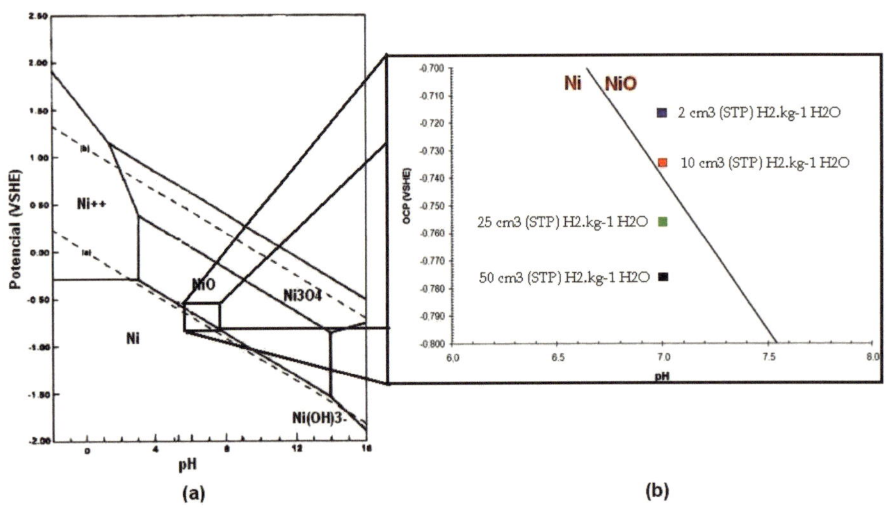

Fig. 11. (a) Potential x pH diagram of nickel at 300°C in pure water (Andresen & Hickling, 2007), (b) detail of the corresponding E_{OCP} values obtained for Alloy 182 weld in PWR environment to the line of Ni/NiO transition as a function of the pH of test solution (pH-7) and the dissolved hydrogen level.

It is observed that at 10 cm^3 H_2 (STP).kg^{-1} H_2O the potential is closer to the Ni/NiO phase transition and at 50 cm^3 H_2 (STP).kg^{-1} H_2O the potential is far away. Studies conducted for the nickel based alloys 600 and X-750 in pure water have shown that the influence of dissolved hydrogen on the SCC susceptibility may be related by the extent that the corrosion potential deviates from of the potential that corresponds to the transition of Ni/NiO ($\Delta ECP_{Ni/NiO}$). A maximum in SCC susceptibility is observed in a narrow region near the Ni/NiO phase transition (Totsuka et al., 2002, Attanasio & Morton, 2003, Andresen et al., 2008).

Table 3 shows the values of $\Delta ECP_{Ni/NiO}$ obtained in PWR solution at 325°C and concentrations of dissolved hydrogen of 2, 10, 25 and 50 cm^3 H_2 (STP).kg^{-1} H_2O. In this table the value of $\Delta ECP_{Ni/NiO}$ for 10 cm^3 H_2 (STP).kg^{-1} H_2O was considered zero because of the proximity of the transition line Ni/NiO. The present result is in accordance with the thermodynamic model proposed by Attanasio and Morton (Attanasio & Morton, 2003). According to this model the concentration of DH at 325°C, which corresponds to this transition, is approximately 10 cm^3 H_2 (STP).kg^{-1} H_2O.

Test Enviroment	E_{OCP} (mV_{SHE})	$\Delta ECP_{Ni/NiO}$ (mV_{SHE})
2(1)	-717	-18
10(1)	-735	0
25(1)	-756	21
50(1)	-776	41

Table 3. E_{OCP} and $\Delta ECP_{Ni/NiO}$ values obtained for Alloy 182 weld in PWR environment at 325°C as a function of the pH of test solution (pH-7) and the dissolved hydrogen level. (1) cm^3 H_2 (STP)/kg H_2O.

The stress-strain curves obtained from SSRT for Alloy 182 weld in the DH levels and in the baseline conditions are presented in Figure 12. Table 4 summarizes the mechanical properties and Table 5 shows the results of time to failure ratio and elongation ratio of the specimens under the five different environments.

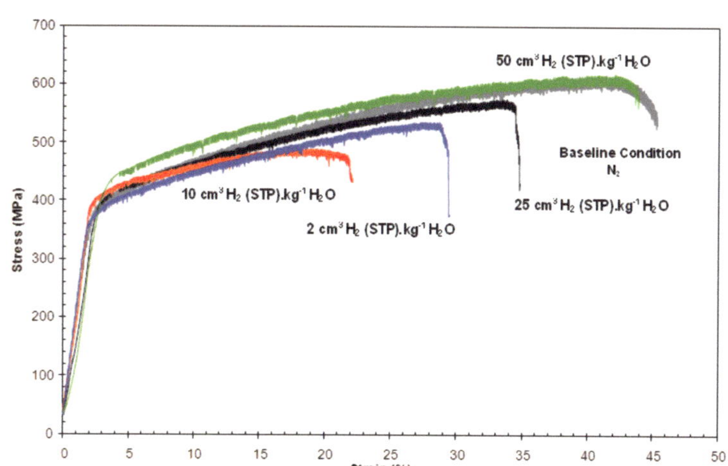

Fig. 12. Stress - Strain curves of Alloy182 weld obtained from SSRT at 325° C, at strain rate of $3x10^{-7}$ s^{-1} in PWR primary water condition with 2, 10, 25 and 50 cm^3 H_2 (STP).kg^{-1} H_2O and baseline (N_2 gas).

Test Environment	Yield Strength (MPa)		Ultimate Tensile Strength UTS (MPa)		Plastic Strain to Failure E (%)	
	Range	Mean	Range	Mean	Range	Mean
Baseline N_2 (gas)	347 - 399	375	610 - 616	612	40 – 45	42
2(1)	385 - 391	388	537 – 575	557	34 - 36	35
10(1)	380 - 390	385	490 – 547	507	20 - 22	21
25(1)	345 - 391	368	524 - 536	530	29 - 32	28
50(1)	380 - 416	398	605 - 618	610	40 -44	42

Table 4. Mechanical Properties obtained from SSRT test at 325°C. (1) cm^3 H_2 (STP)/kg H_2O.

Test Environment	Time to Failure (Hours)		$T_{f\ solution}/T_{f\ N2}$ (%)	$E_{solution}/E_{N2}$ (%)
(1) cm^3 H_2 (STP)/kg H_2O.	Range	Mean		
Baseline N_2 (gas)	367 - 418	391	-	-
2(1)	317 - 336	324	83	83
10(1)	197 - 240	216	55	51
25(1)	266 - 298	278	72	71
50(1)	367 - 401	384	99	100

Table 5. Time to failure and elongation ratios obtained from SSRT test at 325°C.

Note that at 10 cm^3 H_2 (STP).kg^{-1} H_2O there was a reduction in the resistance limit and ductility of the material, the ultimate tensile strength was 21% lower than the baseline. This reduction was attributed to the SCC process that led to the weakness of the material. It can also be seen that the specimens exposed at 10 cm^3 H_2 (STP).kg^{-1} H_2O presented the elongation and time to failure ratios lower than 100% indicating an effect of the environment on the material behavior and a higher susceptibility to SCC. At 50 cm^3 H_2 (STP).kg^{-1} H_2O these values are close to 100% indicating the lower susceptibility to SCC. Figure 13 shows the SEM micrographs of the fracture surfaces of Alloy 182 weld tested in nitrogen gas at 325°C. The fracture surface was completely ductile with extensive shear parts.

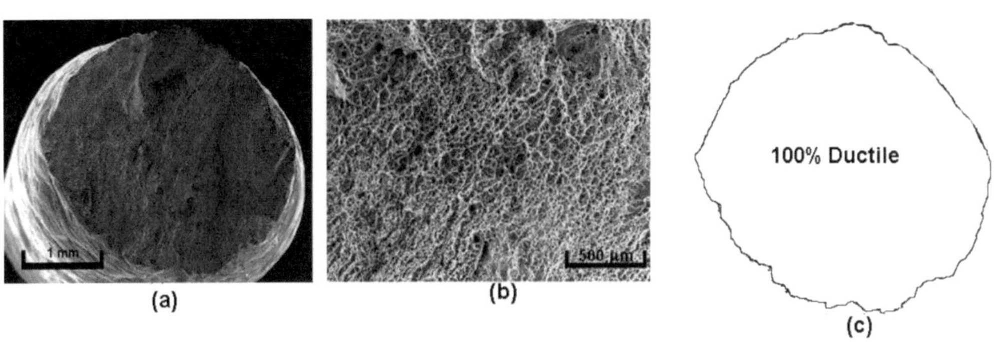

Fig. 13. SEM micrographs of Alloy 182 weld fractured surface tested at 325°C in baseline (N_2 gas) and at strain rate of $3x10^{-7}$ s^{-1} (a) overview (b) detail (c) fraction of fracture.

The respective SEM micrographs of Alloy 182 weld tested at 2, 10 25 and 50 cm^3 H_2 (STP).kg^1 H_2O are shown in Figures 14 to 17. All surfaces exhibit ductile fracture in the middle of the specimen and areas of brittle fracture at the edges, indicating crack initiation by SCC. The fracture mode was intergranular. The intergranular stress corrosion cracking (IGSCC) facets reached an average depth of 836 μm, 1300 μm, 1040 μm and 573 μm for 2, 10, 25 and 50 cm^3 H_2 (STP).kg^{-1} H_2O, respectively. The area of intergranular fracture decreased from 3.33 mm^2 to 0.43 mm^2 when the DH concentration in test solution increased from 10 to 50 cm^3 H_2 (STP).kg^{-1} H_2O. These results are consistent with the stress-strain curves obtained in SSRT, which indicates a higher susceptibility to SCC at 10 cm^3 H_2 (STP).kg^{-1} H_2O.

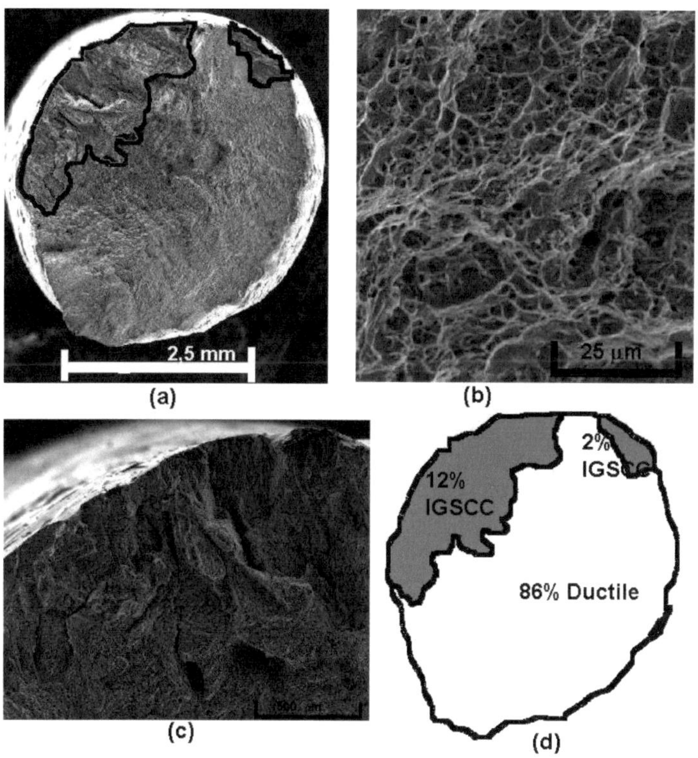

Fig. 14. SEM micrographs of Alloy 182 weld fractured surface of SSRT at 325°C in PWR primary water with 2 cm^3 H_2 (STP).kg^{-1} H_2O and at strain rate of $3x10^{-7}$ s^{-1} (a) overview (b) detail of ductile fracture (c) detail of IGSCC fracture failure (d) fraction of fracture.

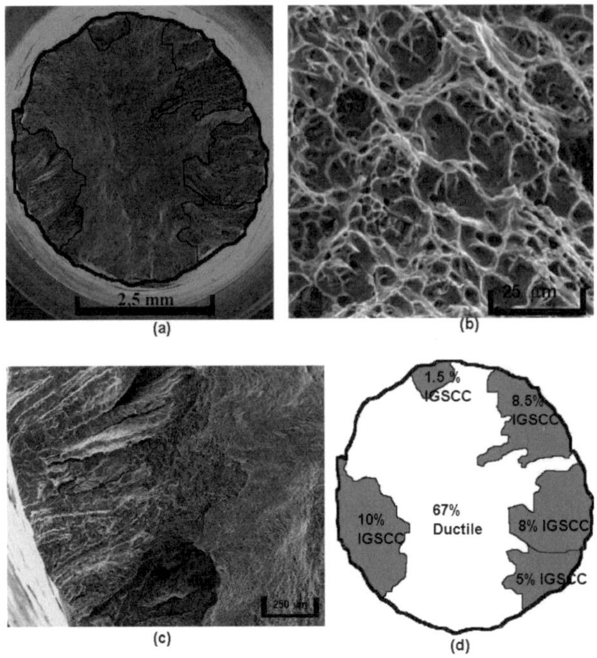

Fig. 15. SEM micrographs of Alloy 182 weld fractured surface of SSRT at 325°C in PWR primary water with 10 cm^3 H_2 (STP).kg^{-1} H_2O and at strain rate of $3x10^{-7}$ s^{-1} (a) overview (b) detail of ductile fracture (c) detail of IGSCC fracture failure (d) fraction of fracture.

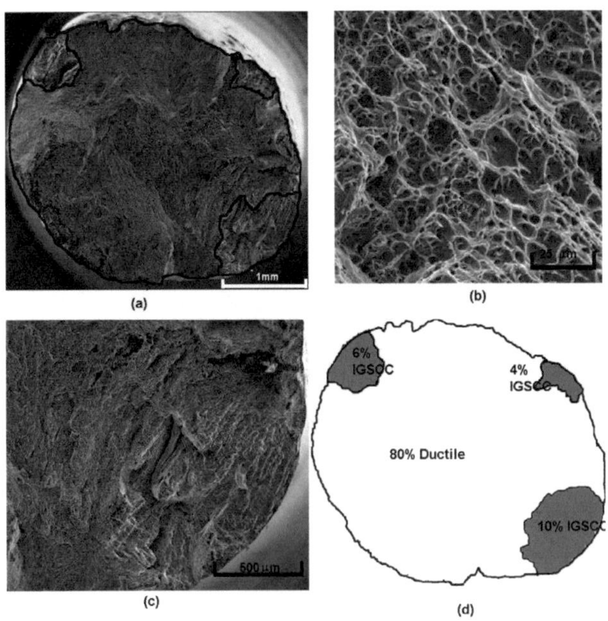

Fig. 16. SEM micrographs of Alloy 182 weld fractured surface of SSRT at 325°C in PWR primary water with 25 cm^3 H_2 (STP).kg^{-1} H_2O and at strain rate of $3x10^{-7}$ s^{-1} (a) overview (b) detail of ductile fracture (c) detail of IGSCC fracture failure (d) fraction of fracture.

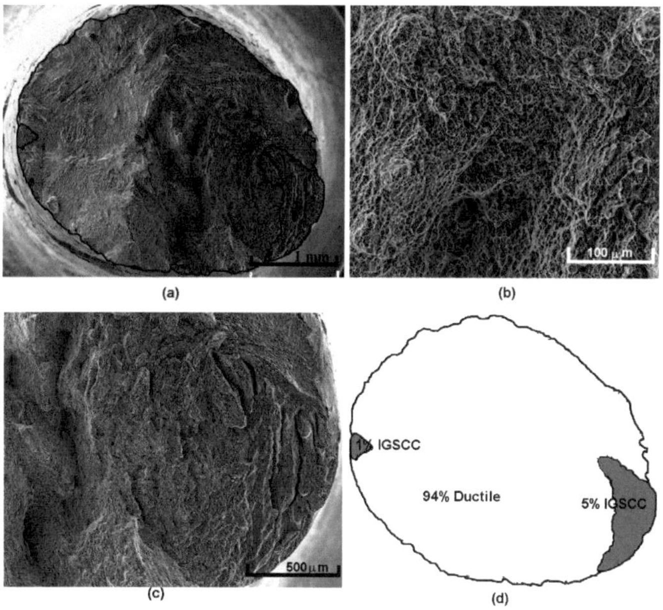

Fig. 17. SEM micrographs of Alloy 182 weld fractured surface of SSRT at 325°C in PWR primary water with 50 cm^3 H_2 (STP).kg^{-1} H_2O and at strain rate of 3x10^{-7} s^{-1} (a) overview (b) detail of ductile fracture (c) detail of IGSCC fracture failure (d) fraction of fracture.

As shown in Figure 18, extensive secondary cracks in the gauge section were observed in the sample that was exposed to 10 cm^3 H_2 (STP).kg^{-1} H_2O in direct contrast with the absence of secondary cracking observed at 50 cm^3 H_2 (STP).kg^{-1} H_2O. According to Brown and Mills (Brown & Mills, 2003) the presence of secondary cracks on the gauge sections of the test samples are also indicative of the weakness of the material attributed to the SCC process, also indicating increased SCC susceptibility of Alloy 182 weld at 10 cm^3 H_2 (STP).kg^{-1} H_2O.

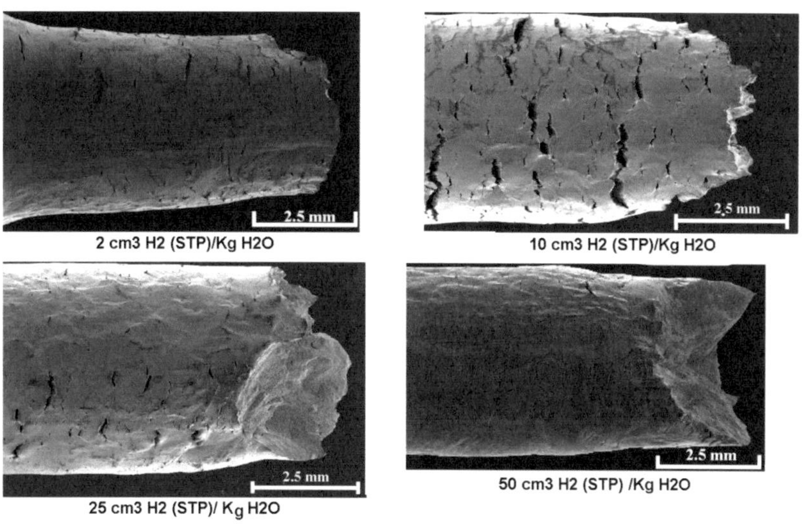

Fig. 18. Gauge section of the specimens tested in the four different environments.

Table 6 shows a summary of the crack growth rate for the four DH conditions calculated using Equation (1) and Figure 19 shows the relationship between dissolved hydrogen and crack growth rate obtained for concentrations of 2, 10, 25 and 50 cm^3 H_2 (STP).kg^{-1} H_2O.

DH cm^3 H_2 (STP) / kg H_2O	Deepest Crack (mm)	A_{IGSCC} (%)	$T_{failure}$ (h)	Crack Growth Rate (mm/s)
2	0.84	14	324	1,3x10^{-7}
10	1.30	33	216	5,0x10^{-7}
25	1.04	20	278	2.1 x 10^{-7}
50	0.57	6	384	2.9 x 10^{-8}

Table 6. SSRT crack growth rate in 325°C PWR primary water.

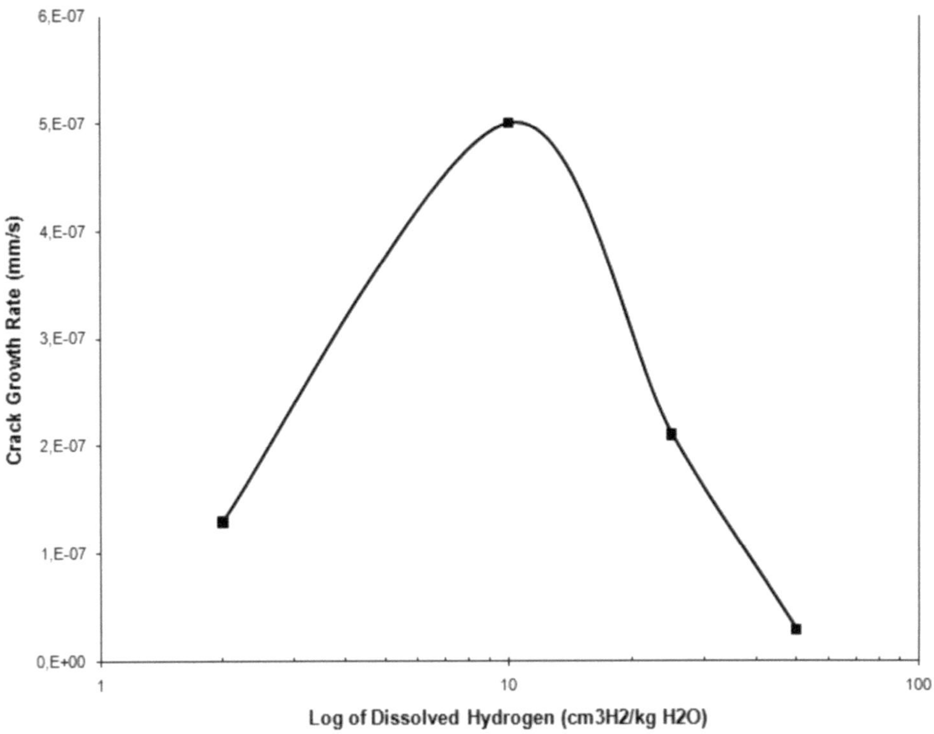

Fig. 19. Relationship between dissolved hydrogen and crack growth rate for Alloy 182 weld in a PWR primary water with obtained for concentrations of 2, 10, 25 and 50 cm^3 H_2 (STP).kg^{-1} H_2O at 325°C.

It is observed that there is a maximum of the crack growth rate (CGR) plot at 10 cm^3 H_2 (STP).kg^{-1} H_2O and that this CGR maximum is 17 times higher than at 50 cm^3 H_2 (STP).kg^{-1} H_2O. The present results are consistent with the $\Delta ECP_{Ni/NiO}$ values obtained in this work. It was observed that at 10 cm^3 H_2 (STP).kg^{-1} H_2O the potential measured is near Ni/NiO phase transition, also indicating a higher susceptibility to SCC. While at 50 cm^3 H_2 (STP).kg^{-1} H_2O the potential measured is well into the nickel metal regime, showing lower SCC susceptibility.

The present results suggest that in the normal range for operating PWRs (25–50 cm^3 H_2 (STP).kg^{-1} H_2O) the influence of hydrogen content on SCC was important. The crack growth rate at 25 cm^3 H_2 (STP).kg^{-1} H_2O was 7 times higher than at 50 cm^3 H_2 (STP).kg^{-1} H_2O. This result is consistent with that reported by Andresen et al., and Moshier, Paraventi for this range of hydrogen level (Andresen et al., 2009, Moshier & Pavarenti, 2005).

Although the crack growth rates obtained in this work have been the result of testing and evaluating the susceptibility to SCC using the technique of SSRT, They are in agreement with studies by Andresen et al., Moshier, Paraventi, and Dozaki et al. (Andresen et al., 2008, Moshier & Pavarenti, 2005, Dozaki, et al., 2010) who used the constant load test (which require a period of time of about six months to be performed).

3. Conclusion

In the present study, a DMW with alloy 182 was made to reproduce a weld joint of a PWR pressurizer nozzle. The susceptibility of Alloy 182 weld to SCC in PWR primary water was studied in four levels of DH, 2, 10, 25 and 50 cm^3 H_2 (STP).kg^{-1} H_2O. From this study the following remarks can be made:

The OCP measurements enabled the quantification of the location of the nickel/nickel oxide (Ni/NiO) phase transition line for the Alloy 182 weld metal in PWR primary water at temperature of 325°C. It was observed that the transition Ni/NiO occurs at potentials close to -740 mV.

Whereas the influence of dissolved hydrogen in the Alloy 182 weld susceptibility to SCC can be described by the extent that the corrosion potential deviates from of the potential that corresponding to the transition of Ni/NiO ($\Delta ECP_{Ni/NiO}$) results indicated that a greater resistance of this material to SCC occurs at 50 cm^3 H_2 (STP).kg^{-1} H_2O, followed by 2, 25 and 10 cm^3 H_2 (STP).kg^{-1} H_2O.

The methodology developed using The SSRT test in assessing the Alloy 182 weld SCC susceptibility reproduced the same order of magnitude of the results obtained in constant load test, demonstrating its feasibility for obtained the CGR of this alloy in PWR primary water in less time and cost.

Within the normal range for operating PWRs (25–50 cm^3 H_2 (STP).kg^{-1} H_2O) the influence of hydrogen content on SCC was significant. The crack growth rate at 25 cm^3 H_2 (STP).kg^{-1} H_2O was 7 times higher than at 50 cm^3 H_2 (STP).kg^{-1} H_2O. It is well known from literature data that the stability of oxides formed on nickel-based alloys at high temperature water is influenced by the closeness to the Ni/NiO transition line. In this study it was observed that at 50 cm^3 H_2 (STP).kg^{-1} H_2O the potential is more cathodic and at this condition the measured potential is away from the Ni/NiO phase boundary. These results suggest less SCC susceptibility in DH content of 50 cm^3 H_2 (STP).kg^{-1} H_2O. In view of that, control of dissolved hydrogen can be an effective countermeasure for the mitigation of SCC in PWR primary water.

4. Acknowledgment

The authors would like to thank Fundação de Amparo à Pesquisa do Estado de Minas Gerais (FAPEMIG), Financiadora de Estudos e Projetos (FINEP), Eletronuclear-Eletrobrás

Termonuclear S.A. and Coordenação de Aperfeiçoamento de Pessoal de Nível Superior (CAPES) for the financial support. They also thank Dr. Rik-Wouter Bosch from SCK-CEN for his help during the performance of electrochemical tests at high temperature.

5. References

Aguilar, J. L, Albarran, L. Martinez & Lopez, H. F. (2003). Effect of Grain Boundary Chemistry on the Intergranular Stress Corrosion Cracking Resistance of Alloy 600 in High Purity Water. *Proceedings of Corrosion 2003 Conference & Expo*, No 03539. San Diego, Ca, March, 2003.

Amzallag, C. J. Boursier, M. C. & Gimond, C. (2002). Stress Corrosion Life Experience of 182 and 82 Welds in French PWRS. *Proceedings Fontevraud 5 th International Symposium*, France September, 2002.

Andresen, P. L.; Emigh, Paul. W. & Morra, Martin. M. (2005). Effects of PWR Primary Chemistry and Deaerated Water on SCC. *Proceedings of Corrosion 2005 Conference & Expo*, No 05592. Houston, Texas, April, 2005.

Andresen, P. L. & Hickilng, J. (2007). Effects of B/Li/pH on CST Growth Rates in Ni-Base Alloys. Materials Reliability Program: (MRP 217), EPRI, Palo Alto, CA: 1015008.

Andresen, P. L.; Young, L. M.; Emigh, Paul. W & Horn. Ron. M. (2002). Stress Corrosion Crack Growth Rate Behavior of Ni Alloys 182 and 600 in High Temperature Water. *Proceedings of Corrosion 2005 Conference & Expo*, No 02510. Denver, Co, April, 2002.

Andresen, P. L.; Hickling, J.; Ahluwalia, A. & J. Wilson. (2008). Effects of Hydrogen on SCC Growth Rate of nickel Alloys in high temperature water. *Corrosion*, Vol. 64, No 9, (September, 2008), pp. (707-720). ISSN 0010-9312.

Andresen, P. L.; Hickling, J.; Ahluwalia, A. & J. Wilson. (2009). Effect of Dissolved Hydrogen on Stress Corrosion Cracking of Nickel Alloys and Weld Metals. *Proceedings of Corrosion 2009 Conference & Expo*, No 09414. Atlanta, GA, March, 2009.

ASM Handbook. (1992). Corrosion in the nuclear power industry. Vol. 13. (Materials Park, OH: ASM International, pp. (927). ISBN 087-170-019-0.

ASM Handbook. (2006). Corrosion Environment and Industries. Vol. 13 C (Materials Park, OH: ASM International, pp. (362-385). ISBN 978-0-87170-709-3.

ASME - American Society of Mechanical Engineers (2004). Boiler and Pressure Vessel Code - Section IX, Welding and Brazing Qualification, 2004. ISBN 079-182-894-8.

ASTM E8. (2000a). Standard Test Methods for Tension Testing of Metallic Materials. In: *Annual book of ASTM Standards. West Conshohocken*, PA: ASTM, 2000a.

ASTM G 129. (1995). Standard Test Methods for Slow Strain Rate Testing to Evaluate the Susceptibility of Metallic Materials to Environmentally Assisted Cracking. In: *Annual book of ASTM Standards*. West Conshohocken, PA: ASTM, 1995.

ASTM G 49. (2000b). Standard Test Methods for Preparation and Use of Direct Tension Stress-Corrosion Test Specimens. In: *Annual book of ASTM Standards*. West Conshohocken, PA: ASTM, 2000b.

Attanasio, S. A. & Morton, D. S. (2003) Measurement of the Nickel Oxide Transition in Ni-Cr-Fe Alloys and Updated Data and Correlations to Quantify the Effect of Aqueous Hydrogen on Primary Water SCC. *Proceedings 11th Int. Conf. Environmental Degradation of Materials in Nuclear Power Systems*, Stevenson, WA, August, 2003.

Alexandreanu, B.; et al. (2007). Environmentally Assisted Cracking in Light Water Reactors. *Argone National Laboratory Annual Report*, NUREG/CR-4667, March, 2007.

Banford, W. & Hall, J. (2003) A Review of Alloy 600 Cracking in Operating Nuclear Plants: Historical Experience and Future Trends. *Proceedings 11th Int. Conf. Environmental Degradation of Materials in Nuclear Power Systems*, Stevenson, WA, August, 2003.

Bosch, R-W.; et al. (2003). LIRES: A European Sponsored Research Project to Develop Light Water Reactor Reference Electrodes. *Proceedings 11th Int. Conf. Environmental Degradation of Materials in Nuclear Power Systems*, Stevenson, WA, August, 2003.

Brown, C. M. & Mills, W. J. (2003). Stress Corrosion Crack Growth Rates for Alloy 82H Welds in High Temperature Water. *Proceedings 11th Int. Conf. Environmental Degradation of Materials in Nuclear Power Systems*, Stevenson, WA, August, 2003.

Davis, J. R. (2006). Hardfacing, Weld Cladding, and Dissimilar Metal Joining. In: *ASM Handbook - Welding, Brazing and Soldering*, Vol. 6, pp. (2044-2061), ISBN 87170-377-7.

Dozaki, K; et al. (2010). Effects of Dissolved Hydrogen Content in PWR Primary Water on PWSCC Initiation Property. *E-Journal of Advanced Maintenance.* Vol. 2, (August, 2010), pp. (65-76). ISSN 1883-9894.

White, G. A. (2004). Crack Growth Rates for Evaluating PWSCC of Alloy 82, 182 and 132 Welds Materials Reliability Program: Materials Reliability Program: (MRP-115). EPRI, Palo Alto, CA: 1006696.

Fallatah, M. G.; Sheikh, K. A.; Khan, Z. & Boah, K. J. (2002). Reliability of Dissimilar Metal Welds subjected to Sulfide Stress Cracking. KFUPM - King Fahd University of Petroleum & Minerals. *Proceedings of 6th Saudi Engineering Conference*, Dhahran, Saudi Arabia, December, Vol. 5, pp. (297-312).

Fukumura, T. & Totsuka, N. (2010). PWSCC Susceptibility of Stainless Steel and Nickel Based Alloy of Dissimilar Metal Butt Welds. *Proceedings of Corrosion 2005 Conference & Expo*, No 10245. San Antonio, TX, March, 2010.

Garbett, K.; Henshaw, J. & Sims, H. E. (2000). Hydrogen and Oxygen Behaviour in PWR Primary Coolant. *International Conference of Water Chemistry of Nuclear Reactor System 8.* ISBN 0727729586. *British Nuclear Energy Society*. Vol. 1, 2000.

Giannuzzi, A.; Hermann, R. & Smith, R. (2004). Recommendations for Testing of Emerging Mitigation Techniques for PWSCC. Materials Reliability Program: (MRP-119). EPRI, Palo Alto, CA: 1009501.

Gomez-Briceño, D. & Serrano, M. (2005). Aleaciones Base Niquel em Condiciones de Primario de Los Reactores Tipo PWR. *Nuclear España: Revista de la Sociedad Nuclear Española*, No. 250, (March, 2005), pp. (17-22), ISSN 1137-2885.

Gorman, J.; Hunt, S.; Pete, R. & White, G. A. (2009). PWR Reactor Vessel Alloy 600 Issues. In: *ASME -American Society of Mechanical Engineering.*

IAEA - International Atomic Energy Agency. (2003). Assesment and Management of Ageing of Major Nuclear Power Plant Components Important to Safety - Primary Piping in PWRs. IAEA - TECDOC 1361, 2003.

Jang, C.; Lee. J.; Kim, S. J. & Jim, E. T. (2008). Mechanical Property Variation Within Inconel 82/182 Dissimilar Metal Weld Between Low Alloy Steel and 316 Stainless Steel. *International Journal of Pressure Vessels and Piping*, Vol. 85, No. 9, (September, 2008), pp. (635-646), ISSN 0308-0161.

King, C. (2005). Primary System Piping Butt Weld Inspection and Evaluation Guidelines Materials Reliability Program: (MRP-139), Eletric Power Research Institute - EPRI, Palo Alto, 2005, CA: 1010087.

Kou, S. (2003). Welding Metallurgy. Second Edition, John-Wiley & Sons, ISBN 0-471-43491-4.

Lima, L. I. L; Schvartzman, M. M. A. M, Figueiredo, C. A. & Bracarense, A. Q. (2011). Stress Corrosion Cracking Behavior of Alloy 182 Weld in Pressurized Water Reactor Primary Water Environment at 325°C. *Corrosion,* Vol. 65, No 085004, (August, 2011), pp. (1-9), ISSN 0010-9312.

Lu, Z.; Takeda, Y. & Shoji, T. (2008). Some Fundamental Aspects of Thermally Activated Process Involved in Stress Corrosion Cracking in High Temperature Aqueous Environments. *Journal of Nuclear Materials,* No. 383, pp. (92–96), ISSN 0022-3115.

Miteva, R & Tayllor, N. G. (2006). General Review of Dissimilar Metal Welds in Piping Systems of Pressurized Water Reactors, Including WWER Designs. *European Comission DG-JRC/IE, Petten,* Netherlands, EUR 22469 EN, 2006.

Morton, D. S.; Attanasio, A. S.; Fish, S. J. & Schurman, K. M. (1999). Influence of Dissolved Hydrogen on Nickel Alloys Stress Corrosion Cracking in High Temperature Water. *Proceedings of Corrosion 1999 Conference & Expo,* No 99447, San Antonio, Texas, April, 1999.

Morton, D. S; Attanasio, S. A. & Young, G. A. (2001). Primary Water SCC Understanding and Characterization Through Fundamental Testing in the Vicinity of the Nickel/Nickel Oxide Phase Transition. LM_01K038.

Nishikawa, Y.; Totsuka, N. & Arioka, K. (2004). Influence of Temperature on PWSCC Initiation and Crack Growth Rate Susceptibility of Alloy 600 Weld Metals. *Proceedings of Corrosion 2004 Conference & Expo,* No 04670, New Orleans, Lousiana, March, 2004.

NACE TMO19 (2004). Slow Strain Rate Test Method for Screening Corrosion Resistant Alloys (CRAs) for Stress Corrosion Cracking in Sour Oifield Service. Houston, TX: Nace, 2004.

Pathania, R. S., Mcilree, R. R. & Hickling, J. (2002). Overview of CST of Alloys 182/82 in PWRs. *Proceedings Fontevraud 5 th International Symposium,* France September, 2002.

Paraventi, D. J. & Moshier, W. C. (2005). The Effect of Cold Work and Dissolved Hydrogen in the Stress Corrosion Cracking of Alloy 82 and Alloy 182 Weld Metal. *Proceedings of 12th International Conference Environmental Degradation of Materials in Nuclear Systems,* Salt Lake City, Utah, August, 2005.

Peng, Q. J., Shoji, T., Yamauchi, H. & Takeda, Y. (2007). Intergranular Environmentally Assisted Cracking of Alloy 182 Weld Metal in Simulated Normal Water Chemistry of Boiling Water Reactor. *Corrosion Science,* No. 49, (January, 2007) pp. (2767-2780), INSS 0010-938X.

Rebak, R. B & Hua, F. H. (2004). The Role of Hydrogen and Creep in Intergranular Stress Corrosion Cracking of Alloy 600 and Alloy 690 in PWR Primary Water Environments - a Review. In: *Environment-Induced Cracking of Materials – Chemistry, Mechanics and Mechanisms,* Sergei A. Shipilov, Russel H. Jones, Jean Marc Olive, Raúl B. Rebak, Elsevier, ISBN 9780080446356.

Rebak, R. B & Szklarska-Smialowska, Z. (1996). The Mechanism of Stress Corrosion Cracking of Alloy 600 in High Temperature Water. *Corrosion Science*, No. 6, (June, 1996), pp. (971 - 988) INSS 0010-938X.

Rebak, R. B, Xia, Z & Szklarska-Smialowska, Z (1993). Effects of Carbides on Susceptibility of Alloy 600 to Stress Corrosion Cracking in High-Temperature Water. *Corrosion*, Vol. 49, No 11, (November, 1993), pp. (1-10), ISSN 0010-9312.

Schaefer, A. (1979). Dissimilar Metal Weld Failure Problems in Large Steam Generators. *Power*, pp. (68 - 69), 1979.

Scott, P. M. (2004). An Overview of Materials Degradation by Stress Corrosion in PWRs. Proceedings of Eurocorr- Annual European Corrosion Conference of the European Federation of Corrosion, ISBN 295168441X, Nice, Acropolis, September, 2004.

Scott, P.M. & Meunier, M. -C. (2007). Review of Stress Corrosion Cracking of Alloys 182 and 82 in PWR Primary Water Service, Materials Reliability Program: (MRP-220), Eletric Power Research Institute - EPRI, Palo Alto, 2007, CA: 1015427.

Schvartzman, M. M. A. M.; Quinan, M. A.; Campos, W. R. C. & Lima, L. I. L. (2009). Avaliação da suscetibilidade à corrosão sob tensão da ZAC do aço inoxidável AISI 316L em ambiente de reator nuclear PWR. *Soldagem & Inspeção*, Vol 14, No 3, (Julho/Setembro, 2009). ISSN 0104-9224.

Sedricks, A. J. (1990). Stress Corrosion Cracking Test Methods. In: *Corrosion Testing Made Easy; Stress Corrosion Cracking Testing Methods*, B.C. Syrett, NACE. ISBN 091-556-740-7.

Seifert, H. P.; et al. (2008). Environmentally Assisted Cracking Behavior in the Transition Region of an Alloy 182/SA 508 Cl.2 Dissimilar Metal Weld Joint in Simulated Boiling Water Reactor Normal Water Chemistry Environment. *Journal of Nuclear Materials*, Vol. 378, pp. (197-290), ISSN 022-3115.

Speidel, M. O. & Magdowski, R (2000). Stress Corrosion Crack Growth in Alloy 600 Exposed to PWR and BWR Environments. *Proceedings of Corrosion 2000 Conference & Expo*, No 00222, Orlando, Fl, April, 2000.

Takiguchi, H.; Ullberg, M. & Uchida, S. (2004). Optimization of Dissolved Hydrogen Concentration for Control of Primary Coolant Radiolysis in Pressurized Water Reactors. *Journal of Nuclear Science and Technology*, Vol. 41, No. 5, pp. (601-609), (May, 2004). ISSN 022-3131.

Totsuka, N.; Sakai, S.; Nakajima, N. & Mitsuda, H. (2000). Influence of Dissolved Hydrogen on Primary Water Stress Corrosion Cracking of Mill Annealed Alloy 600. *Proceedings of Corrosion 2000 Conference & Expo*, No 00212, Orlando, Fl, April, 2000.

Totsuka, N.; Nishikawa, Y. & Nakajima, N. (2002). Influence of Dissolved Hydrogen and Temperature Primary Water Stress Corrosion Cracking of Mill Annealed Alloy 600. *Proceedings of Corrosion 2002 Conference & Expo*, No 02523, Denver, Co April, 2002.

Totsuka, N.; Nishikawa, Y.; Kaneshima, Y. & Arioka, K. (2003). The Effect of Strain Rate on PWSCC Fracture Mode of Alloy 600(UNS N06600) and 304 Stainless Steel (UNS S30400). *Proceedings of Corrosion 2003 Conference & Expo*, No 03538, San Diego, Ca, March, 2003.

Tsai, W.T.; Yu, C.L. & Lee, J. I. (2005). Effect of Heat Treatment on the Sensitization of Alloy 182 Weld. *Scripta Materialia*, No. 53, pp. (505–509), ISSN 1359-6462.

White, G. A.; Nordmann, N. S.; Hickling, J. & Harrington, C. D. (2005). Development of Crack Growth Rate Disposition Curves for Primary Water Stress Corrosion Cracking (PWSCC) of Alloy 82,182 and 132 Weldments. *Proceedings of 12th International Conference Environmental Degradation of Materials in Nuclear Systems*, Salt Lake City, Utah, August, 2005.

7

SMAW Process in Terms of the Amount of Oxygen

Węgrzyn Tomasz
Silesian University of Technology, Faculty of Transport
Poland

1. Introduction

It has long been known criterion for welding due to the hydrogen content in the steel weld. The presence of hydrogen in the welds is translated into deterioration of plastic properties and strength of welded joints. Hydrogen amount has an adverse effect on safety of structures. Hydrogen in welded joints (e.g. of the structure of the vehicle) is an unfavourable element. There is a very similar negative role of nitrogen effect in steel WMD (weld metal deposit), and classification of welding methods in terms of WMD was done in 2000. There is not the same role of oxygen in weld metal deposits. Some authors are convinced, that oxygen could be treated as a negative elements, some others suggest that also very small amount of oxygen in weld is not beneficial. It is connected with austenite transformation info acicular ferrite. That phase has better condition to be formed when there are oxide inclusions inside the austenite grains. Because of that oxygen amount in steel should be better known and correctly classified similarly to the amount of hydrogen and nitrogen in welds.

Proposal of the welding processes classification of low-carbon and low-alloy steel in terms of the amount of oxygen was firstly suggested on ISOPE Conference in Brest in 1999. Since then, a new research has confirmed the valid of this concept. Nevertheless that criterion is still not very popular. Welding with coated electrodes (especially basic, rutile, acid electrodes) is very representative method to analyse that problem. The amount of oxygen in WMD has main influence on the acicular ferrite percentage in it. Metallographic structures, impact toughness and fractograph analysis of weld metal deposit with varied amount of oxygen were carried out (by putting attention to non-metallic inclusions presence in deposit) to analyse mechanical properties of WMD in terms of oxygen content. Also N-S curves were measured for typical deposits with varied amount of oxygen in WMD. Additional inclusions observation and measurements were prepared using a scanning electron microscope equipped with an energy-dispersive X-ray spectrometer.

2. Oxygen effect in welds

Effect of oxygen in weld is not the same like in steel. Amount of oxygen in WMD is normally ten times higher in comparison with steel (normally in range 300-1000 ppm O). It was observed that oxygen amount and further oxide inclusions in steel metal weld deposit have main influence on the transformation austenite→acicular ferrite (AF). The quality, quantity, type and size of oxide inclusions determines the formation of acicular ferrite. Thus the toughness and fatigue properties of the weld metal deposits are affected by the amount of oxygen and the amount of acicular ferrite in the metal weld deposits. The toughness of the WMD is also affected by morphology and density of inclusions. This is the reason, why the amount of oxygen could be treated as the important factor on metallographic structures and impact properties of weld metal deposit. In metallurgy of steel, it is treated that the lowest amount of oxygen gives good toughness properties of steel. The amount of oxygen in weld metal deposit depends mainly on filler materials and methods of welding.

2.1 Oxygen effect in covered electrode deposits

Shielded Metal-Arc Welding (SMAW) process was chosen to assess the effect of oxygen on mechanical properties of WMD. The electrodes contained coatings, constant or variable proportions of standard components in powder form. The principal composition was modified by separate additions of oxidiser (Fe_3O_4) and deoxidisers (FeTi, FeSi and Al in powder form) in electrode coatings. The principal diameter of the electrodes was 4 mm. The standard current was 180 A, and the arc voltage was 22 V. As a result after welding, the amount of oxygen in low-carbon and low-alloy steel metal weld deposits ranged between 200 and 1100 ppm.

A typical low alloy low carbon weld metal deposits after SMAW process had the following chemical composition (table 1):

Chemical element in WMD	Amount of element in WMD
C	bellow 0,06%
Mn	up to 1,4%
Si	up to 0,4%
Al	up to 0,02%
Ti	up to 0,02%
P	max 0,013%
S	max. 0,013%
N	max. 80 ppm
O	200 to 1100 ppm

Table 1. Chemical composition of WMD

For each the following deposits a chemical analysis was carried out. It was shown that in basic, rutile and acid electrode deposits were gettable various oxygen contents (table 2).

Covered electrodes	Oxygen amount in WMD, ppm
acid	in range 800-1100
rutile	in range 600-800
basic	in range 200-500

Table 2. Oxygen amount in various covered electrodes

For each the following deposits (basic, acid, rutile) micrograph tests, impact toughness, fatigue and fractograph tests were precisely carried out.

2.2 WMD properties in terms of oxygen amount

After chemical analysis for each the following deposits micrograph tests, and Charpy V-notch impact toughness test were firstly taken. The main Charpy tests were done mainly at +20°C using 5 specimens from each weld metal. Charpy V-notch impact toughness tests of the selected weld metal at lower temperatures were also done with 5 specimens. On the bases of the results shown in figure 1, the role of oxygen in the SMAW process was analysed.

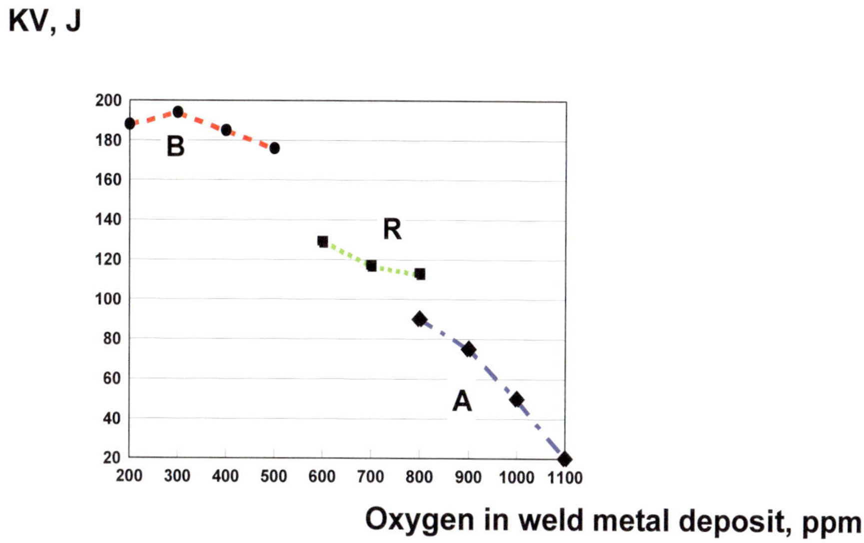

Fig. 1. Impact toughness (at 20°C) of deposits with variable amount of oxygen. Deposits were experimented by: acid electrodes (EA), rutile electrodes (ER) and basic electrodes (EB)

In figure 1 it is well shown that oxygen has an influence on impact toughness properties of metal weld deposit. For each the following deposits (basic, acid, rutile) micrograph tests were carried out. Micrograph structure are presented in figure 2, 3, 4.

Fig. 2. Typical micrograph structure of acid electrode WMD

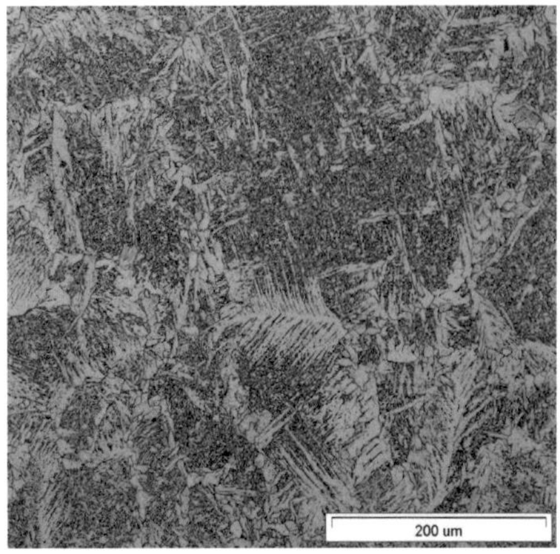

Fig. 3. Typical micrograph structure of rutile electrode WMD

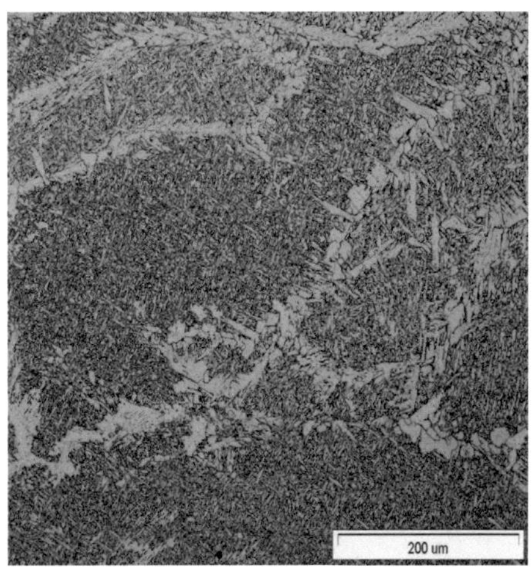

Fig. 4. Typical micrograph structure of basic electrode WMD

Analyzing figures 2, 3, 4 it was easy to deduce that there is varied amount of acicular ferrite, the most beneficial phase in deposits. The difference between amount of AF in welds is shown below in table 3.

Covered electrodes	Acicular ferrite in WMD, %
acid	in range 25-33
rutile	in range 36-44
basic	in range 52-65

Table 2. Oxygen amount in various covered electrodes

Acid electrode deposits were not analysed precisely in that chapter because of unfavourable structure and low impact toughness properties. Fractograph tests were chosen understand more the difference between structure and impact toughness of deposits with various amount of oxygen. Fractograph tests indicates more that amount of acicular ferrite in WMD is connected with the size of inclusions (and their chemical composition), figure 5, 6.

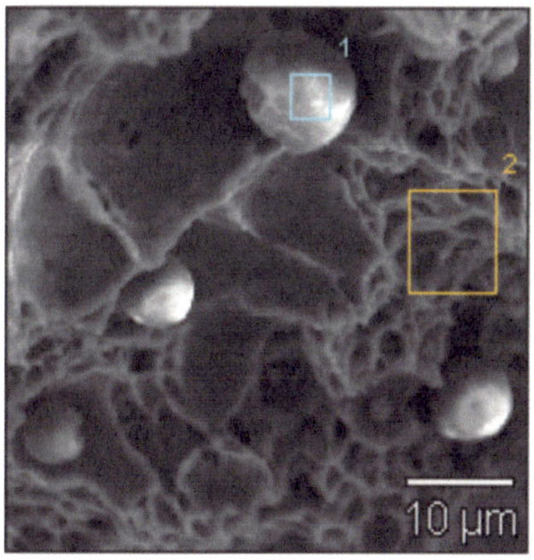

Fig. 5. Very big inclusions of rutile electrode WMD

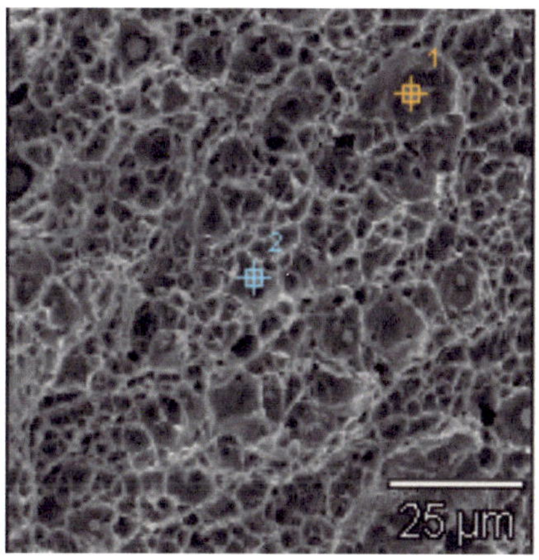

Fig. 6. Typical inclusions of basic electrode WMD

It is possible to observe, that there is not the same size of inclusions in basic and rutile deposits. Size distribution is shown in figures 7, 8, 9.

Only fractography of rutile and basic electrodes were precisely tested. The quality, quantity, type and size of inclusions determines the formation of acicular ferrite. It is possible to deduce that inclusions are heterogeneous nature and that the following oxides TiO_2, TiO, FeO, SiO_2, MnO, CaO, MgO, $MnAl_2O_4$, Al_2O_3 are dominant. Also the size of inclusions could have an influence on forming acicular ferrite and thereby resulting in obtaining better impact toughness properties. Thus the toughness of the weld metal deposits is affected by the amount of oxygen and the amount of acicular ferrite in the metal weld deposits. Basic electrode deposits had about 50-60% of acicular ferrite, rutile electrode deposits had 40-50% of acicular ferrite, and finally acid electrode deposits had less than 40 % of AF. It could be easy to understand relation between impact toughness and AF content in WMD (figure 1).

Last part of that chapter was to compare N-S curves (figures 12, 13) of typical deposits of basic and rutile electrodes giving various amount of oxygen in welds after process. Fatigue tests were generated for two deposits analysed in figures 10 and 11 with low amount of oxygen on the level of 345 ppm (typical for basic electrodes), and 650 ppm (typical for rutile electrodes).

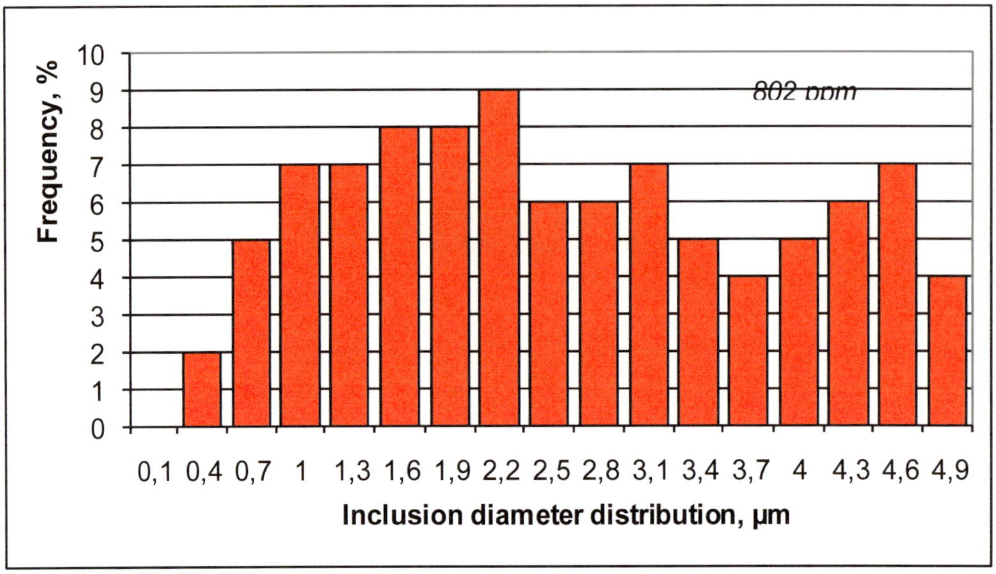

Fig. 7. Typical inclusions distribution of acid electrode WMD

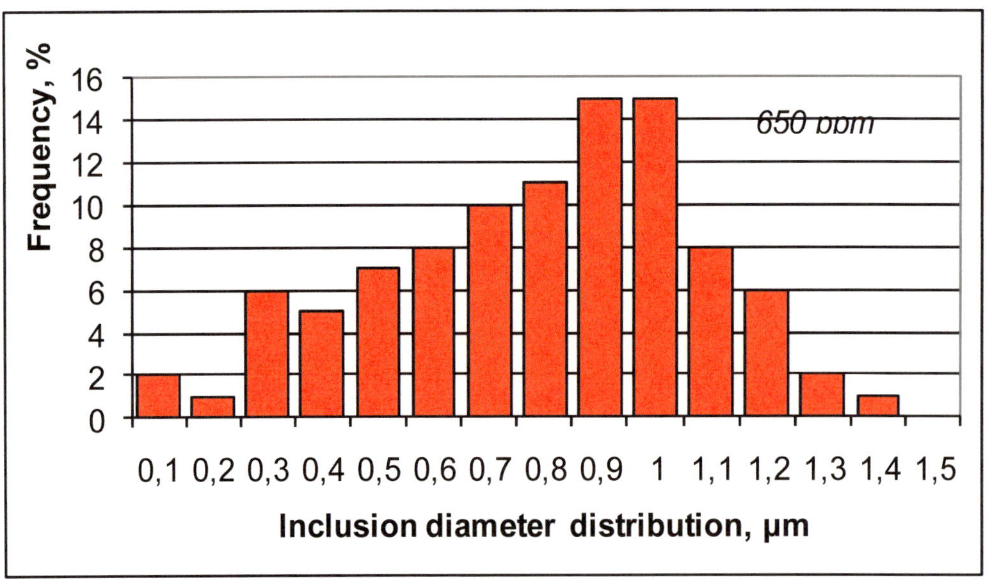

Fig. 8. Typical inclusions distribution of rutile electrode WMD

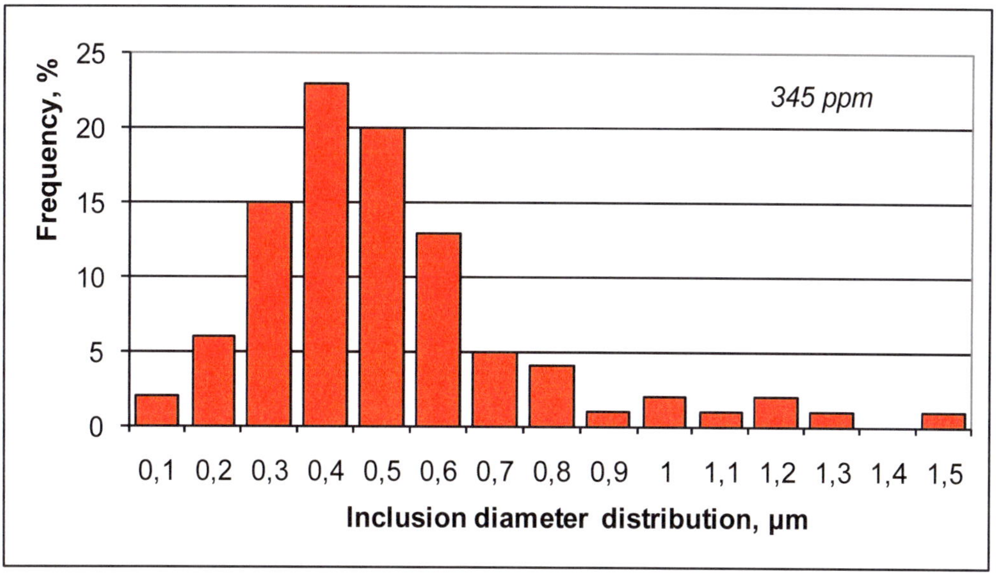

Fig. 9. Typical inclusions distribution of basic electrode WMD

Also chemical composition of inclusions was carefully measured (figures 10, 11).

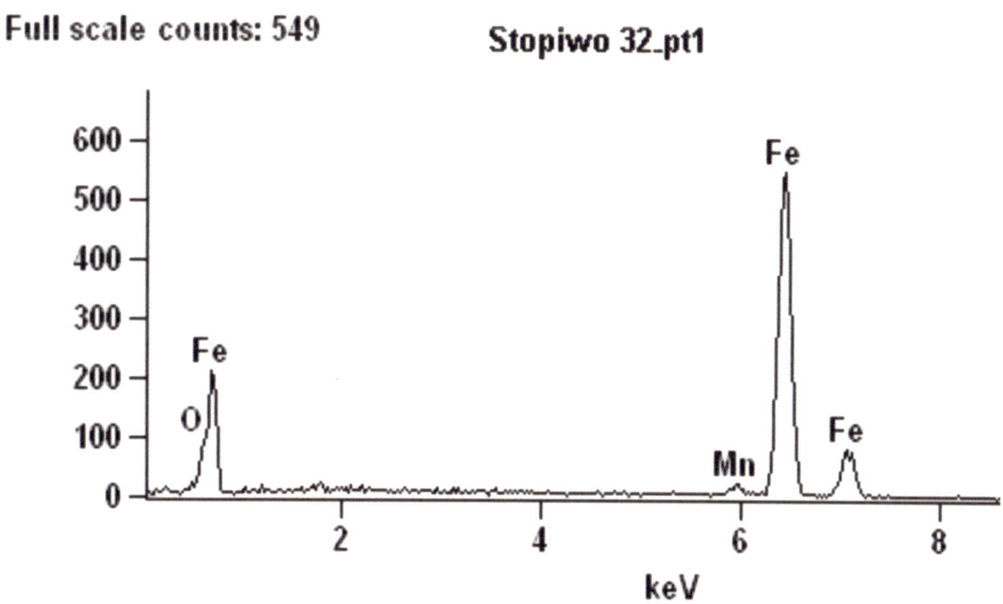

Fig. 10. Elements of oxide inclusions in rutile WMD

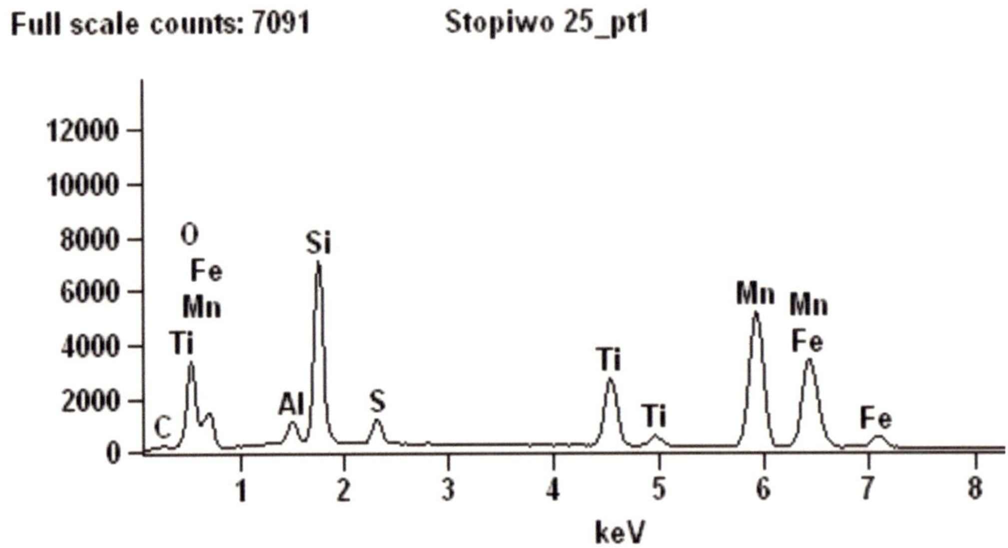

Fig. 11. Elements of oxide inclusions in basic WMD

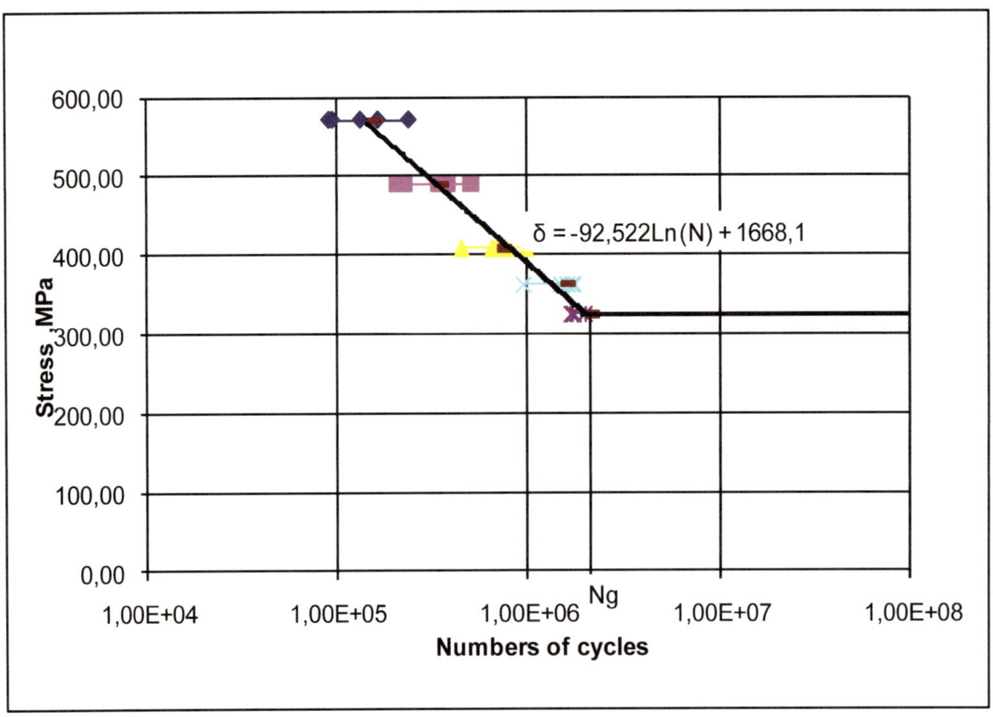

Fig. 12. S-N Fatigue properties for WMD with 345 ppm O

Looking for the S-N curve for the deposit to make an estimate of its fatigue life it easy to deduce, that low amount of O could be treated as beneficial (comparison with WMD having 650 ppm).

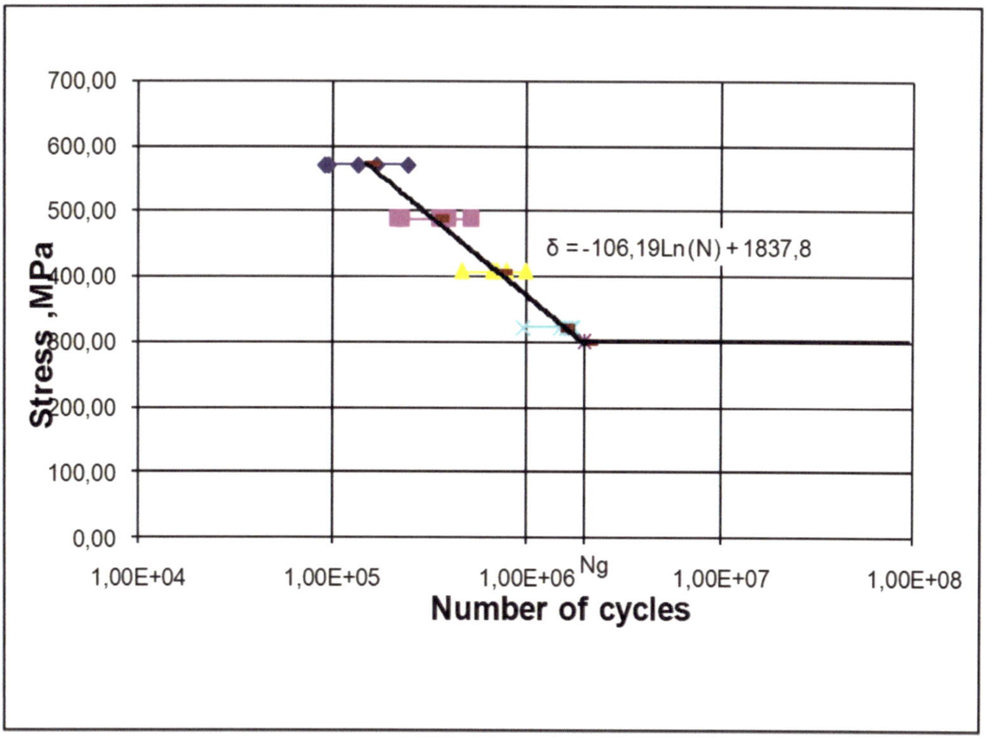

Fig. 13. S-N Fatigue properties for WMD with 650 ppm O

I was able to compare the fatigue values for deposits having various amount of oxygen. Also in this case deposits having lower amount of oxygen could be treated as more beneficial.

Examination of the influence of amounts of oxygen on alloy steel weld metal deposit allows to prove that classification of arc welding processes in terms of the amount of oxygen in weld metal deposits is very important.

Proposal of the classification for electrodes and SMAW welding processes:

- low-oxygen electrodes/process (amount of oxygen in metal weld deposit is in range 250 ÷450 ppm),
- medium-oxygen electrodes/process (amount of oxygen in metal weld deposit is in range 450 ÷650 ppm),
- high-oxygen electrodes/process (amount of oxygen in metal weld deposit is greater than 650 ppm).

3. Main elements in steel WMD

Apart from oxygen it is important to control amount other important elements in weld metal deposit: especially Mn, Ni, Mo, Cr, V). On the study done in chapter 2 it is easy to deduce that amount of oxygen in metal weld deposit should be always in range 250 ÷450 ppm. Nickel, molybdenum, chromium, vanadium are also regarded as the main factors effecting on mechanical properties and metallographic structure of low alloy welds. However there is different influence of all those elements on mechanical properties of welds in comparison with oxygen. The influence of the variable amounts of nickel, molybdenum, chromium, vanadium on impact properties of low alloy metal weld deposit was carefully presented below. Chromium, vanadium, and especially nitrogen are regarded rather as the negative element on impact toughness properties of low alloy basic electrode steel welds in sub zero temperature, meanwhile nickel and molybdenum have the positive influence on impact properties. Authors of the main publications present that the content of nitrogen in low alloy weld metal deposit should not be greater than 80 ppm, and that nickel content should not exceed 3%. It is observed that nickel (from 1% to 2%) in metal weld deposit gives good impact toughness properties of welds. The lowest amount of nitrogen in all weld metal gives the best impact results of metal weld deposit. It was suggested that nitrogen has similar role as carbon in the ferrite. The amount of nitrogen in low-carbon and low-alloy steel is limited, but in high alloy steel welds the amount of nitrogen could be sometimes even augmented to obtain optimal mechanical properties of welds. After the welding process using basic covered electrodes (low oxygen process) there were compered metal weld deposits with the variable amounts of tested elements (Mn, Cr, Mo, V, Ni) in it. After that the chemical analysis, micrograph tests, and Charpy notch impact toughness tests of the deposited metal were presented. The Charpy tests were done mainly at +20°C and -40°C with 5 specimens having been tested from each weld metal. The impact toughness results are given in figures 14-19.

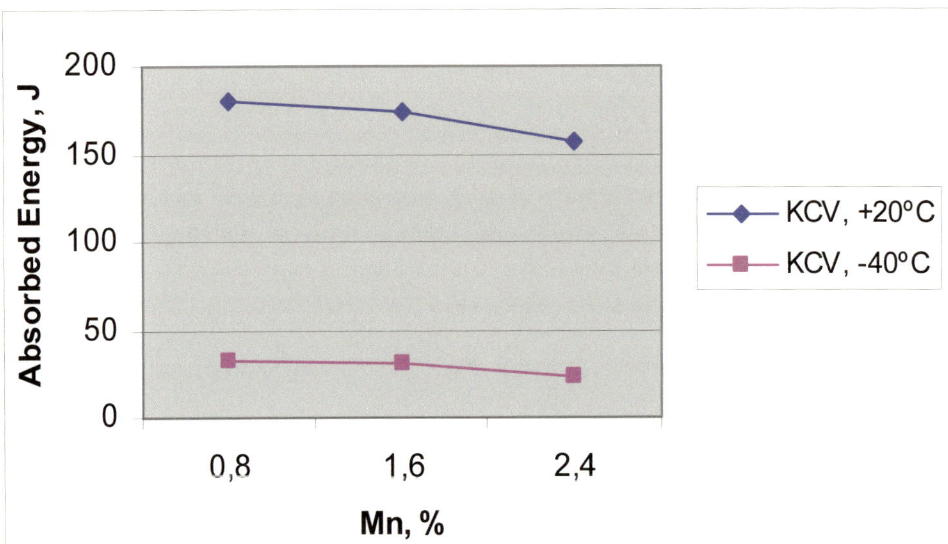

Fig. 14. Relations between the amount of Mn in MWD and the impact toughness of MWD (low-oxygen welding process)

Analysing figure 14 it is possible to deduce that impact toughness of metal weld deposit is not strongly affected by the amount of manganese. Absorbed energy in terms of the amount of vanadium in metal weld deposit is shown in figure 15.

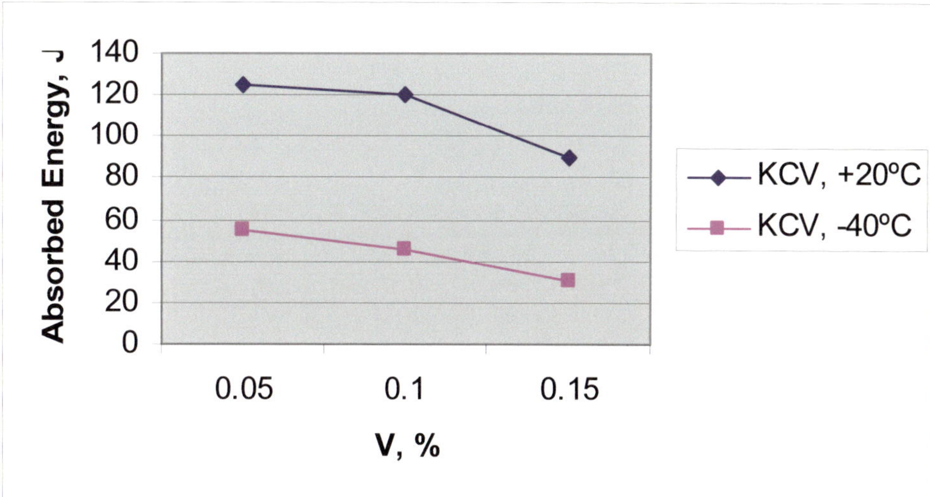

Fig. 15. Relations between the amount of V in MWD and the impact toughness of MWD (low-oxygen welding process)

Analysing figure 15 it is possible to deduce that impact toughness of metal weld deposit is much more affected by the amount of vanadium than manganese. Absorbed energy in terms of the amount of chromium in metal weld deposit is shown in figure 16.

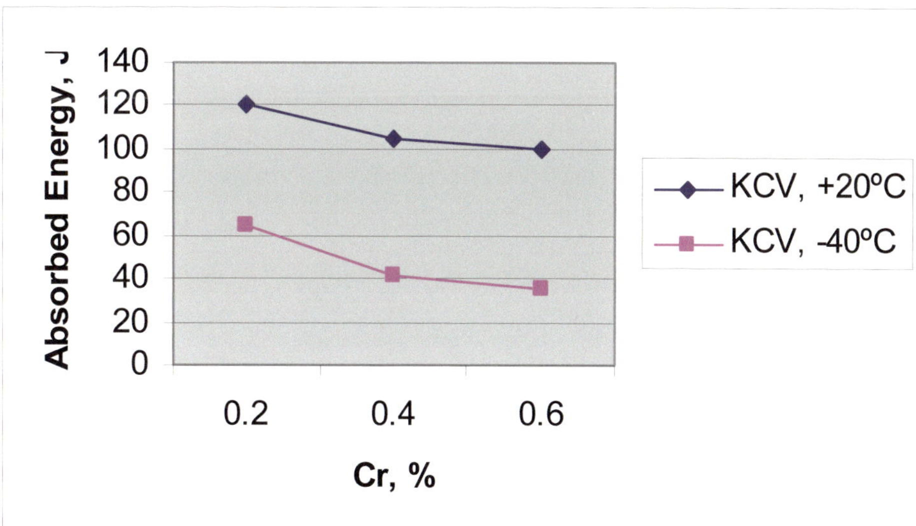

Fig. 16. Relations between the amount of Cr in MWD and the impact toughness of MWD (low-oxygen welding process)

Analysing figure 16 it is possible to observe that impact toughness of weld metal deposit is also much more affected by the amount of chromium than manganese. Absorbed energy in terms of the amount of nickel in metal weld deposit is shown in figure 17.

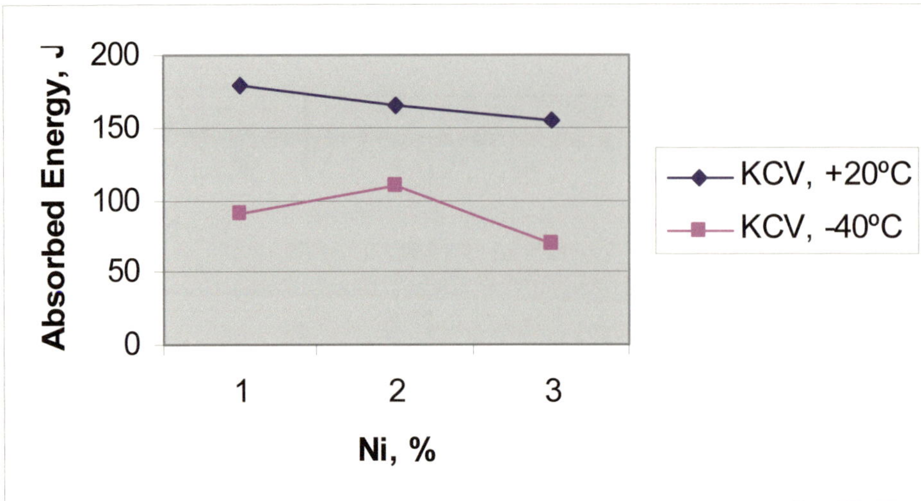

Fig. 17. Relations between the amount of Ni in MWD and the impact toughness of MWD (low-oxygen welding process)

Analysing figure 17 it is possible to deduce that impact toughness of metal weld deposit is very positively affected by the amount of nickel. Absorbed energy in terms of the amount of molybdenum in metal weld deposit is shown in figure 18.

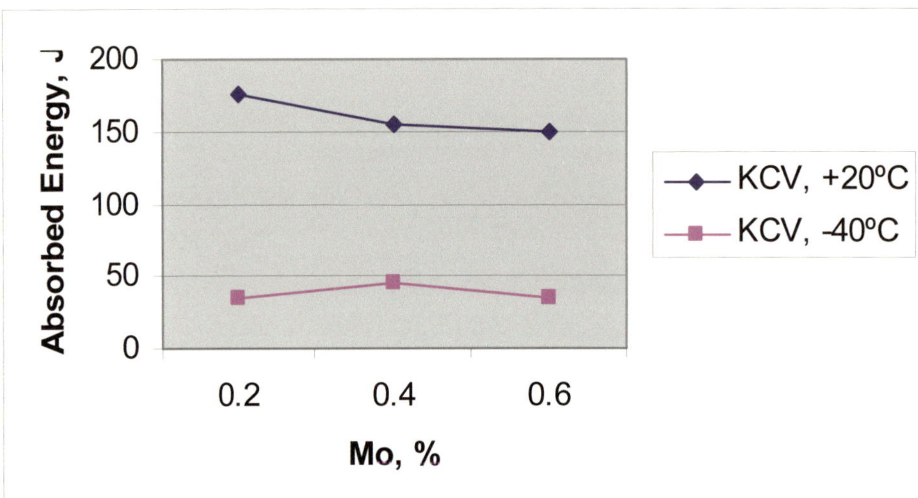

Fig. 18. Relations between the amount of Mo in MWD and the impact toughness of MWD (low-oxygen welding process)

Analysing figure 18 it is possible to observe that impact toughness of weld metal deposit is also very positively affected by the amount of molybdenum. The microstructure and fracture surface of metal weld deposit having various amount of nickel and vanadium was also done. Acicular ferrite and MAC phases (self-tempered martensite, upper and lower bainite, rest austenite, carbides) were analysed and counted for each weld metal deposit. Amount of AF and MAC were on the similar level in deposits with Ni and Mo, also for deposits with V and Cr there were observed rather similar structure. Results of deposits with various structure are shown in figures 19, 20.

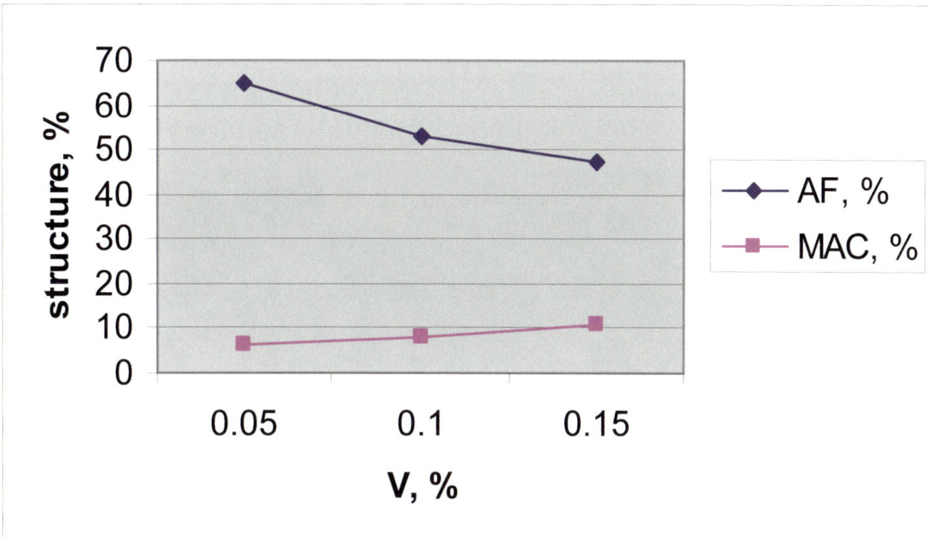

Fig. 19. Metallographic structure with V in MWD (low-oxygen welding process)

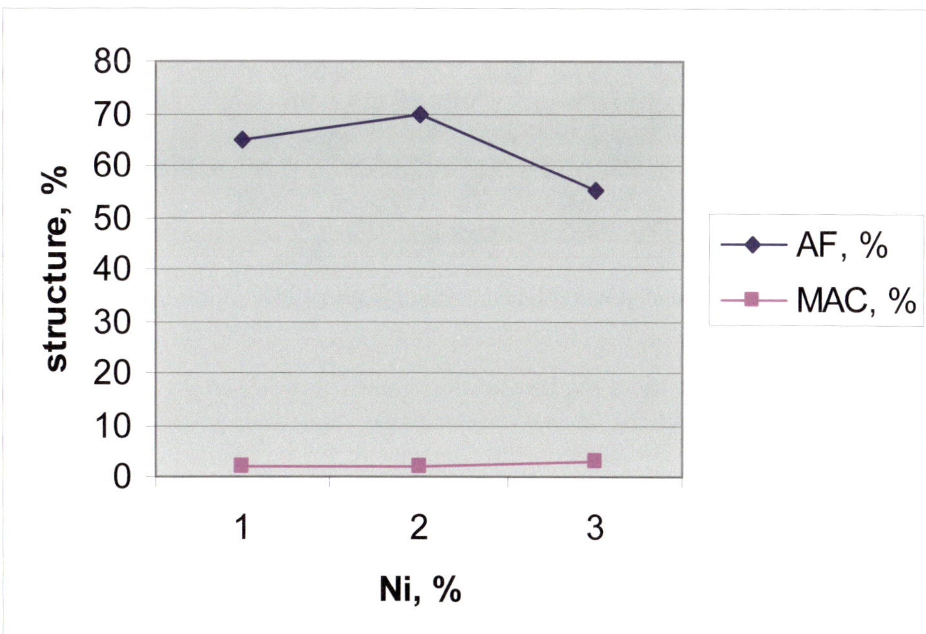

Fig. 20. Metallographic structure with Ni in MWD (low-oxygen welding process)

It was easy to deduce that nickel and molybdenum have positive influence on the structure. That relation was firstly observed in impact toughness tests. Nickel and molybdenum could be treated as the positive elements influencing impact toughness and structure of MWD because of higher amount of acicular ferrite and lower amount of MAC. Chromium and vanadium could be treated as the negative elements influencing impact toughness and structure of MWD. Manganese could be treated as a neutral element influencing impact toughness of MWD. Additional fracture surface observation was done using a scanning electron microscope. The fracture of weld metal deposit having 1.1% Ni is presented in figure 21, and the fracture of metal weld deposit having 0.6% Cr is presented in figure 22.

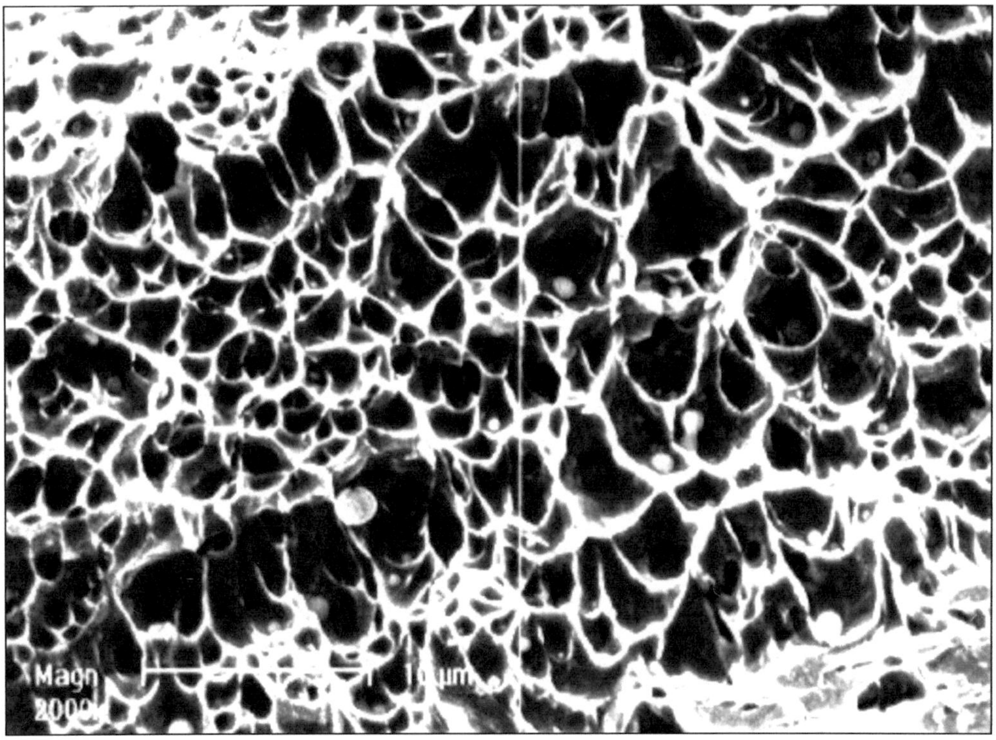

Fig. 21. Fracture surface of metal weld deposit, magnification 2000× (low-oxygen welding process)

The surface is ductile, because of the beneficial influence of nickel on the deposit structure. After microscope observations it was determined that the amount of nickel (or molybdenum) has a great influence on the character of fracture surface. The surface was ductile also for MWD having Mo in it. The character of fracture surface changed from ductile to much brittle in terms of the increscent of the amount of vanadium (or chromium). The typical fracture of metal weld deposit having 0.6% Cr is presented in figure 12.

The surface is less ductile, because of the higher amount of chromium in deposit. The surface was brittle also for MWD having V in it. After microscope observations it was determined that the amount of chromium (or vanadium) has also a great influence on the character of fracture surface. The character of fracture surface changed from ductile to much brittle in terms of the increscent of the amount of chromium. Nickel and molybdenum are positive elements in low alloy metal weld deposits. Chromium and vanadium cannot be treated as positive elements in low alloy metal weld deposits. Manganese could be rather treated as a neutral element influencing impact toughness properties. Nevertheless it is important to remember, that all investigation about Ni, Mn, Cr, V, Mo was done always for log-oxygen welding methods.

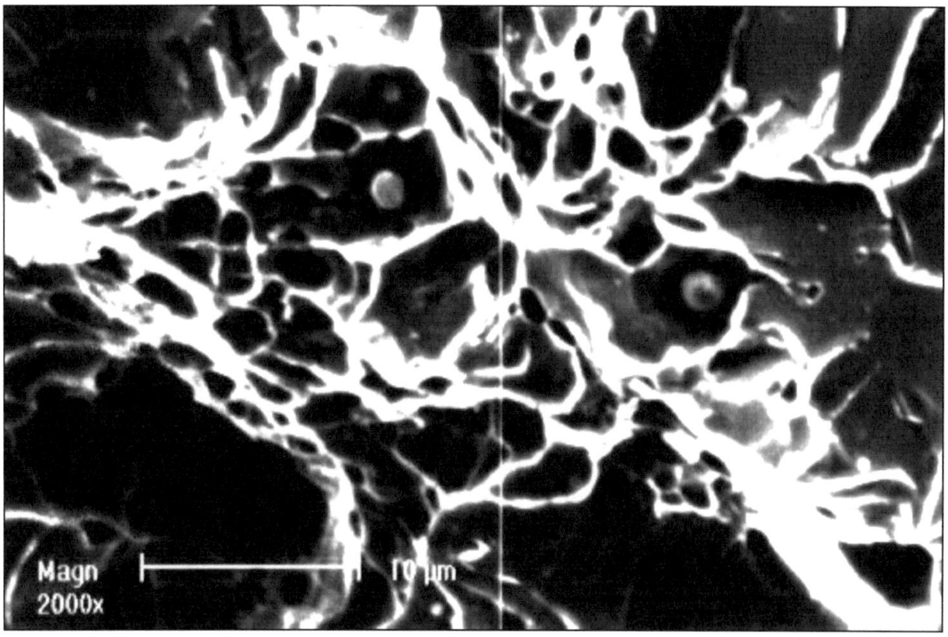

Fig. 22. Fracture surface of metal weld deposit, magnification 2000× (low-oxygen welding process)

4. Conclusion

Actual examination of the influence of amount of oxygen on alloy steel weld metal deposit allows to suggest that classification of filler materials and arc welding processes in terms of the amount of oxygen in WMD was correctly proposed in 1999 y. It is still important to remind a proposal of the classification for electrodes and SMAW processes:

- low-oxygen electrodes/process
  (amount of oxygen in metal weld deposit is in range 250 ÷450 ppm)
- medium-oxygen electrodes/process
  (amount of oxygen in metal weld deposit is in range 450 ÷650 ppm)
- high-oxygen electrodes/process
  (amount of oxygen in metal weld deposit is greater than 650 ppm)

5. Acknowledgment

Purpose: Main goal of this chapter is to remind proposals for welding processes classification in terms of oxygen in WMD, and confirmation of that concept after further 12 years of new research and results. Welds were obtained with various oxygen content. There were investigated properties of WMD, especially metallographic structure, toughness and fatigue strength of welds with various oxygen amount. The connection between the properties of welds with the content of oxygen in WMD were carried out.

Findings: Demonstrated that oxygen content in WMD has an important influence on metallographic structure, especially on the percentage of acicular ferrite in weld. The preferred structure improves the mechanical properties of welded joints. The research results indicate that it should be limited oxygen content in steel welds. Subsequent researchers could find more precisely the most beneficial oxygen amount in the welds in terms of the amount of acicular ferrite in welds.

Practical implications: To obtain welds with the best properties should be chosen suggested low-oxygen process. It is therefore suggestion to use much more basic electrodes than rutile for steel welding.

Originality/value: Proposal of welding methods in terms of oxygen content in welds was given 12 years ago, however still it is not very popular. New researches and results could prove that it is very important and original proposal.

6. References

Sylvain St-Laurent and Gilles L'Esperance.: Effects of Chemistry and size distribution of inclusions on the nucleation of acicular ferrite of C-Mn steel shielded-metal-arc-welding weldments, IIW Doc II-A-900-93, 1993

Węgrzyn T., Mirosławski J., Silva A., Pinto D., Miros M.: Oxide inclusions in steel welds of car body. Materials Science Forum 2010, vol. 6, pp 585-591

Węgrzyn T., Szopa R., Miros M.: Non-metallic inclusions in the weld metal deposit of shielded electrodes used for welding of low-carbon and low-alloy steel. Welding International, vol. 23, Issue 1, 2009, 54 - 59

Wegrzyn T.: Oxygen and nitrogen in metal weld deposits using arc welding processes. copyright by Publishing House of the Warsaw University of Technology, Warszawa 1999, 1-140

Wegrzyn T.: The Classification of Metal Weld Deposits in Terms of the Amount of Oxygen. Conference of International Society of Offshore and Polar Engineers ISOPE´99, Brest, France 1999, Copyright by International Society of Offshore and Polar Engineers, ISBN 1-880653-43-5, vol. IV Cupertino - California, USA 1999, 212-216

8

Advanced Austenitic Heat-Resistant Steels for Ultra-Super-Critical (USC) Fossil Power Plants

Chengyu Chi[1,2], Hongyao Yu[2] and Xishan Xie[2]
[1]*School of Metallurgical and Ecological Engineering, University of Science and Technology, Beijing,*
[2]*School of Materials Science and Engineering, University of Science and Technology, Beijing, China*

1. Introduction

In recent years, construction of fossil-fired power plants with higher thermal efficiency has been speed up all over the world for meeting the large requirement of electricity with the rapid development of economy and continuous raising of people's living standard. With the promoting of steam parameters to USC(Ultra-Super-Critical) level even higher, high temperature materials with improved creep resistant strength, steam corrosion and oxidation resistance over 600°C such as newly developed ferritic heat resistant steels, advanced austenitic heat resistant steels and Ni-base superalloys are required. Indeed the research and development of these high quality materials has become a key factor for USC power plants construction.

Nowadays, only advanced austenitic heat resistant steels are most suitable for USC power plants as the highest temperature components materials in the view of high temperature performance and cost. Mainly, there are three kinds of newly developed austenitic heat-resistant steels such as TP347H, Super304H and HR3C have been used as superheater/reheater tubes extensively all over the world(Viswanathan et al., 2005; Xie et al., 2010; Iseda et al., 2008; Hughes et al., 2003). In this chapter, thermodynamic calculation, SEM, TEM and Three Dimensional Atom Probe(3DAP) technology were used to analyze the microstructure evolution of these advanced austenitic heat-resistant steels during aging at 650°C till 10,000hrs. The relationship of microstructure with age hardening effect and strengthening mechanism of these austenitic steels will be discussed. According to experimental results, the advice of further improvement of these austenitic heat-resistant steels will be also proposed.

2. Development of fossil-fired power plants

Thermal power generation boilers generate electricity by consuming fossil fuels such as coal, oil and liquid natural gas. It is generally recognized that fossil fuels and coal in particular will remain the primary energy source for electric power generation for many years(Gibbons, 2009). In this situation, for reducing the levels of CO_2 in the atmosphere and saving fossil energy, it is necessary to improve the efficiency of power generation. Especially in many coal resource rich countries such as U.S., China and India, where fossil-fired power generation occupies the most important part in the structure of electricity supply. It is urgent to develop ultra-supercritial, high-efficiency coal-fired power plants to assure electricity supply safety and fossil resources effective use(Viswanathan & Bakker, 2001; Lin et al, 2009). United State was the first country to develop high efficiency USC steam generator technology in 1957, but it was ceased for a number of problems including superheater/reheater tube materials broke down(Masuyama, 2001). However, since the energy crisis broke out in the 1970s and the requirement of electricity with fast world economy growing was continually increasing, research to increase the efficiency of conventional fossil-fired power plants has been pursued worldwide. The need to reduce CO_2, SO_X and other environmentally hazardous gases emissions has recently provided another incentive to increase efficiency of fossil-fired power plants since the early of 1980s.

As well known, the efficiency of conventional fossil power plants is strongly affected by the steam temperature and pressure to reduce coal consumption and protect the environment. During the last fifty years steam parameters of fossil-fired power plants in the world have been gradually raised in program. Fig.1 shows the development of steam conditions of power plants in the world(Chen et al, 2007). It can be seen that the steam parameters of power plant is speed-up increasing and still aiming at higher level sponsored by new projects. For examples, the AD700 Project in Europe aims to build a demonstration plant operating with a main steam temperature of 700°C in 2014 (or later time), and its estimated efficiency under these conditions will improve from 35% to nearly 46%, a 11% increase(Blum et al, 2004). Similarly, project sponsored by the U.S. Department of Energy has set 760°C for advanced ultra-supercritical(A-USC) project as the goal by 2020(Viswanathan et al, 2006). In Japan, efforts to retrofit older units to enable operation at higher steam temperatures with higher efficiency. It parallels along with the Sunshine Program of setting 700°C as the goal of steam temperature(Masuyama, 2005).

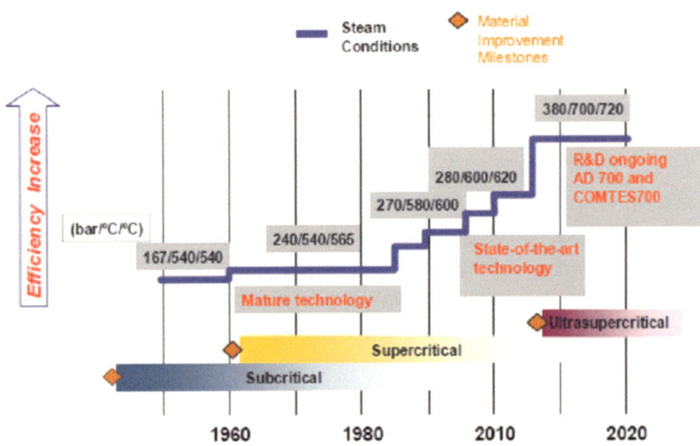

Fig. 1. Development of steam conditions in the world

It is also can be seen from Fig.1 that on every large steam parameters growing step there is always a material improvement. It means that invention of new materials with high performance promotes the improvement of fossil power plant technology. Although the steam temperature will reach about 700°C in the next 30 years (advanced ultra-supercritical technology), parameters of most power plants in the world recently still keep in the temperature range of 600-620°C with the pressure of 27-28MPa, which belongs to USC power plant level. Nowadays, some high efficiency ultra supercritical(USC) power plants with 600°C steam temperature have been built up and commercially operated all over the world, such as in USA, Germany, Denmark, Japan and China, which result not only in reduced fuel costs, but also reduced waste emission. It demonstrates that development of USC power plants is an effective way to achieve the aim of energy saving and environmental protection.

China also has a long history of fossil power plants development, and coal-fired power plants have occupied the main part in the structure of electricity supply and this situation will still keep for a long time. Recently Chinese government has announced to the world that CO_2 emission per GDP unit in 2020 would be decreased 40-45% in comparison with it in 2005. Chinese fossil power units have faced to a tough task to increase thermal efficiency and to decrease the emission of CO_2, SO_X and NO_X by means of shutting down small coal-fired power plants and developing ultra supercritical power units with high thermal efficiency. Fig.2 shows steam parameters evolution of Chinese coal-fired power plants(Liu et al, 2011). It can be seen that steam parameters of coal-fired power plants in China is increasing quickly. The first commercial operated USC unit with 26.25MPa/600°C/600°C had been started since Nov.28 in 2006 at Zhejiang province in China. The thermal efficiency reaches 45.4% and the coal consumption decreases to 283.2g/kWh for 1,000MW unit at Yuhuan USC power plant, indicating that development of USC power plants can meet the requirement of economy development. Up to now, there are sixteen 1,000MW USC power plants and twenty-one 600MW USC power plants in operation or under construction, and installed USC units are 72 in sum, which need a large amount of high performance heat-resistant steels and alloys. With the promoting of steam parameters of advanced USC power plants to 700°C/40MPa even higher(shown in Fig.2), high temperature materials with requiring good performance should be a big issue to be faced.

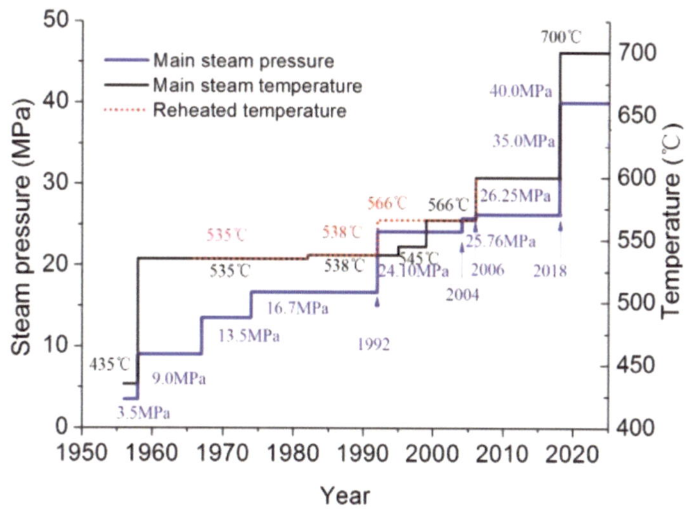

Fig. 2. Steam parameters evolution of Chinese USC power plants

There are many different components in USC power plants, such as water wall, superheater/reheater tube, header and main steam pipe need to use materials with excellent high temperature strength. However, superheater/reheater tubes are operated at the highest temperature range among these components. Therefore, the key issue to assure increase of steam temperature and pressure in boiler is the materials that will be used for superheater/reheater tubes, which must provide high creep rupture strength and high corrosion/oxidation resistance both at high-temperature conditions. Fig.3 shows the relationship of the allowable stresses and temperature for different type boiler materials(Viswanathan, 2004). Stress rupture data of these materials are dramatically decreased with the increasing of temperatures. It can be seem from Fig.3 that ferritic steels and advanced 9-12%Cr heat-resisting steels which has been widely used in conventional boilers can not be used for USC boilers with steam temperature higher than 600°C, because of its abruptly decreased allowable stress and relatively poor corrosion/oxidation resistance. Although Ni-base superalloys can meet the requirement both of high temperature strength and corrosion/oxidation resistance, however it is hard to be accepted because of its high price. Nowadays, only advanced austenitic heat-resistant steels with good high temperature performances and relatively lower cost are suitable to produce superheater/reheater tubes for 600°C USC power plants.

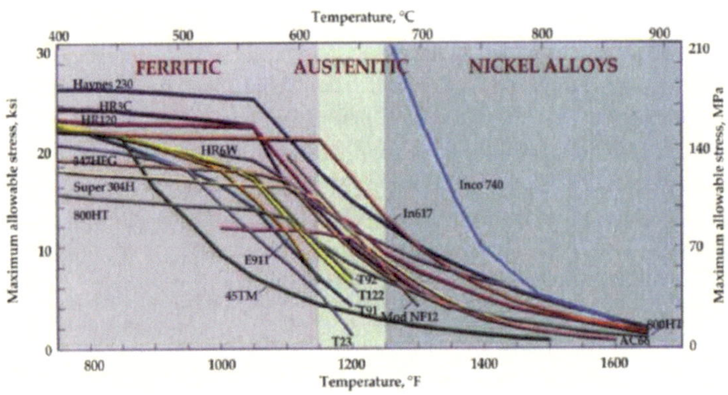

Fig. 3. Boiler materials for ultra-super critical coal-fired power plants

3. Development of austenitic heat-resistant steels

The service life of coal-fired power plants needs to reach 30~40 years. The material used for superheater/reheater tubes must be reliable over very long times at high temperatures and in severe environments. The main enabling technology is the development of stronger high-temperature materials capable of operating under high stresses at high temperatures. As mention in above paragraph, only new type austenitic heat resistant steels can be used as superheater/reheater tubes for 600°C USC power plants on the view of high temperature performance. New type austenitic heat resistant steels are origin of 18Cr-9Ni austenitic stainless steel which was early chosen for high temperature components in the first USC power plant in U.S. However, it broke down because of its poor high temperature performance and the unit must be shut down. At the efforts to pursuing for high temperature strength and good oxidation resistance by adding or optimizing alloying elements and increasing Cr, Ni content, conventional Cr-Ni austenitic steel has developed to modified 18Cr-9Ni type or 25Cr-20Ni type austenitic heat resistant steels(Yoshikawa et al, 1988; Sawaragi et al, 1992; Sourmail & Bhadeshia, 2005).

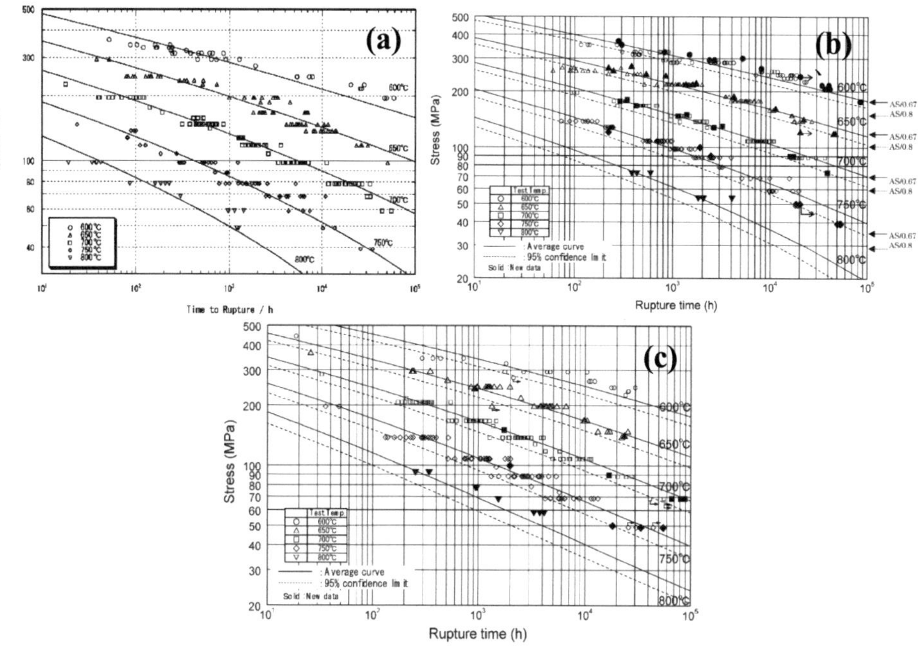

Fig. 4. Creep rupture data for TP347H(a), Super304H(b) and HR3C(c)

In order to improve high temperature creep rupture strength, not only solid solution strengthening elements such as W and Mo, but also precipitation strengthening elements such as Nb, Ti, V, are added to 18Cr-9Ni base austenitic steels to form carbon-nitrid MX for age hardening. Additionally, the element which can segregate at grain boundaries such as B is also added to increase grain boundary strengthening. The details information on improving austenitic heat resistant steel strength by alloying addition can refer paper(Masuyama, 2001). However, with the steam parameter increasing continually, 18Cr-9Ni type austenitic heat resistant steel can not burden the severe environmental corrosion and oxidation. Therefore, the Cr content has to increase to 25% and newly 25Cr-20Ni type austenitic heat resistant steel was invented. In recent 30 years researching several series of innovative austenitic heat resistant steels have been developed or in developing. Their chemical compositions are listed in Table.1. These austenitic heat resistant steels are all precipitation strengthening type steels by adding Nb element. Some steels also added W, Mo, V, N and B elements. Cu element is even added to some of these steels to further increase its creep rupture strength. Three typical newly developed austenitic heat resistant steels, TP347H, Super304H and HR3C, have been successfully serviced as superheater/reheater tubes in 600°C USC power plants all over the world. Fig.4(a), (b) and (c) show creep rupture data for TP347H, Super304H and HR3C steels, respectively(Iseda et al., 2008). In comparison with these three creep rupture data, it can be seen that creep strength of Super304H at 650°C is much higher than TP347H, and its long-term creep strength is even slightly higher than HR3C. It shows that Super304H has better strengthening effect and more stable microstructure during high temperature long time service. So in order to deeply understand strengthening effect and its mechanism of these newly developed austenitic heat resistant steels, the microstructure evolution of these three typical austenitic heat resistant steels during 650°C aging will be analyzed and comparatively discussed.

Steel	C	Si	Mn	Ni	Cr	Fe	Mo	W	V	Nb	Ti	Others
TP347H	0.08	0.6	1.6	10.0	18.0	Bal.	-	-	-	0.8	-	-
TempaloyA-1	0.12	0.6	1.6	10.0	18.0	Bal.	-	-	-	0.1	0.08	-
Super304H	0.1	0.2	0.8	9.0	18.0	Bal.	-	-	-	0.4	-	3.0Cu, 0.2N, 0.003B
XA704	0.03	0.3	1.5	9.0	18.0	Bal.	-	2.0	0.3	0.35	-	0.2N
SAVE 25	0.1	0.1	1.0	18.0	23.0	Bal.	-	1.5	-	0.45	-	3.0Cu, 0.2N
Sanicro 25	0.08	0.2	0.5	25.0	22.0	Bal.	-	3.0	-	0.3	-	3. 0Cu, 0.2N
TempaloyA-3	0.05	0.4	1.5	15.0	22.0	Bal.	-	-	-	0.7	-	0.15N, 0.002B
HR3C	0.06	0.4	1.2	20.0	25.0	Bal.	-	-	-	0.45	-	0.2N
NF709	0.02	0.5	1.0	25.0	22.0	Bal.	1.5	-	-	0.2	0.1	0.2N, 0.004B

Table 1. Nominal chemical compositions of typical austenitic heat resistant steels used or under development for superheater/reheater tubes in 600°C USC power plants(wt%)

4. Typical advanced austenitic heat-resistant steels for superheater /reheater tubes

In this part the character of microstructure and strengthening mechanism in three austenitic heat resistant steels (TP347H, Super304H and HR3C) for 600°C USC power plants have been compared according to the results of microstructure analyses and thermodynamic calculation. The microstructure of steels aging at 650°C was observed by Scanning Electron Microscopy (SEM), Transmission Electron Microscopy (TEM) and Three Dimensional Atom Probe (3DAP) have been also used for detail analyses. The thermodynamic calculation results were obtained by using Thermo-Calc software.

4.1 TP347H heat-resistant steel

TP347H steel is a kind of traditional austenitic heat-resistant steels, which is used as superheater/reheater tube material. This kind of 18%Cr-9%Ni steel contains with about 0.7%Nb and it keeps higher allowable stress and creep rupture strength than TP304, TP321H and TP316H(Viswanathan, 2009). The chemical composition of TP347H steel for our research is as follows(in mass%): 0.07C, 0.35Si, 1.35Mn, 0.016P, 0.003S, 11.22Ni, 18.08Cr, 0.68Nb, bal. Fe. Steel was solution treated at high temperature of 1150-1200°C, then water quenched to prevent new phase precipitation during cooling. In order to investigate the change of precipitated phases with aging time, aging treatments were conducted from 1,000h to 10,000h at 650°C.

4.1.1 Long-term age hardening effect

The changes of tensile strength of TP347H steel after long-term aging at 650°C are shown in Fig.5. The tensile strength increases quickly at the initial stage and has a peak value at

650°C/1,000h. The tensile strength stably keeps in a high level from 1,000h till 10,00h during 650°C long-term aging. It is clear that TP347H characterizes with stable mechanical properties and age strengthening effect developed by strengthening phase precipitation.

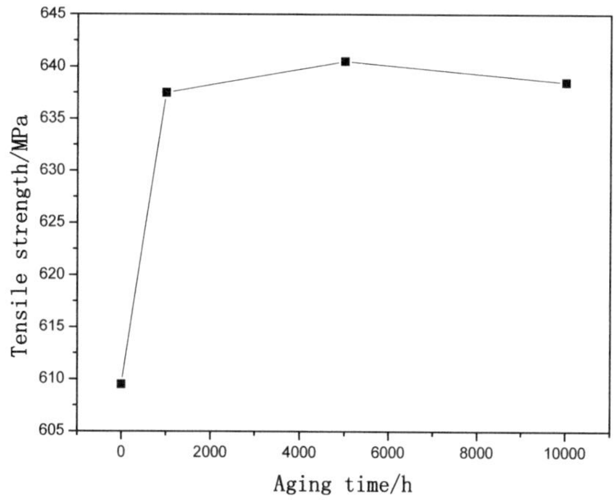

Fig. 5. Tensile strength of TP347H steel at 650°C long-term aging

4.1.2 Micro-structure analyses

Micro-structure characterization of TP347H was analyzed firstly by means of scanning electron microscope (SEM). Fig.6 shows SEM images and EDS spectrum corresponding to the precipitates of TP347H steel at solid solution treatment condition and after 1,000h and 5,000h long-term aging at 650°C. Fig.6(a) and (b) are the SEM images at solid solution treatment condition. Some primary NbC inclusions randomly exist in the grains and also occasionally at grain boundaries and their size is about 1-3μm, shown in Fig.6(a) and (b). Fig.6(c) and (d) are the SEM images after 1,000h and 5,000h long-term aging at 650°C, respectively. Fig.6(e) is the EDS analysis result of the nano-size phase precipitated in grains shown by the arrow. This EDS spectrum shows that nano-size precipitate is rich in Nb and C, which indicates that the nano-size precipitate is MX type NbC phase. After long-term aging at 650°C, the amount of precipitates in grains increases intensively and there are also some Cr-rich precipitates($M_{23}C_6$ carbide) at grain boundaries, shown in Figure.6(c) and (d). The main precipitated phase in grains is Nb-rich(NbC) precipitates.

Transmission electron microscope (TEM) has been used for surveying nano-size particles. Fig.7 shows TEM images, diffraction patterns and EDS spectrum of TP347H austenitic heat resistant steel at solid solution treatment condition. There are some primary NbC carbide particles as inclusions and some undissolved MX phase in grains confirmed by diffraction pattern, shown in Fig.7(a) and (b). The size of primary NbC carbide is about 1-3μm which can directly form during solidification. The primary NbC inclusion is not good for strengthening during long-term creep, because of its large size and the cracks may happen nearby the inclusions and grow up quickly. The size of undissolved MX phase is about 300-600nm and most of them are spherical. Fig.7(c) and (d) show the EDS analysis of matrix and the undissolved MX phase, respectively. According to the diffraction pattern and EDS results, it can be known that the nano-size undissolved MX particle is Nb rich NbC phase.

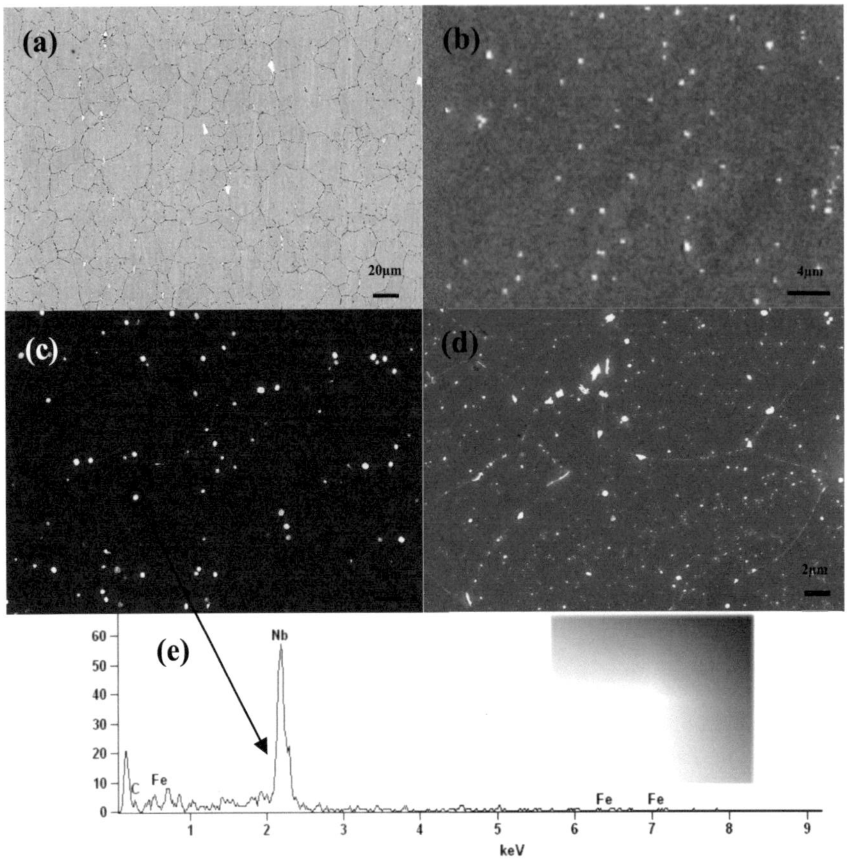

Fig. 6. SEM images of TP347H austenitic heat resistant steel after long-term aging at 650°C: solution treatment condition(a)(b), 650°C aging for 1,000h (c), 5,000h(d), and EDS spectrum corresponding to the precipitates in the grain(e)

The TEM micrographs of TP347H austenitic heat resistant steel after 1,000h long-term aging at 650°C is shown in Fig.8. During the initial aging, very fine nano-size MX phase with NaCl crystal structure precipitate in grains, as shown in Fig.8(b). Its size is about 28nm only. The morphology of MX is not spherical but in quadrate shape. $M_{23}C_6$ carbide can precipitate at grain boundaries when a solution-treated stainless steel is aged isothermally or slowly cooled within the temperature range of 500-900°C(Tanaka et al, 2001). The Cr-rich phase $M_{23}C_6$ carbide precipitates at grain boundaries of this steel, shown in Fig.8 (c). The corresponding EDS spectrum is shown in Fig.8(d). It shows that this Cr-rich precipitates at grain boundaries is $Cr_{23}C_6$ phase. The morphology of $Cr_{23}C_6$ is globular and keeps in chain-like distribution at grain boundaries after 650°C/1,000h aging. The size of grain boundary carbide $Cr_{23}C_6$ is about 100nm, which is much bigger than MX precipitates in grains. $Cr_{23}C_6$ carbide is very often found in the early stage of precipitation because it nucleates easily at grain boundaries. Incoherent or coherent twin boundaries and sometimes intragranular sites are all the nucleation sites of $Cr_{23}C_6$ carbide, but the most favorable sites for $Cr_{23}C_6$ carbide precipitation are grain boundaries(Hong et al, 2001). In this research, it is confirmed that $M_{23}C_6$ carbide mainly precipitates at grain boundaries.

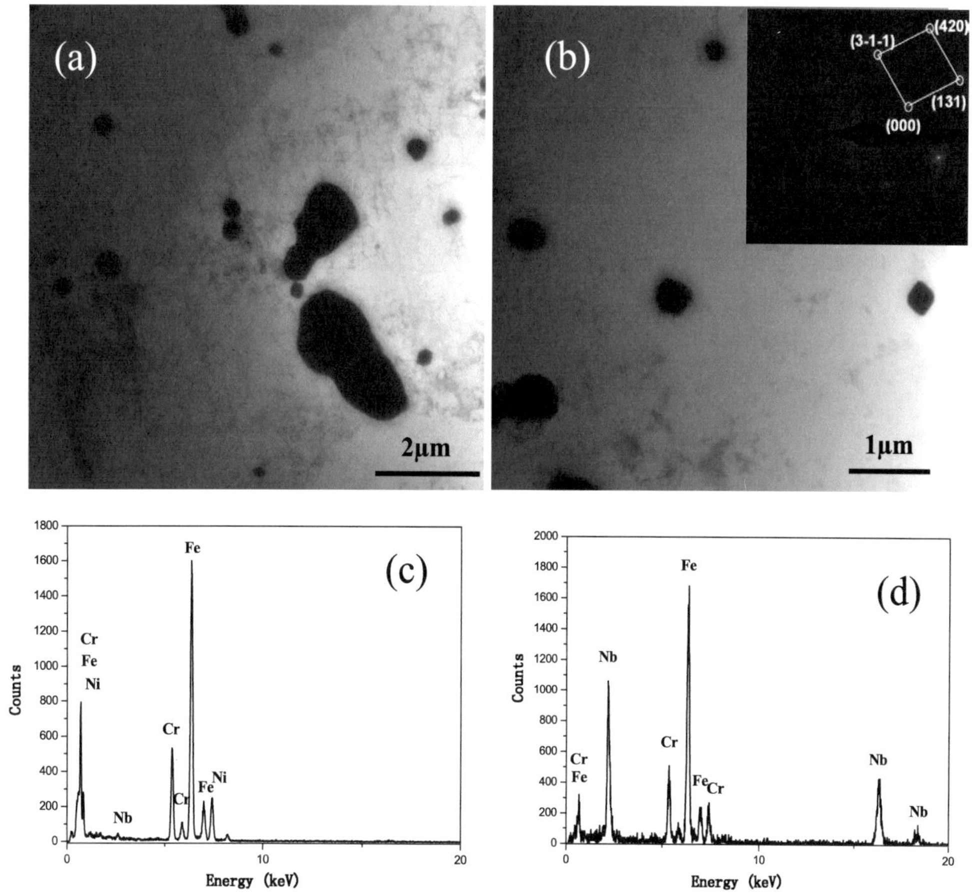

Fig. 7. TEM images of TP347H heat resistant steel after solution treatment (a) with diffraction pattern corresponding to the undissolved MX phase (b), EDS spectrum of the austenitic matrix (c) and the undissolved MX phase in grains (d)

After further long-term aging, the microstructures of TP347H austenitic heat resistant steel have changed. Figure.9 shows TEM images and diffraction patterns of precipitates in TP347H steel after 5,000h and 10,000h long-term aging at 650°C. Compared Fig.9(a) and (c) with Fig.8(b), it confirms that MX phase grows slowly. After 5,000h aging its size keeps in 35nm and even aging 10,000h its size still keeps in 50nm. Fig.9(d) shows the diffraction patterns of this nano-size particles and γ-matrix. It can be seen that the orientation relationship between MX phase and γ-matrix is $<\text{-}121>_{MX}//<1\text{-}1\text{-}2>_{\gamma}$. The lattice parameter of MX phase (a_{MX}) is 0.4421nm which is bigger than the lattice parameter of matrix phase (a_{γ}) 0.3635nm. Comparing Fig.8(a) and Fig.9(b), the density of the nano-size MX particle is increasing with aging time. The undissolved MX phase still exit after 10,000h long-term aging, shown in Fig.9(b). It can be seen clearly that there are two types of MX phase exited in the grains during long-term aging. One is the undissolved MX phase and its size keeps 300-600nm all the time. The other one is nano-size MX phase which precipitate during long-term aging and its growth rate is very slow. There is a very small amount of undissolved MX phase in the grains compared with precipitated MX particles.

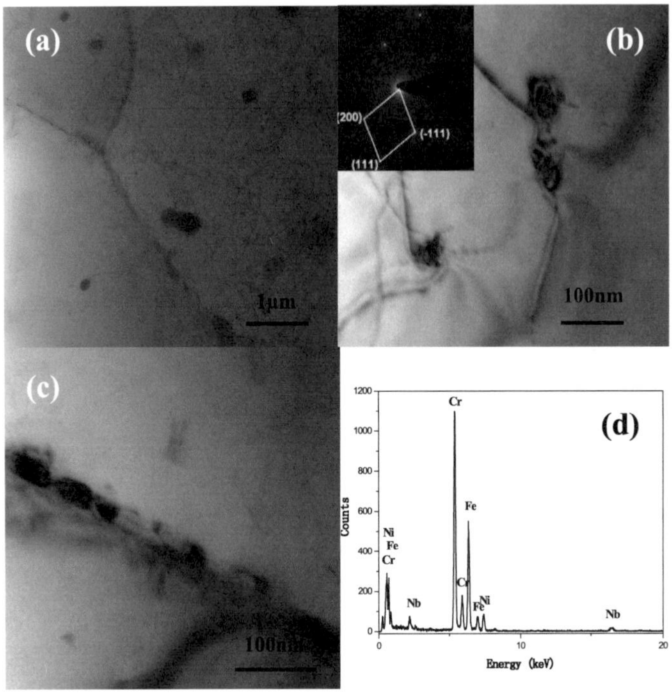

Fig. 8. TEM images of TP347H austenitic heat resistant steel after 1,000h long-term aging at 650°C (a) with diffraction pattern corresponding to precipitation in grain (b), precipitation at grain boundary(c), and corresponding EDS spectrum(d)

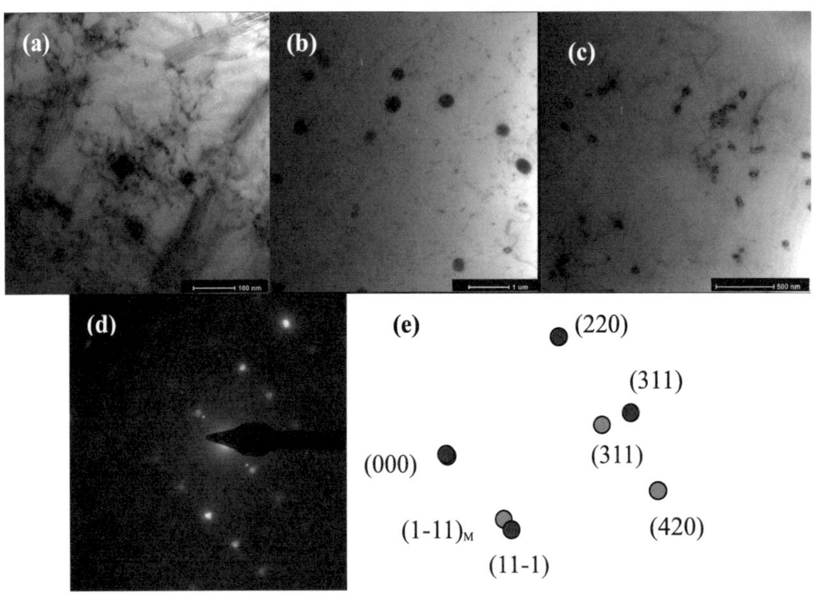

Fig. 9. TEM images and diffraction patterns of precipitations in TP347H steel after long-term aging at 650°C (a) 5,000h; (b) and (c) 10,000h; (d) and (e) diffraction patterns of γ-matrix and precipitation in(c)

4.1.3 The relationship between thermal equilibrium phases and temperature in TP347H

The fraction, size, morphology and distribution of carbides/nitrides have a great impact on the strengthening effect of age-hardening steels. According to Fig.10(a) and (b), MX, $M_{23}C_6$ and Sigma phase are the equilibrium phases at 650°C for TP347H austenitic heat resistant steel. Fig.10(b) shows $M_{23}C_6$ carbide solutes in austenite matrix at 840°C. At this temperature, the rest equilibrium phases are austenite and MX phase. The amount of MX phase decreases with increasing temperature. MX phase solutes in austenite matrix at 1340°C. It is clearly that MX phase is a very stable phase in this steel. It explains that bigger MX phase exits because MX phase can precipitate and grow up during hot working and undissolve at solution treatment condition. This phenomenon is also reported by T. Sourmail and R. Ayer(Sourmail, 2001; Ayer et al, 1992).

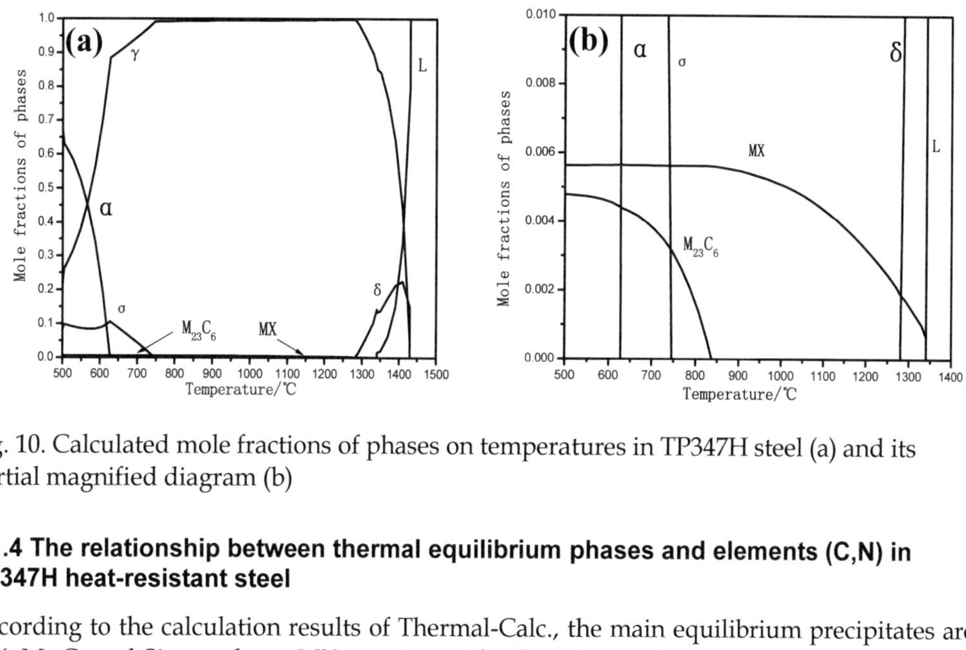

Fig. 10. Calculated mole fractions of phases on temperatures in TP347H steel (a) and its partial magnified diagram (b)

4.1.4 The relationship between thermal equilibrium phases and elements (C,N) in TP347H heat-resistant steel

According to the calculation results of Thermal-Calc., the main equilibrium precipitates are MX, $M_{23}C_6$ and Sigma phase. MX contains with Nb and C and $M_{23}C_6$ carbide contains with Cr, Fe and C. It also can confirm that the precipitation of carbides, especially nano-size MX in grains, plays the most important strengthening effect on creep rupture strengths according to experimental observation described in above paragraph. In order to optimize chemical composition of TP347H austenitic heat resistant steel with high creep rupture strength. Thermal-calc software has been used to predict the effect of MX containing elements C and Nb on the fraction of MX strengthening phase. The results are shown in Fig.11(a) and (b), respectively.

Fig.11(a) shows the effect of C content on mole fractions of MX phase. From 500°C to 730°C the content of C increases from 0.02% to 0.06%, the mole fraction of MX phase increases intensively. The C content increases 0.02%, the solution temperature of MX phase increases 60-80°C. However, when the content of C increases to 0.1%, the mole fractions of MX phase do not increase obviously. It can be seen that the fractions of MX phase increases with the C content till 0.06%C at service condition (600-650°C). When the C content increases from 0.06% to 0.1%, the mole fractions of MX phase keep in a same level. Thus, the C content should be controlled in the level of 0.06%-0.07% C.

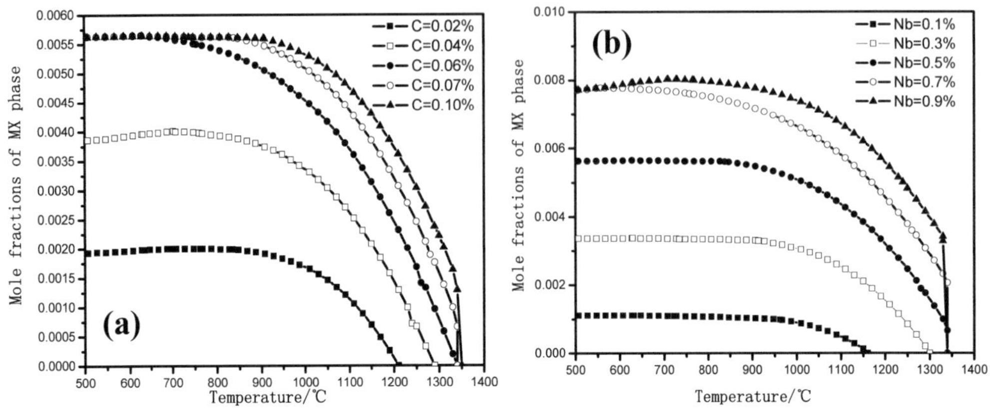

Fig. 11. Effect of C(a) and Nb(b) content on mole fractions of MX phase in TP347H

Fig.11(b) shows the effect of Nb content on mole fractions of MX phase. With the content of Nb increasing from 0.1% to 0.7%, the mole fractions of MX phase increase obviously. However, when the content of Nb increases to 0.9%, the mole fractions of MX phase do not increase obviously. Thus, the content of Nb should be controlled at about 0.7% for good strengthening effect of MX precipitation.

It is reported that the brittle σ phase can precipitate after long-term service in 18Cr-9Ni austenitic steels and it will cause mechanical property degredation(Minami et al, 1986). This research gave some results on σ phase. Fig.12 shows the effect of C content on the mole fractions of σ phase. The mole fractions of σ phase decrease obviously with the content of C increasing from 0.02% to 0.06%. Especially, when the content of C increases to 0.10%, the mole fractions of σ phase decrease quickly. Thus, the content of C should be controlled at high level to decrease the mole fractions of σ phase for good strengthening effect.

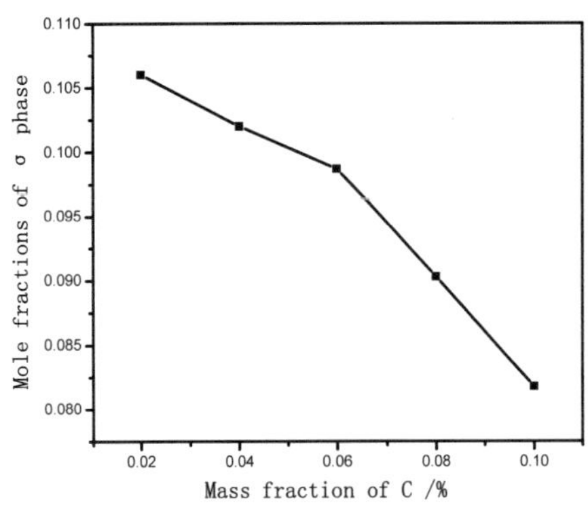

Fig. 12. Effect of C content on mole fractions of σ phase in TP347H steel at 650°C

The yield strength could increase from 229MPa to 238MPa with a certain amount of N element was reported in the literature(Ayer et al, 1992). The elongation of this steel with a certain amount of N element also increases. If this steel contains a certain amount of N, the complex carbonitride(MX) which contains with Nb, Cr, N and C forms and it can promote strengthening effect. This is why the mechanical properties of TP347H containing with N are increased. Nitrogen element can also play a good role in solid solution strengthening. Thus, to add a certain amount of N in TP347H steel will get better mechanical properties.

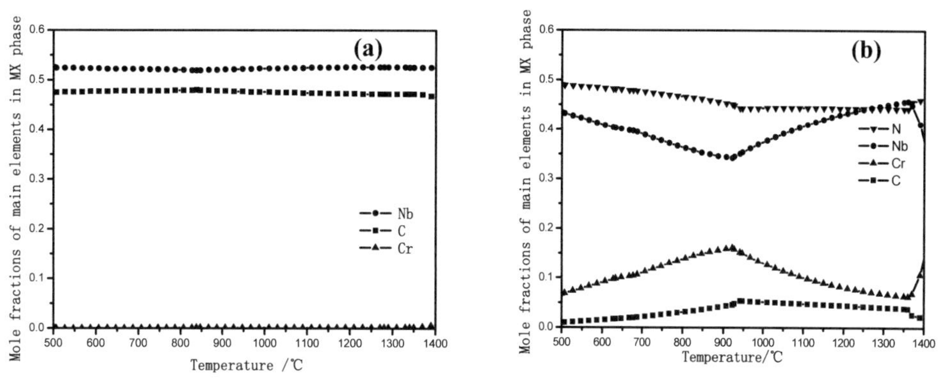

Fig. 13. Effect of N content on mole fractions of elements in MX phase for TP347H steel N= 0% (a) and N= 0.2%(b)

Fig.13 shows the effect of N content on mole fractions of elements in MX phase for TP347H steel. Fig.13(a) is the experimental steel without N and Fig.13(b) is the steel with 0.2%N. When there is no N in steel, Nb and C are the main elements in MX phase and their atomic ratio is 1:1. At this time MX phase is the simple niobium carbide(NbC). While the steel containing with 0.2%N, MX phase contains not only Nb and C but also Cr and N, which is a complex carbon-nitride. The mechanical properties can be increased by adding N in TP347H.

4.2 Super304H heat-resistant steel

Super304H steel belongs to 18%Cr-9%Ni system austenitic stainless steel, which has been added copper (Cu), niobium (Nb) and nitrogen (N) for precipitation strengthening. It is resulted in a good combination of elevated temperature creep strength and corrosion resistance(Sawaragi & Hirano., 1992; Sawaragi et al., 1994). Especially it has more than 20% higher stress rupture strength at 650°C than TP347H which is conventionally used as superheater/reheater tube material(Muramatsu, 1999). The resistance to oxidation and corrosion of Super304H is superior to TP321H at high temperatures(Igarashi, 2004). It makes Super304H heat resistant steel to be an very effective austenitic heat resistant steel and has been widely accepted for the application of superheater/reheater tubes in boilers of USC power plants all over the world(Senba et al., 2002). The application of Super304H in USC power plants confirms its good high temperature performance and stability, which is important to be used in USC boilers(Igarashi et al., 2005; Komai et al., 2007). Cu addition is its distinct character compared with other austenitic heat resistant steels, which also causes its good performance. This part attempts to provide the detail precipitation behavior and strengthening mechanism of Cu-rich phase in Super304H steel during long-term aging process at 650°C.

The chemical composition of investigated Super304H steel is as follows(in mass%): 0.08C, 0.23Si, 0.80Mn, 0.027P, 0.001S, 9.5Ni, 18.51Cr, 2.81Cu,0.51Nb, 0.11N, 0.0034B, bal. Fe. Its heat treatment is same with TP347H that is also solid solution treated at 1150°C high temperature. In order to examine long-term precipitation behavior, aging treatments at 650°C were conducted from 1h to 10,000h.

4.2.1 Long-term age hardening effect

Fig.14 shows the micro-hardness change in grains of Super304H aged at 650°C. After high temperature solid solution treatment, the value of micro-hardness is only 185HV. However, micro-hardness increases rapidly just from the very beginning of aging till 1,000h and at that time the highest micro-hardness value has gained at about 245HV. This value of micro-hardness is much higher than the micro-hardness of solid solution treatment condition. The micro-hardness of Super304H steadily keeps at high value about 240HV till 8,000h. It shows a very effective hardening effect during 650°C aging.

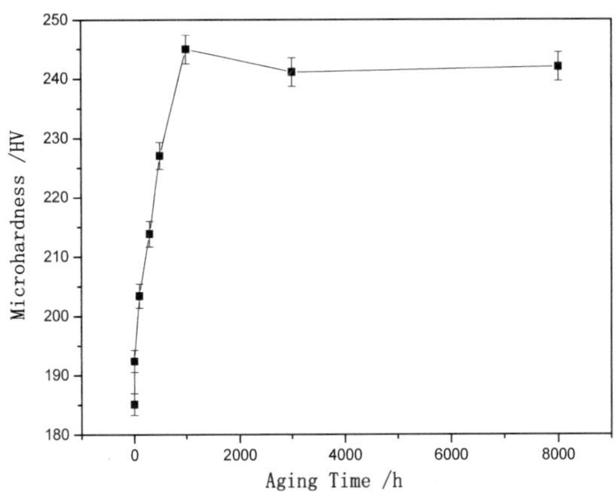

Fig. 14. The changes of micro-hardness in grain of Super304H after 650°C long time aging

4.2.2 Tensile property after long-term aging

The room temperature ultimate tensile strength of Super304H after 650°C long time aging keeps a similar tendency as micro-hardness change, as shown in Fig.15. The ultimate tensile strength also increases rapidly just from the very beginning of aging. Its value keeps around 700MPa till 10,000hrs after reaching the highest value. Although the room temperature ultimate tensile strength is increasing with aging time, the plasticity of aged steels does not drop sharply. The results of micro-hardness and ultimate tensile strength both indicate that some precipitates are forming to make an excellent hardening effect during 650°C long time aging of Super304H.

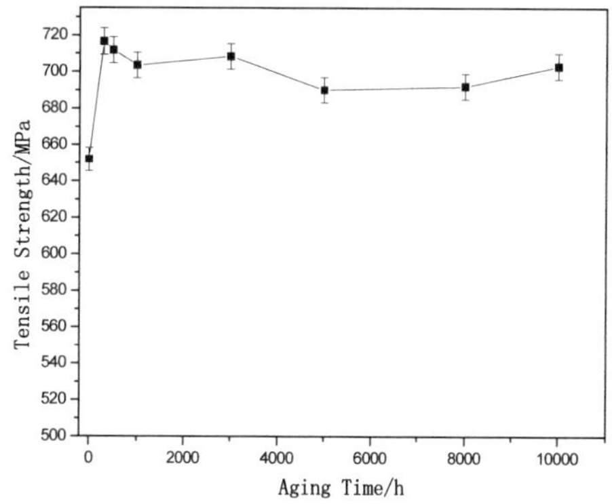

Fig. 15. The changes of ultimate tensile strength of Super304H after 650°C long time aging

4.2.3 Micro-structure analyses by SEM and TEM

In order to study this unique age hardening effect, microstructure changes of aged samples have been detailed analyses by SEM and TEM. Typical SEM images of Super304H after aging for different times are shown in Fig.16. Fig.16(a) is the SEM image of Super304H just after solid solution treatment. Its microstructure characterizes with equiaxed austenite grains and with a few coarse and fine particles also. The coarse particles, which are in the size range of 1 to 3μm, as inclusions directly formed during solidification, randomly distributed in grains and partially at grain boundaries. There are also some spherical undissolved small particles distributed in grains with the size of about 0.1μm. Energy-dispersive spectroscopy(EDS) analysis reveals that two kinds of particles both mainly contain with niobium, they are niobium carbonitride directly formed during solidification and undissolved Nb(C,N) carbonitride respectively, which are similar to the two kinds of MX type phases existed in TP347H austenitic steel at solid solution condition.

Fig.16(b), (c) and (d) represents the SEM images of Super304H after exposure at 650°C for 500, 5,000 and 10,000hrs, respectively. There are almost no precipitates at grain boundaries except larger niobium carbonitride just after solid solution treatment(see Fig.16(a)). The grain boundary precipitates contain with high content of Cr identified by EDS and also a few of particles in the grains are $M_{23}C_6$ carbide as shown in Fig.16(b) after 650°C aging for 500h. Fig.16(c) shows that the size of $M_{23}C_6$ at grain boundaries is about 0.5μm and has contacted to each other to form chain-like precipitates after 650°C/5,000h ageing. Fig.16(d) shows the fine precipitates in grains which are rich in Nb detected by EDS. These precipitates are suggested as Nb rich MX phase. The size of MX precipitates is about 70nm when aged at 650°C for 10,000h. SEM images can not find very fine precipitates except primary MX phase, undissolved small size MX phase after solid solution treatment and nano-size secondary MX phase precipitate in grains and $M_{23}C_6$ phase at grain boundaries during long-term aging. This result is similar to TP347H steel. However, the high temperature performance of Super304H is higher than TP347H steel, which means there must be some other precipitated strengthening phase in Super304H that is different from the precipitates in TP347H. For the purpose of detail study on precipitation hardening behavior of Super304H at 650°C aging, TEM and 3DAP have been used to do more detail observation.

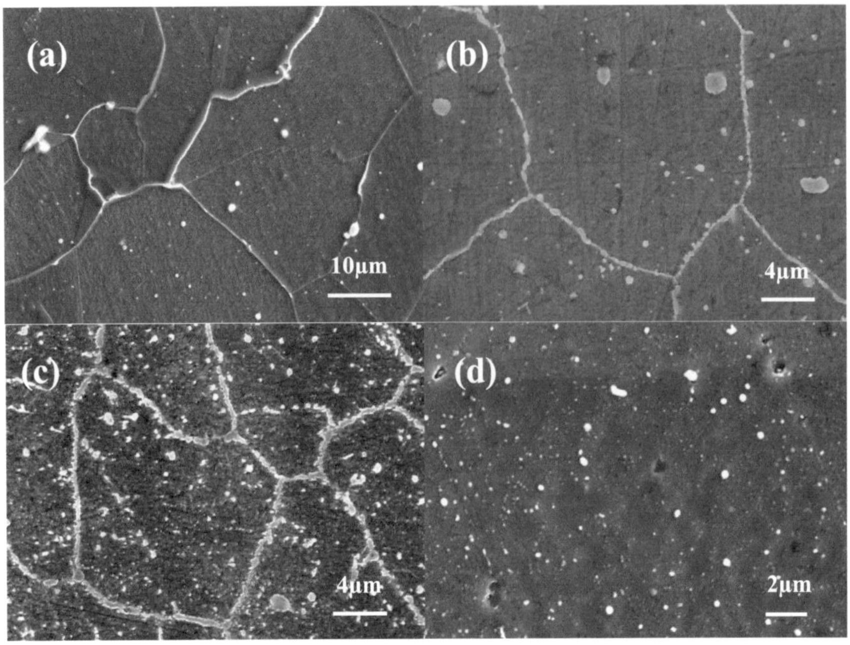

Fig. 16. SEM images of Suepr304H after solid solution treatment(a), aging at 650°C for different times 500h(b), 5,000h(c) and high magnified image in grain aging for 10,000h(d)

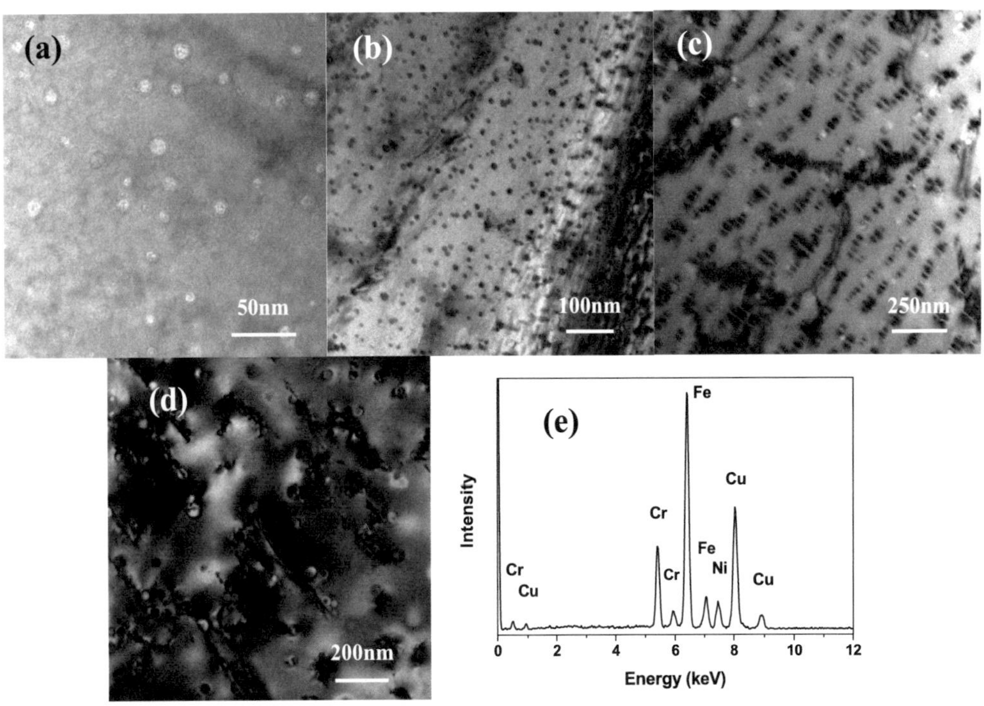

Fig. 17. TEM images of Super304H heat resistant steel aged at 650°C for 500h (a), 1,000h (b), 5,000h (c), 10,000h (d) and EDS spectrum for Cu-rich phase(e).

Fig.17 shows the TEM images of Super304H aging at 650°C for different times. Fig.17(a) is the TEM image of the sample aged at 650°C for 500h. It shows that there are dense distribution of nano-size spherical precipitates in grains which contain high content of Cu determined by EDS spectrum as shown in Fig.17(e). It confirms that these nano-size precipitates are Cu-rich phase. Cu-rich phase has been clearly found by TEM during long-term aging at 650°C for 1,000h as shown in Fig.17(b). The average size of Cu-rich phase is about 10nm only, and it distributes homogeneously with a higher density than MX phase. Because the size of Cu-rich phase is very fine and the lattice parameter of Cu(3.6153Å) is close to γ-matrix(3.5698Å), the diffraction pattern of Cu-rich phase is hard to be separated from γ-matrix. However, the images of spherical precipitate particles are separated into two parts by no-contrast line shown in Fig.17(c). It shows that Cu-rich phase is coherent with γ-matrix even aged at 650°C for 5,000h. After aging for 10,000h, the Cu-rich phase still keeps nano-size(as shown in Fig.17(d)). At this time, it can be clearly seen from Fig.17(d) that dislocations contact with Cu-rich phase particles and are effectively blocked by Cu-rich phase. It indicates that these fine Cu-rich phase particles distribute uniformly with high density in grains are the main strengthening phase which can cause strong age hardening effect and supply excellent high temperature strength for Super304H steel, which has been reported by Yu(Yu et al., 2010).

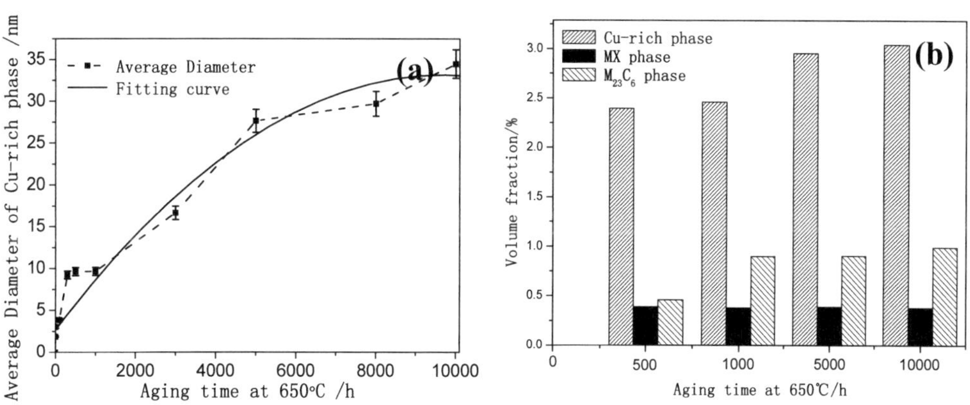

Fig. 18. The change of Cu-rich phase average diameters (a) and precipitates volume fraction (b) in Super304H at 650°C with aging time

The Cu-rich phase is growing during 650°C long time aging till 10,000hrs, but the growth rate is very slow. The average diameters change of Cu-rich phase has been determined by TEM as shown in Fig.18(a). This curve confirms that Cu-rich phase grows slowly at 650°C long time aging. The average size of Cu-rich phase still keeps about 34nm at 650°C aging for 10,000h. Experimental results confirm that Cu-rich phase and MX phase precipitate in grains, while $M_{23}C_6$ phase mainly precipitates at grain boundaries. The volume fractions of these phases change with aging time is shown in Fig.18(b). The volume fraction of MX is the sum of undissolved MX at solid solution treatment condition and secondary MX precipitates during long-term aging. It is stable and increases slightly during aging time. The volume fraction of Cu-rich phase increases gradually with aging time and its fraction is the highest in these three phases. It is clear that nano-size Cu-rich phase with the highest volume fraction and homogeneously distribution plays a very important role for strengthening effect in Super304H steel. However, Cu-rich precipitates are so fine that hardly to be clearly detected by TEM. So three dimensional atom probe(3DAP) has been used to study the precipitation behavior of Cu-rich phase.

4.2.4 Cu-rich phase investigation by means of 3DAP technology

3DAP can effectively detect the nano-size precipitates by analyzing atoms reconstruction. Fig.19 shows the elements mapping of Super304H after solid solution treatment. The analyzed volume is 10nm×10nm×70nm. In these figures, one point represents an atom. From Fig.19 it is confirmed that all atoms have solved in austenitic matrix and homogeneously distributed at solid solution condition. There are no any traces of precipitate formation after high temperature solid solution treatment.

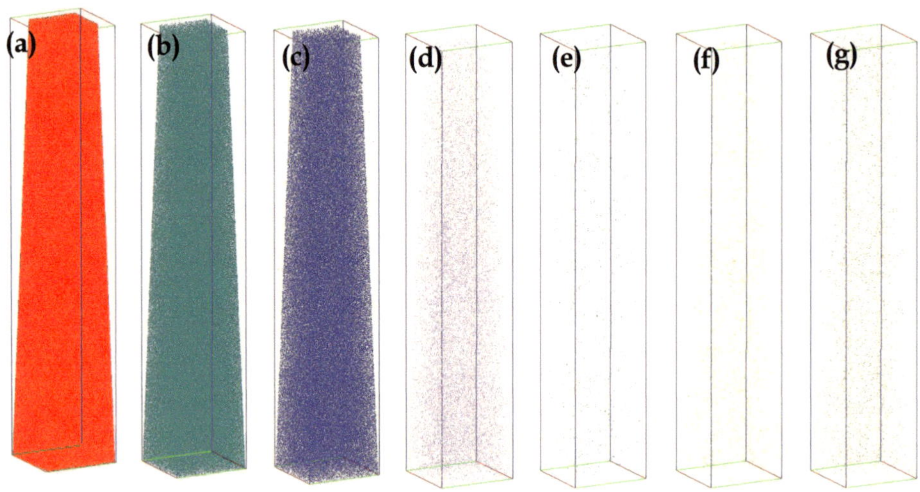

Fig. 19. Atomic mapping of the Super304H after solution treatment: Fe(a), Cr(b),Ni(c),Cu(d),Nb(e),C(f) and N(g)(size of selected box is 10nm×10nm×70nm)

Fig.20 shows the Cu atomic mapping of Super304H aged for different times. Fig.20(a) shows that Cu atoms continuously concentrate in clusters(about 1.5nm in radius) just at the beginning of aging treatment, which homogeneous distribute in austenitic matrix after 650°C aging for 1h only. The average size of Cu-rich phase particles is increasing while the density of its distribution is decreasing with aging time(see Fig.20(b) and (c)). The density of Cu atoms in austenitic matrix is also slightly decreasing with aging time. It means that Cu atoms diffuse from matrix and concentrate into Cu-rich areas. This diffusion control process is continuing during Cu-rich phase growing process. When aged till to 1,000h the one Cu-rich particle is larger than above mentioned analyzed volume and is partially intercepted. However, this size is still very small and radius of this Cu-rich particle is about 3nm as shown in Fig.20(d). The Cu-rich phase particle density calculated according to the reference(Chi et al., 2010) is still very high and keeps at the level of 0.07×10^{24}n/m^3. These important results show that Cu-rich phase can keep a level of nano-size and high density distribution at 650°C aging till 1,000 hrs.

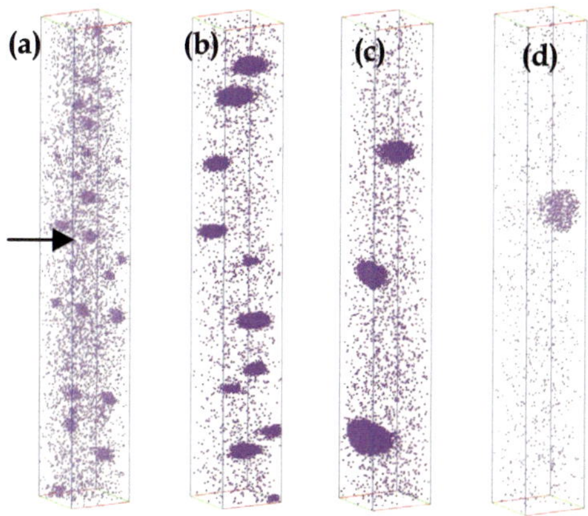

Fig. 20. Cu atomic mapping of Super304H aged for 1h(a), 100h(b), 500h(c) and 1,000h(d) at 650°C (size of selected box is 10nm×10nm×70nm)

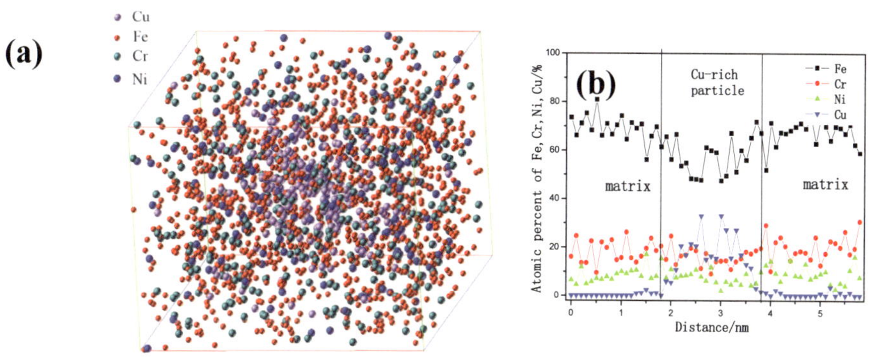

Fig. 21. 3DAP reconstruction map of a Cu-rich segregation area selected from Fig.20(a) as shown by the arrow (size of selected box is 3.5nm×3nm×2nm) (a); and a concentration depth profile through a Cu-rich segregation area and adjacent γ-matrix of the sample aged at 650 °C for 1h(b)

In order to detect the Cu atom composition concentration in Cu-rich phase, just one particle selected from 1h aged specimen(as shown in the Fig.20(a) by the arrow) was analyzed and the concentration profile of Fe, Cr, Ni and Cu through the Cu-rich phase particle is shown in Fig.21. Fig.21(a) shows that the Cu-rich particle is composed of Cu ,Fe, Cr and Ni, which indicates a certain degree of Cu atom concentration when aged at 650°C for 1h only. The concentration depth profile through the Cu-rich segregation area is shown in Fig.21(b). It can be seen that the composition profiles are asymmetric across adjacent γ-matrix and Cu-rich particle two interfaces. Although the boundary between Cu-rich segregated particle and austenitic matrix is fluctuant, Cu compositions undergo a sharp transition across the Cu-rich particle and austenitic matrix interface. The Cu atom concentration is increasing from the edge to the centre of Cu-rich segregation area and reaches the highest degree in the centre of Cu-rich segregation area, while the others composed atoms such as Fe, Cr and Ni appear inverse tendency.

The concentration depth profile through a Cu-rich particle selected from the specimens with different aging times and the proportion of main composition at the centre part of Cu-rich phase particle changing with aging time is shown in Fig.22. It shows that the Cu-rich particle is mainly composed of Cu and also a part of Fe, Cr and Ni when aged for 5h(see Fig.22(a)). The Cu concentration in Cu-rich particle is gradually increasing with aging time, while others composed atoms appear inverse tendency(as shown in Fig.22(b) and (c)). Fig.22(d) shows that Cu content in the Cu-rich segregation area is almost lower than 20at% at early stage of precipitation when Fe is still the mainly composed composition, and then Cu content is increasing continually with aging time and reaches almost 90at% at the centre of Cu-rich particle when aging for 500h. These results represent that Cu atoms gradually concentrate to Cu-rich particles and the other elements(such as Fe, Cr, Ni etc) diffuse away from Cu-rich particles into γ -matrix with the increasing of aging time at 650°C. It may suggest that Cu will be the only main composition in Cu-rich phase when aging for very long time.

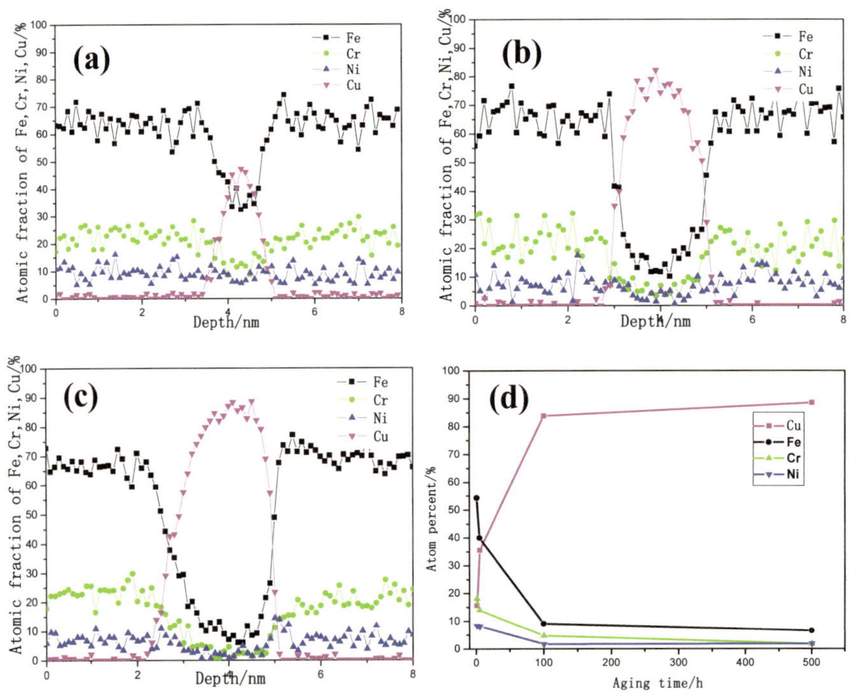

Fig. 22. Concentration depth profile through a Cu-rich particle selected from the specimens aged at 650°C for different age times, (a)5h, (b)100h, (c)500h and proportion of main composition at the centre part of Cu-rich phase particle changing with aging time(d)

Fig. 23 shows compared results of the atomic reconstruction in a selected area that cut through a Cu-rich particle in Super304H after solid solution treatment and aged for 500h. The depth in the vertical direction is so small that only reflects a few planes of atoms, so it can clearly show the composition through the whole Cu-rich particle. Fig.23(a) shows that all elements distributed homogeneously after solid solution treatment. With the lasting of aging time, much more Cu atoms concentrate into the Cu-rich phase and the other elements such as Fe, Cr, Ni are rejected from Cu-rich phase into austenitic matrix. It can be seen from Fig.23(b) that Cu atoms are the main atoms in the centre of Cu-rich particle(circled by dashed line), and the other elements are in a very few numbers at 650°C/500h aging. At this time the interphase boundary between Cu-rich phase and austenitic matrix is much clearer than the boundary at the condition of 650°C aging for 1h only. These results clearly indicate that Cu atoms gradually concentrate into Cu-rich particle and the other elements such as Fe, Cr, Ni diffuse away from Cu-rich phase particle to γ-matrix with the increasing of aging time at 650°C. According to the experimental results, it can suggest that Cu atoms concentrate to form Cu-rich segregation clusters just at the beginning of 650°C aging, the change from Cu-rich segregated clusters to Cu-rich phase by Cu atom diffusing into the Cu-rich particles and other atoms diffuse to γ-matrix.

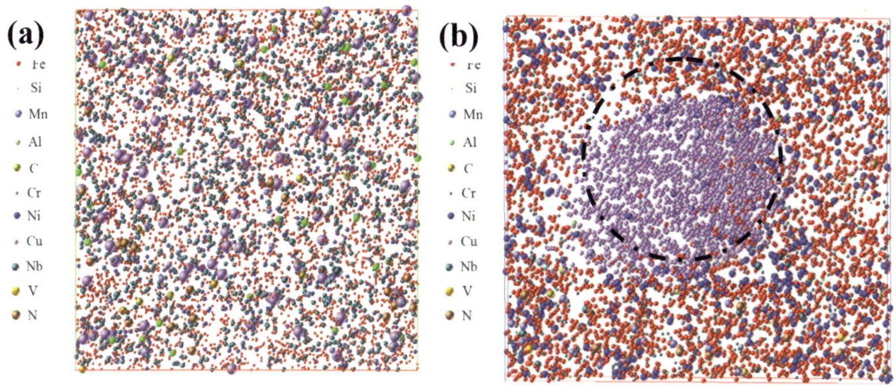

Fig. 23. All elements atomic mapping of Super304H after solid solution treatment(size of selected box is 10nm×10nm×1nm)(a) and aged at 650°C for 500h (size of selected box is 10nm×9nm×1nm) (b)

4.2.5 The relationship between Cu-rich phase formation and Cu content in Super304H heat-resistant steel

As described in above paragraph, Cu-rich phase is the main strengthening phase in Super304H steel, and Cu atom gradually becomes the main composed element in Cu-rich phase. It means that content of Cu added in Super304H can develop the precipitation of Cu-rich phase, resulting in a strong strengthening effect. Thermal-calc software can be used to predict the formation of Cu-rich phase with different Cu content.

The equilibrium diagrams of Super304H steels with different Cu content calculated by Thermal-calc software are shown in Fig.24. Fig.24(a) is the equilibrium diagram of Super304H steel with 1% Cu. It shows that Cu atoms all dissolve in γ-matrix and Cu-rich phase can not form at 650°C. When the addition of Cu increases to 3%, Cu-rich phase can form at 650°C. Its mole fraction is higher than MX phase which is another strengthening phase precipitates in grains. The mole fraction of Cu-rich phase at equilibrium condition is increasing with the increase of Cu content, and will be more than $M_{23}C_6$ phase as shown in Figure.24(c). Although more Cu addition will induce more Cu-rich phase precipitate which is good for high temperature strength. However, the steel with too much Cu content do not have the best performance in the view point of high temperature ductility(Tan et al., 2010).

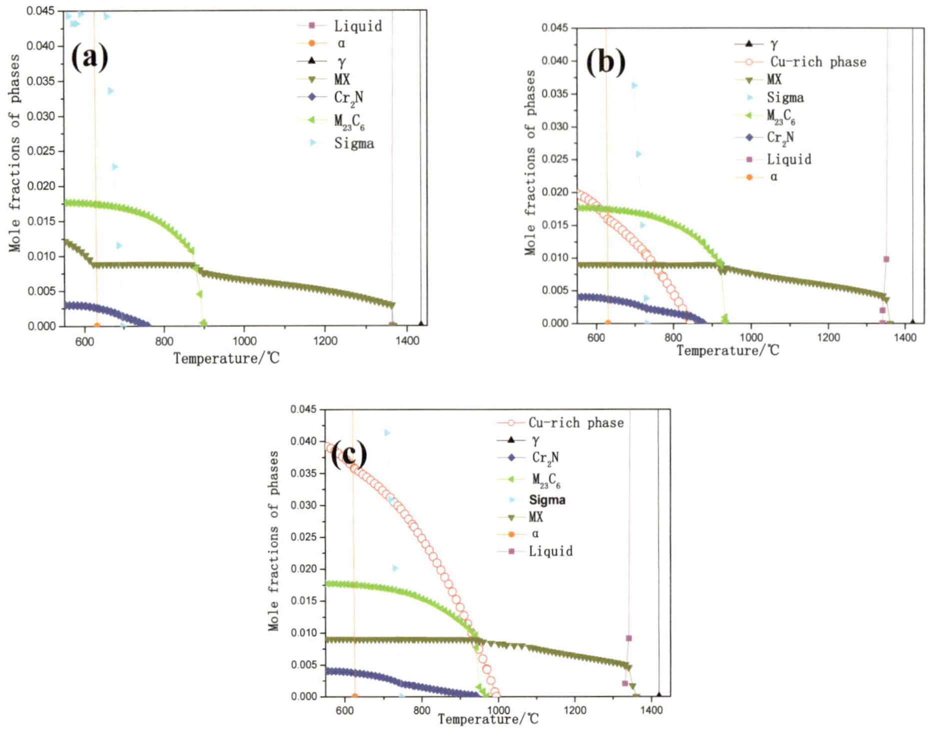

Fig. 24. Equilibrium diagrams of Super304H steels with different Cu content calculated by Thermal-calc software: (a)1%Cu, (b)3%Cu, (c)5%Cu

4.3 HR3C heat-resistant steel

HR3C steel is a kind of austenitic heat-resistant steels. HR3C steel containing with higher chromium(25%Cr) and nickel(20%Ni) is developed to improve corrosion and oxidation resistant properties both for high temperature application. Because of good structure stability of HR3C steel, it characterizes with good high temperature mechanical properties, corrosion/oxidation resistance and weldability for long-term service. It has been also used as superheater/reheater tubes for ultra-supercritical power plant boilers, especially in final stage of reheater tubes.

The chemical composition of HR3C steel for this study is as follows(in mass%): 0.06C, 0.55Si, 0.97Mn, 0.022P, 0.001S, 19.82Ni, 25.23Cr, 0.46Nb, 0.20N, bal. Fe. Its heat treatment is solid solution treated at high temperature 1200-1250°C. According to the service environment of HR3C steel—the highest temperature of reheater tubes, the aging temperature of this steel was selected at 650°C and 700°C. Long-time aging treatments reach 3,000h-5,000h in order to analyze micro-structure evolutions of HR3C steel.

4.3.1 Long-term mechanical properties

The Vickers micro-hardness of all specimens with aging time at 700°C have been shown in Fig.25. At initial aging stage the Vickers micro-hardness increases rapidly with aging time.

At 1,000h, the Vickers micro-hardness of the steel has got the maximum. After that, with further aging till 5,000h the hardness still keeps at a high level of 255HV. It is clear that age hardening effect of HR3C steel develops by strengthening phase precipitation.

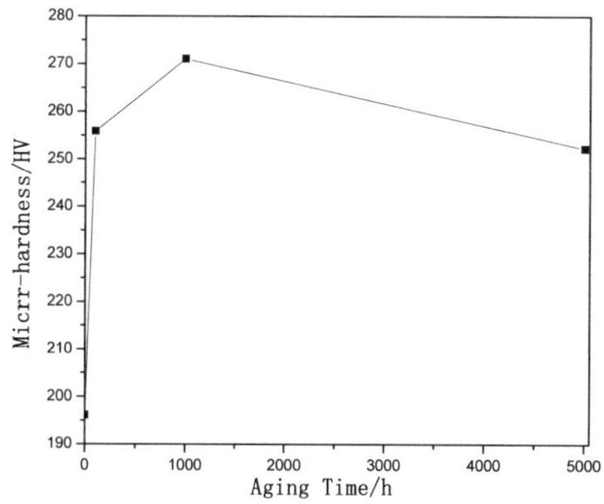

Fig. 25. Micro-hardness of HR3C steel at 700°C long-term aging

The impact property of HR3C heat-resistant steel at 700°C is shown in Fig. 26. At the initial aging, impact property decreases dramatically with aging time. The value of impact toughness decreases to 7J at 700°C/300h from 220J at solution condition. But it can keep in a stable level for long time aging till 3,000h. Therefore, the impact property of HR3C steel needs to be improved after long-term aging at 700°C.

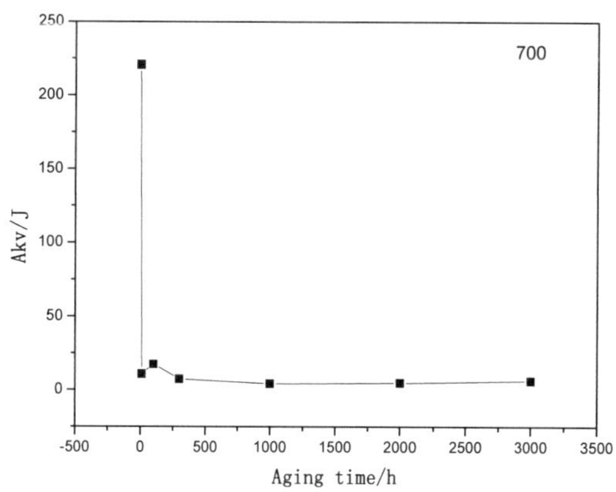

Fig. 26. The impact property changes with aging time at 700°C in HR3C

4.3.2 Micro-structure analyses

Fig. 27 shows SEM images and EDS of HR3C steel after 500h and 3,000h long-term aging at 650°C. Fig.27 (a) and (b) are the SEM images after 500h and 3,000h aging at 650°C respectively. Fig. 27 (c), (d) and (e) are EDS results of γ-matrix and precipitated particles in grains, and at grain boundaries respectively. There are several kinds of precipitates in grains and at grain boundaries after long-term aging. $M_{23}C_6$ carbide containing with Cr precipitates at grain boundaries and NbCrN phase containing with high Nb and Cr precipitates in grains, as shown in Fig. 27(c), (d) and (e).

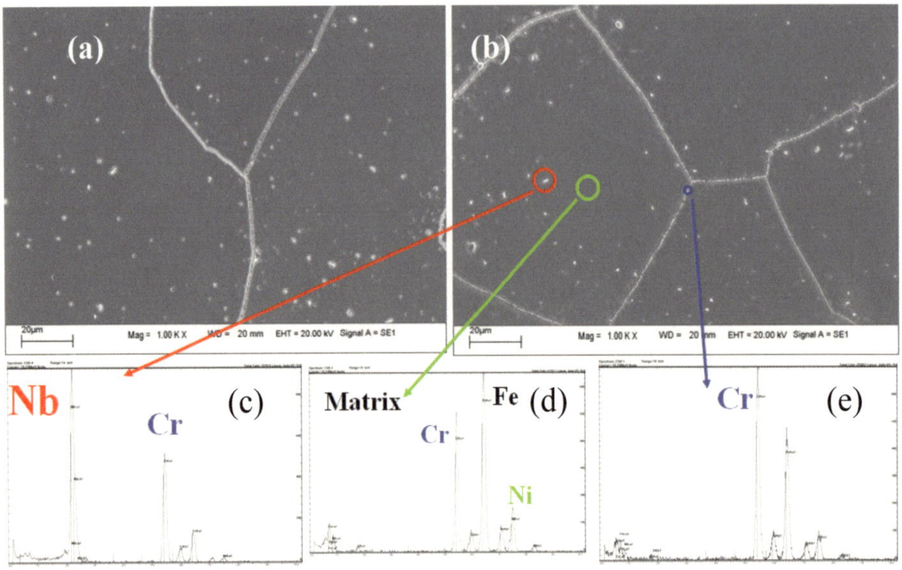

Fig. 27. SEM images of HR3C after aging at 650°C for (a) 500h and (b) 3,000h, EDS results of (c) the particle in grains, (d) matrix and (e) the particles at grain boundaries

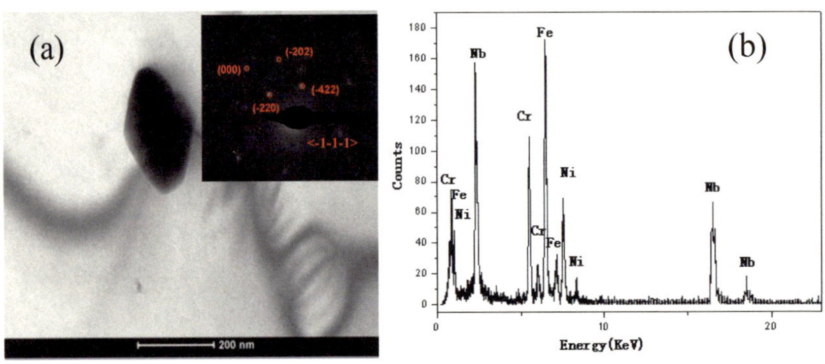

Fig. 28. TEM image of HR3C at solution treatment(a) precipitation with diffraction patterns and (b) EDS

Fig. 28 shows TEM images, diffractions pattern and EDS results of HR3C steel after solid solution treatment condition. MX phase exits in grains at solid solution treatment. Its crystallographic structure is NaCl type(a= 0.4331nm). This kind of MX phase contains with Nb and its size is about 300nm.

Fig. 29 shows TEM images, EDS and diffraction of $M_{23}C_6$ carbides after 3,000h aging at 650°C in HR3C. $M_{23}C_6$ phase can also precipitate in grains(but mainly at grain boundaries) after long-term aging. Its crystallographic structure is complicated face-center cube. Its lattice parameter is about three times than that of austenite matrix.

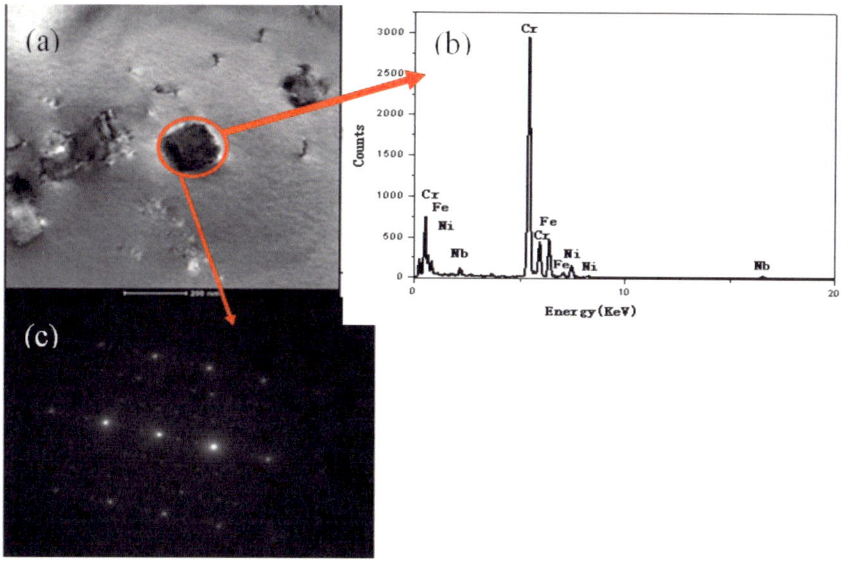

Fig. 29. TEM images of $M_{23}C_6$ carbides after aging 3,000h at 650°C in HR3C(a) $M_{23}C_6$ carbides (b) EDS and (c) diffraction patterns

4.3.3 The relationship between thermal equilibrium phases and the contents of Nb, N and C in HR3C heat-resistant steel

According to the results and the reported literature(Iseda et al., 2008), the main precipitated phases in HR3C steel are NbCrN phase, $M_{23}C_6$ carbides and σ phase. NbCrN phase and $M_{23}C_6$ carbides play an important role on strengthening effect. However, the brittle σ phase is harmful for mechanical properties at high temperatures. The relationship between thermal equilibrium phases (NbCrN phase, $M_{23}C_6$ carbides and σ phase) and elements (C, N and Nb) in HR3C heat-resistant steel are evaluated by thermodynamic calculation.

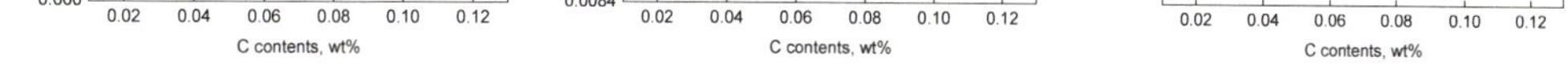

Fig. 30. Effect of C content on mole fractions of $M_{23}C_6$ phase (a), NbCrN phase (b) and σ phase (c) in HR3C steel at 650°C

Fig. 30 shows the effect of C content on mole fractions of $M_{23}C_6$ phase, NbCrN phase and σ phase in HR3C steel, respectively. The mole fractions of $M_{23}C_6$ phase increase in a straight line tendency with C content increasing, shown in Fig. 30(a). For every increasement of 0.02%C content, the mole fractions of $M_{23}C_6$ phase increase 0.5%. Fig.30(b) shows that the mole fractions of NbCrN phase keep in a stable value with C content increasing. The mole fractions of σ phase decrease in a straight line tendency with C content increasing, shown in Fig. 30(c). With the content of C increasing from 0.02% to 0.12%, the mole fractions of σ phase decrease from 0.23 to 0.19.

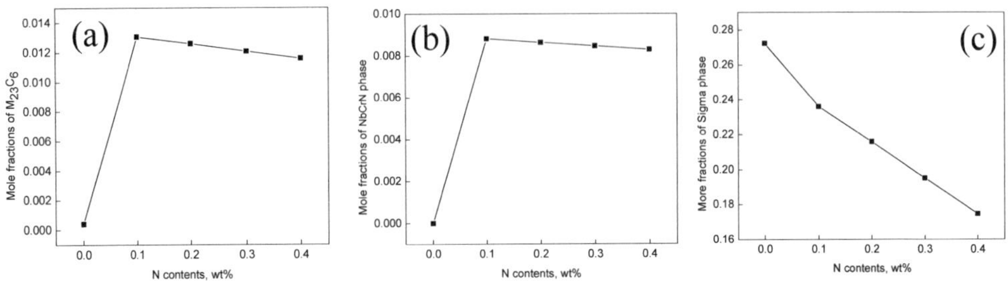

Fig. 31. Effect of N content on mole fractions of $M_{23}C_6$ phase (a), NbCrN phase (b) and σ phase (c) in HR3C steel at 650°C

Fig. 31 shows the effect of N content on mole fractions of $M_{23}C_6$ phase, NbCrN phase and σ phase in HR3C steel, respectively. It is obviously to know that the tendency between the mole fractions of $M_{23}C_6$ phase and N content is similar to that of the mole fractions of NbCrN phase. The mole fractions of $M_{23}C_6$ phase and NbCrN phase both increase quickly with N content increasing from 0 to 0.1%. But the mole fractions of $M_{23}C_6$ phase and NbCrN phase both decrease very slowly with N content increasing from 0.1% to 0.4%, shown in Fig.31(a) and (b). The mole fractions of σ phase decrease very quickly with the increasing of N content. When the N content increases to 0.4%, the mole fractions of σ phase decreases from 0.27 to 0.17, shown obviously in Fig.31(c).

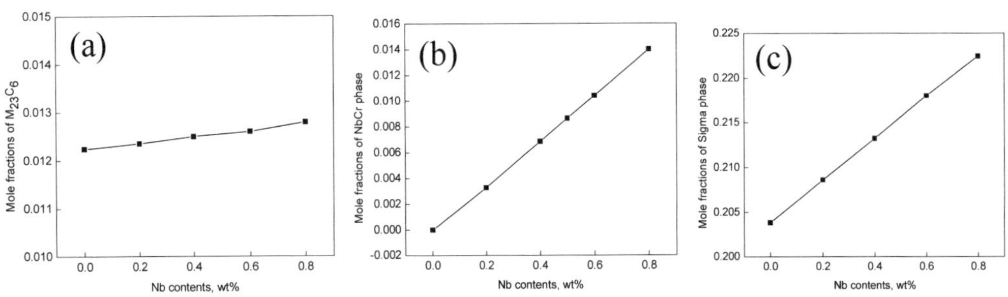

Fig. 32. Effect of Nb content on mole fractions of $M_{23}C_6$ phase (a), NbCrN phase (b) and σ phase (c) in HR3C steel at 650°C

Fig. 32 shows the effect of Nb content on mole fractions of $M_{23}C_6$ phase, NbCrN phase and σ phase in HR3C steel at 650°C. The mole fractions of $M_{23}C_6$ phase increase very slowly with Nb content increasing, shown in Fig.32(a). The mole fractions of the strengthening phase NbCrN and the brittle σ phase both increase with the increasing of Nb content. For every increasement of 0.2%Nb content, the mole fractions of NbCrN phase and σ phase increase 0.4% and 0.5%, respectively shown in Fig.32(b) and (c). Thus, the Nb content should be controlled in a correct level for reducing Sigma phase in order to get good strengthening effect and to avoid embrittlement.

From these results obtained by experimental observation and thermodynamic calculation, advanced austenitic heat resistant steels are strengthened by second phase precipitation. High Cr and Ni containing HR3C heat resistant steel is the candidate material for high terminal superheater/reheater components in USC fossil power plants with higher steam parameters on the view point of corrosion and oxidation resistance. However, it is a little insufficient in high temperature stress rupture strength. Cu addition in austenitic heat resistant steel is effective to develop nano-size Cu-rich phase precipitation during high temperature long-term service, which causes an excellent high temperature strengthening effect and also good microstructure stability. Cu-rich phase precipitates in Super304H heat resistant steel is a good example. In order to improve the strength of 25Cr-20Ni type austenitic heat resistant steel, the addition of Cu to induce Cu-rich phase precipitation is strongly suggested. So the further development of austenitic heat resistant steels may contain high content of Cr and Ni to meet the requirement of corrosion/oxidation resistance and coupled with nano-size MX, Cu-rich phase in the grains for second phase precipitation strengthening and to add B for stabilization $M_{23}C_6$ phase precipitates at grain boundaries.

5. Conclusion

The development of advanced austenitic heat resistant steels is promoted by the progress of USC fossil power plants, which urgently needs new materials with high temperature strength and good corrosion/oxidation resistant properties for superheater/reheater components. In this chapter three advanced austenitic heat resistant steels widely used for 600°C USC power plants, TP347H, Super304H and HR3C, have been investigated by SEM, TEM and 3DAP technologies coupled by thermodynamic calculation. Results show that high temperature strength of these three heat resistant steels is mainly dependent on fine second phase precipitation during long-term aging. The MX phase mainly precipitates in the grains and $M_{23}C_6$ phase mainly precipitates at grain boundaries. The Cu-rich phase precipitates in grains characterize with outstanding precipitation strengthening effect for Super304H in comparison with TP347H and HR3C. This is also the main reason why Super304H gains superior high temperature strength. According our research results we conclude as follow:

1. In TP347H austenitic heat resistant steel, some primary NbC as inclusions with 1~3μm in size randomly distribute in grains and sometimes also at grain boundaries at solid solution treatment condition. There are also some undissolved NbC particles with about 200nm in size in the grains. After long-term aging at 650°C, the Nb rich MX type phase intensively precipitates in grains and there are also some $Cr_{23}C_6$ precipitates at grain boundaries. These very fine high Nb containing MX phase precipitates during long time aging and keeps in nano-size contribute to good strengthening effect on TP347H steel. The calculated results show that the amount of MX phase increases with increasing of C/N and Nb contents. The amount of σ phase decreases with the increasing of C content. Adding 0.2%N to this steel, MX phase contains N, Nb, Cr with a certain amount of C, which is a complex carbon nitride and makes excellent strengthening effect.
2. In Super304H austenitic heat resistant steel, there are about 3% Cu-rich phase and 0.38% MX in volume fraction precipitate in grains. Elaborate analyses conducted by unique 3DAP technology show that Cu atoms are soluted in matrix after high temperature solid solution treatment and have formed austenitic matrix with high saturation of Cu. Cu-rich segregated clusters in 2nm size contained about 20at% Cu can

be quickly formed just after 650°C aging for 1h. The average size of Cu-rich segregation clusters are increasing slowly with aging time. The concentrations of Cu atom in Cu-rich segregation clusters are continuously increasing and simultaneously the atoms of Fe, Cr and Ni are defusing to γ-matrix. In result of that Cu becomes the main composition of Cu-rich phase. However, there are still some other elements in the Cu-rich phase after long time aging. In fact, Cu-rich phase is not a pure Cu phase. The precipitation character of Cu-rich phase is different from ε-Cu phase precipitation in ferritic steels. It is defined that when the Cu-rich particle contains more than 50at% Cu to be called Cu-rich phase. The copper element has almost concentrated to 90at% in the center part of Cu-rich phase after 500h aging. It is clearly shown that the formation of Cu-rich phase from Cu-rich segregated clusters has been completed.

Cu-rich phase grows slowly during 650°C aging and can keep in nano-size for very long aging. The average size of Cu-rich phase is only about 35nm at 650°C aging even for 10,000h. The density of Cu-rich phase particles is very high and can reach to $0.38\times10^{14}/m^2$ after long-term aging at 650°C for 10,000h. Cu rich phase characterizes with fine nano-size, high density distribution and low growth rate and it is also the largest volume fraction among all the precipitates in Super304H. Homogeneously distributed fine Cu-rich phase can effectively block dislocation moving for excellent strengthening effect. It is clearly confirmed that this kind of nano-size Cu-rich phase is the most important precipitation strengthening phase in Super304H.

3. In HR3C austenitic heat resistant steel, the main equilibrium phases are NbCrN, $M_{23}C_6$, MX, σ and Cr_2N according to Thermal-Calc results. The amounts of NbCrN and σ phase increase with the increasing of Nb content. There are a lot of phases precipitate in the grains and at grain boundaries during 650°C long time aging by experimental observation. Among these phase the main dispersive precipitates are NbCrN, MX and $M_{23}C_6$ phase. NbCrN phase and MX phase precipitated in grains. $M_{23}C_6$ carbides precipitated mainly at grain boundaries and it can also precipitated partially in grains. During long-term high temperature service a brittle phase (σ phase) may precipitate which will degrade mechanical properties .

6. Acknowledgment

This project is supported by China National Natural Science Foundation under the grant No:50931003. Authors would like also to thank Profs Bangxin Zhou and Wenqing Liu who work in Instrumental Analysis & Research Center, Shanghai University for their guidance on 3DAP experiments and valuable discussion on the results.

7. References

Ayer, R.; Klein, C. F. & Marzinsky, C, N. (1992). Instabilities in stabilized austenitic stainless steels. *Metall Mater Trans,* 1992, Vol.23A, (May 1991), pp.2455-2467

Blum, R.; Vanstone, R, W and Messelier-Gouze, C. (2004). Materials Development for Boilers and Steam Turbines Operating at 700°C, *Proceedings to the Fourth International Conference on Advances in Materials Technology for Fossil Power Plants* , Hilton Head Island, South Carolina, U.S., October 25-28, 2004

Chen, Q, R.; Stamatelopolous, G, N. Helmrich, A. Heinemann, J. Maile, K. & Klenk, A. (2007). Materials Qualification for 700°C Power Plants, Fifth International Conference on Advances in Materials Technology for Fossil Power Plants, U.S., October 3-5, 2007

Chi, C, Y.; Dong, J, X. Liu, W, Q. & Xie, X, S. (2010). 3DAP investigation of precipitation behavior of Cu-rich phase in Super304H heat resistant steel. *Acta Metallurgica Sinica,* Vol.46, No.9, (Sept 2010), pp.1141-1146

Gibbons, T, B. (2009). Superalloys in modern power generation applications. *Materials Science and Technology,* Vol.25, No.2, (July 2008), pp.129-135

Hong, H, U.; Rho, B, S. & Nam, S, W. (2001). Correlation of $M_{23}C_6$ precipitation morphology with grain boundary characteristics in austenitic stainless steel. *Materials Science and Engineering A,* 2001, Vol.318, (Jan 2001), pp. 285–292

Hughes, A.; Dooley, B. & Paterson, S. (2003). Oxide Exfoliation of 347HFG in High Temperature Boilers, *7th International Conference and Exhibition on Operating Pressure Equipment,* Sidney, Australia, April 2-4, 2003

Igarashi, M. (2004). Development of 18-8 steel (Super304H) having high elevated temperature strength for fossil fired boilers. *CAMP-ISIJ,* 2004, Vol.17, pp. 336-340

Igarashi, M.; Okada, H. & Semba, H. (2005). Development of 18-8 steel (Super304H) having high elevated temperature strength for fossil fired boilers, *Proceedings 9th Workshop on the Innovative Structural Material for Infrastructure in 21st Century,* NIMS, Tsukuba, 2005

Iseda, A.; Okada, H. Semba, H. & Igarashi, M. (2007). Long term creep properties and microstructure of SUPER304H, TP347HFG and HR3C for A-USC boilers. *Energy Mater,* Vol.2, No.4, (August 2008), pp. 199-206

Komai, N.; Igarashi, M. Minami, Y. Mimura, H. Masuyama, F. Prage, M. & Boyles, P, R. (2007). Field test results of newly developed austenitic steels in the Eddystone Unit No.1 boiler, *Proceedings of CREEP8 Eighth International Conference on Creep and Fatigue at Elevated Temperatures,* San Antonio, Texas, U.S., July 22-26, 2007

Lin, F. S.; Cheng, S. C. & Xie, X. S. (2008). Ultrasupercritical power plant development and high temperature materials applications in China. *Energy Mater,* Vol.3, No.4, (October 2009), pp. 201-207

Liu, Z, D.; Xie, X, S. Cheng, S, C. Lin, F, S. Wang, Q. J. & Xu, S. Q. (2011). The research and development of advanced boiler steels used for USC power plants, *4th Symposium on Heat Resistant Steels and Alloys for High Efficiency USC Power Plants 2011,* Beijing, China, April 11-13, 2011

Minami, Y.; Kimura, H. & Ihara, Y. (1986). Microstructural changes in austenitic stainless steels during long-term aging. *Materials Science and Technolgoy,* Vol.2 (August 1986), pp. 795-806

Muramatsu, K. (1999). Advanced heat resistant steel for power generation, The university Press, Cambridge

Masuyama, F. (2001). History of Power Plants and Progress in Heat Resistant Steels. *ISIJ Int,* Vol.41, No.6, (December 2000), pp. 612-625

Masuyama, F. (2005). Alloy development and material issues with increasing steam temperature, *Proceedings to the Fourth International Conference on Advances in Materials Technology for Fossil Power Plants ,* Hilton Head Island, South Carolina, U.S., October 25-28, 2004

Sawaragi, Y. & Hirano, S. (1992). The Development of a new 18-8 austenitic stainless steel (0.1C-18Cr-9Ni-3Cu-Nb,N) with high elevated temperature strength for fossil fired boilers. Jono, M. & Inone, T. ed. Mechanical Behavior of Materials, Vol.4, London: Pergamon Press, pp. 589-594

Sawaragi, Y.; Ogawa, K. Kato, S. Natori, A. & Hirano, S.(1992). Development of the economical 18-8 stainless steel (Super304H) having high elevated temperature strength for fossil fired boilers. *The Sumitomo Search,* 1992, No.48, (Jan 1992), pp. 50-58

Sawaragi, Y.; Otsuka, N. Senba, H. & Yamamoto, S. (1994). Properties of a new 18-8 austenitic steel tube(Super304H) for fossil sired boilers after service exposure with high elevated temperature strength. *The Sumitomo Search,* 1994, No.56, (Oct 1994), pp.34-43

Sourmail, T. (2001). Precipitation in creep resistant austenitic stainless steels. *Materials Science and Technology,* 2001, Vol.17, (Jan 2001), pp. 1-14

Senba, H.; Sawaragi, Y. Ogawa, K. Natori, A. & Kan, T. (2002). The development of high efficiency 18-8 type Super304H tube used in USC power plants. *Materia Jpn,* 2002, Vol.41, pp. 120-125

Sourmail, T. & Bhadeshia, H, K, D, H. (2005). Microstructural evolution in two variants of NF709 at 1023 and 1073K. *Metall Mater Trans,* 2005, Vol.36A,(Feb 2004), pp.23-34

Tanaka, H.; Muruta, M. Abe, F. & Irie, H. (2001). Microstructural evolution and change in hardness in type 304H stainless steel during long-term creep. *Materials Science and Engineering A,* 2001, Vol.(319–321), pp. 788–791

Tan, S. P.; Wang, Z, H. Cheng, S, C. Liu, Z, D. Han, J, C. & Fu, W, T. (2010), Effect of Cu on Aging Precipitation Behavior of Cu-Rich Phase in Fe-Cr-Ni Alloy. *Journal of Iron and Steel Research, International,* 2010, Vol.17, No.5, (Mar 2009) pp.63-68

Viswanathan, R. & Bakker, W. (2001). Materials for Ultrasupercritical coal power plants-Boiler Materials: Part 1. *J Mater Eng Perform,* Vol.10, No.1, (February 2001), pp. 81-95

Viswanathan, R. (2004). Materials technology for coal-fired power plants. *Advanced Materials & Processes,* (August 2004), pp. 73-76

Viswanathan, R.; Henry, J, F. Tanzosh, J. Stanko, G. Shingledecker, J. Vitalis, B. & Purgert, R. (2005). U.S. Program on Materials Technology for Ultra-Supercritical Coal Power Plants. *J Mater Eng Perform,* Vol.14, No.3, (June 2005), pp. 281-292

Viswanathan, R.; Coleman, K. & Rao, U. (2006). Materials for ultra-supercritical coal-fired power plant boilers. *International Journal of Pressure Vessels and Piping,* 2006, Vol. 83, pp. 778-783

Viswanathan, R.; Shingledecker, J. Hawk, J. & Goodstine, S.(2009) Advanced Materials for Use Ultrasupercritical Coal Power Plants, *2009 Symposium on Advanced Power Plant Heat Resistant Steels and Alloys,* Shanghai, China. Oct, 2009

Xie, X, S.; Chi, C, Y. Yu, H, Y. Yu, Q, Y. Dong, J, X. Chen, M, Z. & Zhao, S, Q.(2010). Structure Stability Study on Fossil Power Plant Advanced Heat-Resistant Steels and Alloys in China. *Advances in Materials Technology for Fossil Power Plants Proceedings from the Sixth International Conference,* Santa Fe, New Mexico, USA. August, 2010

Yoshikawa, K.; Teranishi, H. Tokimasa, K. Fujikawa, H. Miura, M. & Kubota, K. (1988). Fabrication and properties of corrosion resistant TP347H stainless steel. *J. Mater. Eng,* Vol.10, No.1, (1988), pp. 69-84

Yu, H, Y.; Dong, J, X. & Xie, X, S. (2010). 650°C long-term structure stability study on 18Cr-9Ni-3CuNbN heat-resistant steel. *Mater. Sci. Forum,* Vol. 654-656, (2010), pp.118-121

Part 4

Fatigue in Steels: From the Basic to New Concepts

9

Metal Fatigue and Basic Theoretical Models: A Review

S. Bhat and R. Patibandla
*School of Mechanical and Building Sciences
Vellore Institute of Technology, Tamil Nadu,
India*

1. Introduction

1.1 History of metal fatigue

Preliminary understanding about fatigue failure of metals developed in 19th century during industrial revolution in Europe when heavy duty locomotives, boilers etc. failed under cyclic loads. It was William Albert who in 1837 first published an article on fatigue that established a correlation between cyclic load and durability of the metal. Two years later in 1839, Jean-Victor Poncelet, designer of cast iron axles for mill wheels, officially used the term *fatigue* for the first time in a book on mechanics. In 1842, one of the worst rail disasters of 19th century occurred near Versailles in which a locomotive broke an axle. Examination of broken axle by William John Macquorn Rankine from British railway vehicles showed that it had failed by brittle cracking across its diameter. Some pioneering work followed from August Wöhler during 1860-1870 [1] when he investigated failure mechanism of locomotive axles by applying controlled load cycles. He introduced the concept of rotating-bending fatigue test that subsequently led to the development of stress-rpm (S-N) diagram for estimating fatigue life and endurance or fatigue limit of metal, the fatigue limit representing the stress level below which the component would have infinite or very high fatigue life. In 1886, Johann Bauschinger wrote the first paper on cyclic stress-strain behavior of materials. By the end of 19th century, Gerber and Goodman investigated the influence of mean stress on fatigue parameters and proposed simplified theories for fatigue life. Based on these theories, designers and engineers started to implement fatigue analysis in product development and were able to predict product life better than ever before. At the beginning of the 20th century, J. A. Ewing demonstrated the origin of fatigue failure in microscopic cracks. In 1910, O.H. Baskin defined the shape of a typical S-N curve by using Wöhler's test data and proposed a log-log relationship. L. Bairstow followed by studying cyclic hardening and softening of metals under cyclic loads. Birth of fracture mechanics took place with the work of Alan A. Griffith in 1920 who investigated cracks in brittle glass. This promoted understanding of fatigue since concepts of fracture mechanics are essentially involved in fatigue crack characteristics. However, despite these developments, fatigue and fracture analysis was still not regularly practiced or implemented by the designers.

Importance of the subject was finally realised when serious accidents took place around World War II in 20th century that spurred full fledged research work on the subject.

Requirement of war necessitated fabrication of ships quickly and on a large scale. These ships were coined by US as Liberty Ships. Their frames were welded instead of time consuming riveting. Soon there were incidents of ships cracking in cold waters of Atlantic Ocean. As a matter of fact, some ships virtually broke into several parts due to origination of fatigue cracks by sea waves followed by their rapid propagation in cold environment. Sub-zero temperatures drastically diminished the ductility of welds and parent metals thereby making them brittle. As fracture energy of brittle metals is much less than the ductile ones resulting in reduced critical crack sizes in them, fracture took place at the loads that were otherwise found safe in ambient environment. Similarly Comet Jet Airliners fractured and exploded in mid air. A jetliner flying at an altitude of 10,000 m functions like a pressurized balloon with the wall of its fuselage under high tensile stresses. Since the aircraft structure was not designed against fatigue, cyclic aerodynamic loads resulted in nucleation and propagation of cracks through fuselage of the aircraft resulting in its rupture in flight.

1.1.1 Fatigue failure assessment

Several studies have been undertaken to quantify fatigue failures. A world-wide survey of aircraft accidents was reported in 1981 [2] in which the extent to which metal fatigue was responsible for aircraft failures was assessed. A total of 306 fatal accidents were identified since 1934 with metal fatigue as the related cause. These accidents resulted in 1803 deaths. They covered civil and to a limited extent military aircrafts also. Failure of wings and engines were the most frequent causes of fixed-wing accidents while for helicopters, failures of main and tail rotors were found to be the common causes. About 18 fatal accidents per year were attributed to metal fatigue which was an alarming figure.

Fatigue technology has vastly improved since then with the subject gaining more and more importance. Lot of attention is now-a-days being directed towards fatigue and fracture studies of modern and sophisticated technological machines and structures like high speed aircrafts, nuclear vessels, space shuttles, launch vehicles, ships, submarines, pressure vessels, high speed trains etc. which can be devastating in the event of their fatigue failure.

1.2 Fundamentals of fatigue

Metals when subjected to repeated cyclic load exhibit damage by fatigue. The magnitude of stress in each cycle is not sufficient to cause failure with a single cycle. Large number of cycles are therefore needed for failure by fatigue. Fatigue manifests in the form of initiation or nucleation of a crack followed by its growth till the critical crack size of the parent metal under the operating load is reached leading to rupture. Behaviour of metal under cyclic load differs from that under monotonic load. New cracks can nucleate during cyclic load that does not happen under static monotonic load. Importantly, fatigue crack nucleates and grows at stress levels far below the monotonic tensile strength of the metal. The crack advances continuously by very small amounts, its growth rate decided by the magnitude of load and geometry of the component. Also the nucleated crack may not grow at all or may propagate extremely slowly resulting in high fatigue life of the component if the applied stress is less than the metal fatigue limit. However, maintaining that condition in actual working components with design constraints and discontinuities calls for limited service loads which may be an impediment. Therefore, fatigue cracks in most cases are permissible but with proper knowledge of fracture mechanics about the allowable or critical crack size. On the other hand, only two possibilities exist in cracked structure under monotonic load.

The crack can be either safe or unsafe. The component under cyclic load works satisfactorily for years, albeit with hidden crack growth, but ruptures suddenly without any pre-warning. Such characteristics make cyclic load more dangerous than monotonic load.

1.2.1 Fractograph of fatigue surface

The surface having fractured by fatigue is characterised by two types of markings termed as beachmarks and striations [3]. Both these features indicate the position of the crack tip at some point of time and appear as concentric ridges that expand away from the crack initiation site frequently in a circular or semicircular pattern. Beachmarks (sometimes also called clamshell marks) are of macroscopic dimensions, Fig. 1a, and may be observed with unaided eye. These markings are found in components that experience interrupted crack propagation e.g. a machine that operates only during normal work-shift hours. Each beachmark band represents a period of time over which the crack growth occurs. On the other hand, fatigue striations are microscopic in size, Fig. 1b, and can be viewed with an Electron Microscope. A striation forms a part of beachmark and represents the distance by which the crack advances during the single load cycle. Striation width increases with increasing stress range and vice-versa. Although both beachmarks and striations have similar appearances, they are nevertheless different, both in origin and size. There may be literally thousands of striations within a single beachmark. Presence of beachmarks and striations on a fractured surface confirms fatigue as the cause of failure. At the same time, absence of either or both does not exclude fatigue as the cause of failure.

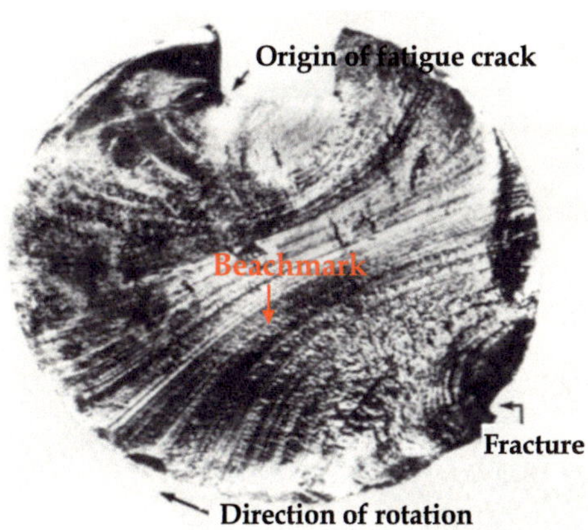

(a)

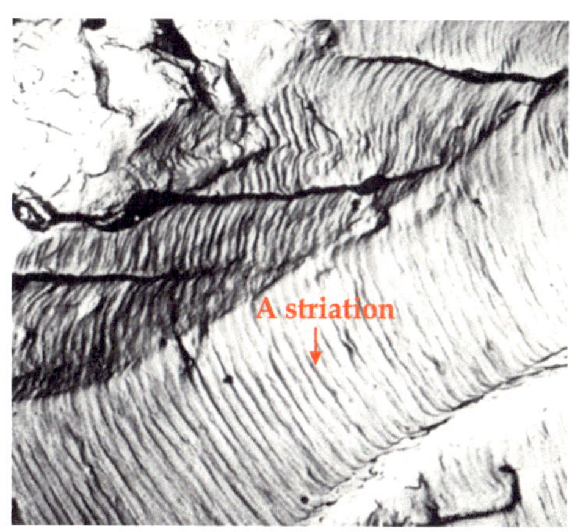

(b)

Fig. 1. Fatigue fracture surface of a steel shaft [3, pp. 216-217]

1.2.2 Types of fatigue load

Load cycles can be of constant amplitude or variable amplitude type. Rotating machines usually operate under pre-decided constant amplitude load cycles whereas aircrafts and ships are subjected to variable amplitude load cycles due to unpredictable and fluctuating wind or sea gusts. Some common types of load cycles are illustrated in Fig. 2. Important fatigue terms are presented.

1.2.3 Classification of fatigue

A. *Load based*: When the stress level is low and the deformation is primarily elastic, the fatigue is called as high cycle type. Number of cycles needed for fracture in this type is high. The account of this regime in terms of stress is more useful. When the stress level is high enough for plastic deformation to occur, the fatigue is known as low cycle type. Number of cycles needed for fracture in such a case is low. The account of this regime in terms of stress is less useful and the strain in the material offers an adequate description. Low cycle fatigue is also termed as strain based fatigue. Direction of load too influences fatigue. Multi-axial loads result in different fatigue characteristics than uni-axial loads. Fatigue under pure mechanical loads is rate independent.

B. *Environment based*: Fatigue characteristics are affected by operating temperature and aqueous and corrosive environments. Fatigue under high temperature is rate dependant.

1.2.4 Factors influencing fatigue life

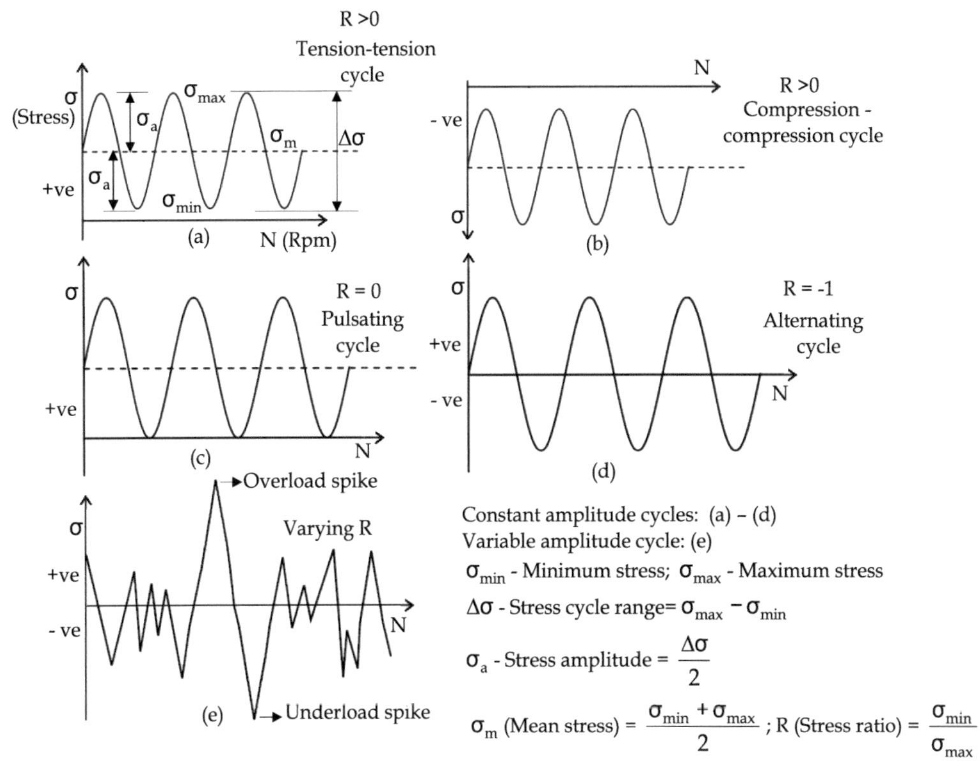

Fig. 2. Types of fatigue cycles

A. *Metal microstructure*: Metal with large grains have low yield strength and reduced fatigue limit and vice-versa. However, at higher temperatures, the coarse grained metal is seen to show better fatigue properties. Barriers to crack growth in the form of precipitates, impurities, grain boundaries, etc. improve fatigue properties. Phase transformations occurring during cyclic loading can also influence the fatigue life.

B. *Manufacturing process*: Fatigue properties are better in the direction of forging, extrusion and rolling and are lower in the transverse direction. Some specific processes like shot peening, cold rolling etc. and other hardening/heat treatment methods that induce compressive residual stresses reduce the chances of crack initiation and enhance the fatigue properties. Tensile residual stresses on the other hand promote crack initiation. Other manufacturing processes like forming, drawing, forging, extrusion, rolling, machining, punching etc., that produce rough surfaces, decrease fatigue life. A rough surface possesses more crack initiation sites due to unevenness and asperities. Polished and ground surfaces on the other hand have excellent high fatigue life due to minimum asperities.

C. *Component geometry*: Discontinuities such as holes, notches and joints, that are the source of stress risers, facilitate crack initiation. Fatigue life of a notched component is less than that of an un-notched one when subjected to similar loads.

D. *Type of environment*: Aqueous and corrosive environments promote crack initiation and increase crack growth rate although crack tip blunting and closure due to accumulation of environmental products at crack tip may dip crack growth rate to some extent. But the overall effect of such environments is to enhance crack growth rate. Under high temperature too, fatigue resistance in most metals generally diminishes with increase in crack growth rate due to the effect of creep.

E. *Loading condition*: Multi-axial loads reduces fatigue life in comparison with uni-axial loads except in the case of pure torsional loading. Mean stress also influences fatigue life. Positive tensile mean stress reduces fatigue life whereas negative mean stress may increase it. The influence of mean stress is more significant in low strain or high cycle fatigue regime.

1.2.5 Fatigue life prediction

1.2.5.1 Constant amplitude load

As discussed earlier in Section 1.1, Wöhler's S-N curves, that were based on experimental fatigue data, were first used for life prediction of metallic structures. A rotating bending test machine is used to obtain S-N curve. The number of cycles, N_f , needed for the specimen to fail when subjected to alternating cycles (R=-1) with maximum stress, σ_{max} , or stress amplitude, σ_a , are recorded. N_f represents fatigue life under σ_{max} . Reduction in σ_{max} enhances N_f and vice versa. Some steels exhibit fatigue or endurance limit whereas non-ferrous metals generally don't. S-N curves have been in use for more than a century now and are still being used by conventional designers. Refer Fig. 3 for S-N curves of various metals. These curves however have certain limitations. They only provide fatigue life without giving any indication about cycles needed for crack initiation and propagation. They also don't take into account the effects of specimen size and geometry i.e. the data generated on a small sized specimen may not be exactly valid for large components in actual use. Also, the component designed on fatigue limit may still fail. S-N curves therefore don't impart sufficient confidence about failure free performance of the component. Consequently, concepts about fatigue damage are required some fatigue damage rules and theories which look into the characteristics of fatigue curves are discussed as follows:

A. *Linear damage rules* (LDR): Refer Fig. 4a and 4b. Basquin in 1910 [4] presented a stress based law, $\sigma_a = \frac{\Delta\sigma}{2} = \sigma_f'(2N_f)^b$, where σ_f' is the fatigue strength coefficient, $2N_f$ is the number of reversals to failure or N_f full cycles and b is the fatigue strength exponent. The stress based approach is mostly applicable in high cycle regime. Coffin and Manson [5, 6] expressed LDR for low cycle regime in terms of plastic strain range as $\frac{\Delta\varepsilon^p}{2} = \varepsilon_f'(2N_f)^c$ where $\frac{\Delta\varepsilon^p}{2}$ is the plastic strain amplitude, ε_f' is the fatigue ductility coefficient and c is the fatigue ductility exponent. For cumulative damage under stress cycles with varying magnitudes, Miner [7] in 1945 first expressed Palmgren's concept [8] in the mathematical form, $D = \sum r_i = \sum \frac{n_i}{N_{fi}}$, where D denotes the quantum of damage, r_i the cycle ratio and n_i and N_{fi} the applied cycles of given stress level and total cycles needed for failure respectively under ith constant amplitude loading cycle. The measure of damage is the cycle ratio with basic assumptions of constant work absorption per cycle. Miner's damage vs cycle ratio plot or D-r curve is a diagonal straight line, that is independent of loading levels. The main deficiencies with LDR are the load level independence, load sequence independence and lack of load-interaction accountability. Life prediction based on linear rule is often un-satisfactory

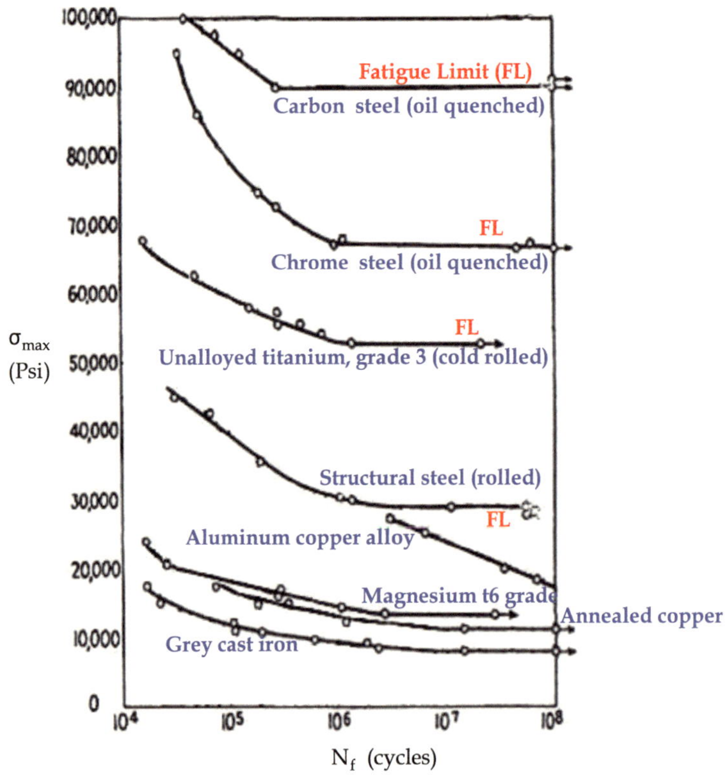

Fig. 3. S-N curves of various metals

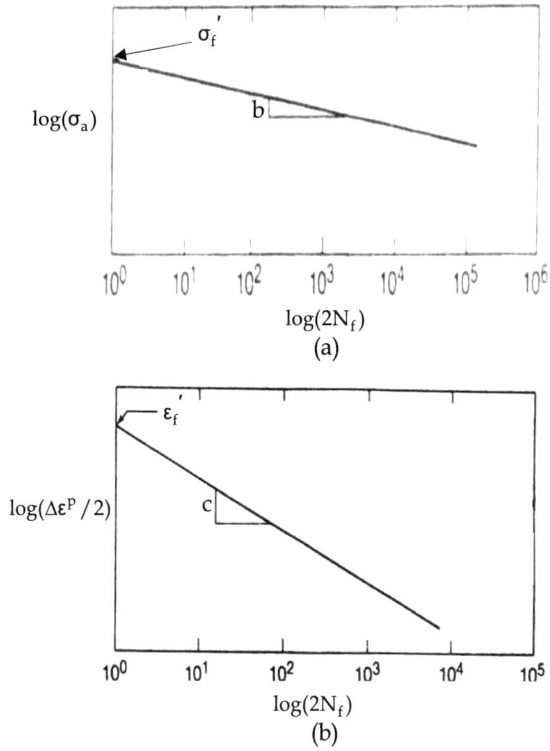

Fig. 4. Stress and strain amplitudes vs number of cycle reversals to failure

B. *Marco-Starkey theory*: Richart and Newmark [9] in 1948 introduced the concept of D-r curves being different at different stress levels. Marco and Starkey [10] followed by proposing the first non-linear load dependant damage theory in 1954 as the power relationship, $D = \sum r_i^{x_i}$, where x_i is a variable quantity related to ith loading. The law results in $\sum r_i > 1$ for low to high load sequence (L-H) and $\sum r_i < 1$ for high to low load sequence (H-L).

C. *Endurance limit reduction theory*: Concept of change in endurance limit due to pre-stress was found to exert an important influence over cumulative fatigue damage. Kommers [11] and Bennett [12] investigated the effect of fatigue pre-stressing on endurance properties using a two-level step loading method. Their experimental results suggested that the reduction in endurance strength could be used as damage measure. Such models are non-linear and are able to account for load sequence effect. But these models fail to include load interaction effects.

D. *Load interaction effect theory*: Corten-Dolon [13] and Freudenthal-Heller models [14] are based on these theories. Both of them are based on modification of S-N diagram, the results being clockwise rotation of original S-N curve around a reference point on the curve. In the former model, a point corresponding to the highest load point is selected as the reference point while in the latter this reference is chosen as the stress level corresponding to fatigue life of $10^3 - 10^4$ cycles. Fig. 5 shows a schematic of fatigue life for two level L-H and H-L stressing with Corten-Dolon model. Actual values of N_f are indicated.

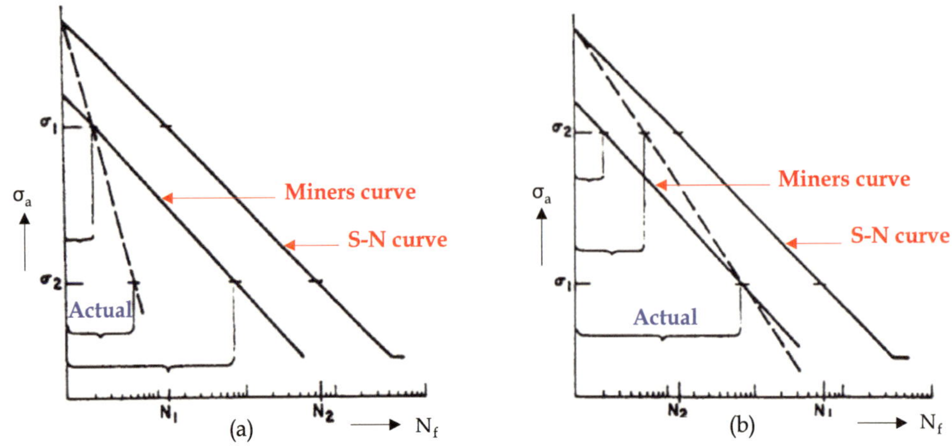

Fig. 5. Fatigue behaviour by rotation method for (a) L-H and (b) H-L load sequences

E. *Two-stage or double linear damage rule* (DLDR): Using Langers concept [15], Grover [16] considered cycle ratios for two separate stages in fatigue damage process. They are the damage in stage I due to crack initiation, $N_I = \alpha N_f$, and damage in stage II due to crack propagation, $N_{II} = (1-\alpha)N_f$, where α is life fraction factor for the initiation stage. In other model by Manson [17], $N_I = N_f - PN_f^{0.6}$ and $N_{II} = PN_f^{0.6}$ where P is the coefficient of stage II fatigue life.

F. *Crack growth based theory*: Important theories are subsequently elucidated in small crack and large crack sections

1.2.5.2 Variable amplitude and complex loads

In order to assess the safe life of a part operating under variable amplitude load cycles, the following steps are taken:- i) Reduce complex to series of simple cyclic load loading using a technique such as rainflow analysis ii) Create a histogram of cyclic stresses from the rain-flow analysis to form a fatigue damage spectrum iii) For each stress level, calculate the degree of cumulative damage from the S-N curve and iv) Combine the individual contributions using an algorithm such as Miner's rule discussed previously.

1.2.6 Design approaches

The component subjected to cyclic load should ideally be without cracks and its surface should be ground and polished for high fatigue life. The following principles are adopted in the design based on fatigue considerations:

A. *Safe-life design*: The underlying assumption in this approach is that the service load is well known and fixed e.g. parts of rotating machinery, engine valves etc. The maximum stress in the fatigue cycle should preferably be less than the fatigue limit of the metal to ensure high fatigue life. If the metal does not exhibit a definite fatigue limit, then the allowable stress level corresponding to desirable life cycles, say 10^6 or 10^7 is chosen from S-N curve as the limiting/design value. In some cases, as specified by ASME code, the design is based on maximum shear stress theory of failure. Alternating stress intensity which is equal to twice the maximum shear stress or difference in principal stresses is used as a design factor in conjunction with ASME fatigue curves.

B. *Fail-safe design*: It is employed when service load is of random nature and contains overload or underload spikes e.g. in aircrafts and ships. Since cracks are unavoidable, the fatigue crack is allowed to grow but the structure is designed in such a way that the crack does not become critical. Crack arrestors are implanted at various positions in the structure. In other words, this concept is based on introducing alternative load bearing members such that failure of one could be tolerated by load redistribution to the remaining members. Structures that don't permit multi-piece parts can't be designed with this principle.

C. *Damage tolerance*: Cracks are allowed in this approach as well but stringent periodic inspections are undertaken to check whether the crack size is below the permissible or critical value. Such inspections are regularly carried out in nuclear and aviation industries. A method is available [18] that provides a rational basis for periodic inspection based on crack growth, taking into account variability of various parameters.

1.3 Important fatigue concepts

A. *Crack tip zones*: Region ahead of crack tip undergoes plastic deformation due to high stress concentration induced by the presence of crack. Strains are very high at the crack tip that cause the tip to blunt thereby reducing the local stress tri-axiality. Refer Fig. 6. The region in immediate vicinity of the crack tip of size, δ^*, in immediate vicinity of the crack tip is called as the process zone. The fracture process resulting in crack advancement by breaking of atomic bonds in the case of brittle metal and coalescence of voids in the case of ductile metal takes place in this zone. The process zone is governed by large strain analysis and Hutchinson, Rice and Rosengren (HRR) elastic–plastic and other stress solutions, that are based upon small strain or deformation plasticity concepts without considering the effect of blunting, are therefore not valid in it. Area ahead of process zone is divided into three regions: Region I or the cyclic plastic zone of size, Δr, where plastic deformation takes place during loading and unloading half-cycles. Region II, between the monotonic plastic zone of size, r_m, and the cyclic plastic zone, where plastic deformation occurs only during the loading part of the cycle and metal is elastic during the unloading part. Region III or the elastic zone beyond the monotonic plastic zone where cyclic strains are fully elastic both during loading and unloading parts of the cycle.

B. *Crack closure*: Refer Fig. 7. During unloading part of a load cycle, the opened crack faces touch each other earlier at position A instead of B. On further reduction of stress from A, the touched surfaces start exerting compressive load, the effect of which is overcome by a part of the next cycle. In other words, the part of energy of new cycle is utilized in overcoming the compressive effects induced by the previous cycle. In the process, the applied cyclic stress intensity parameter, ΔK, reduces to effective stress intensity parameter, ΔK_{eff}, that is equal to $K_{max} - K_{cl}$. The closure ratio, U, is defined as $U = \frac{\Delta K_{eff}}{\Delta K}$. Closure is induced by factors like crack tip plasticity, surface oxides, surface roughness and operating viscous fluids (if any) and is desirable since it retards crack growth rate.

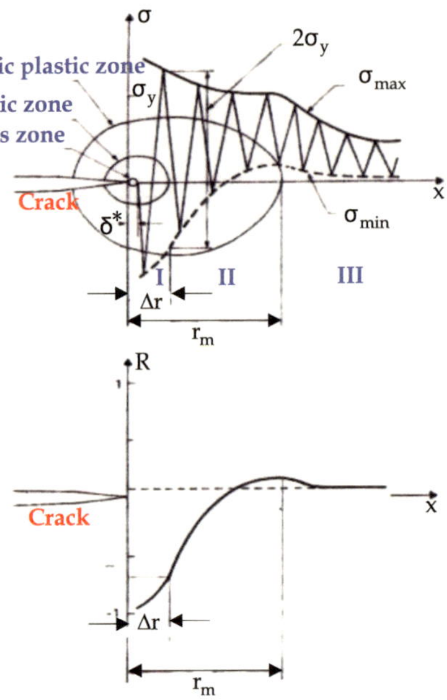

Fig. 6. Fatigue crack tip zones

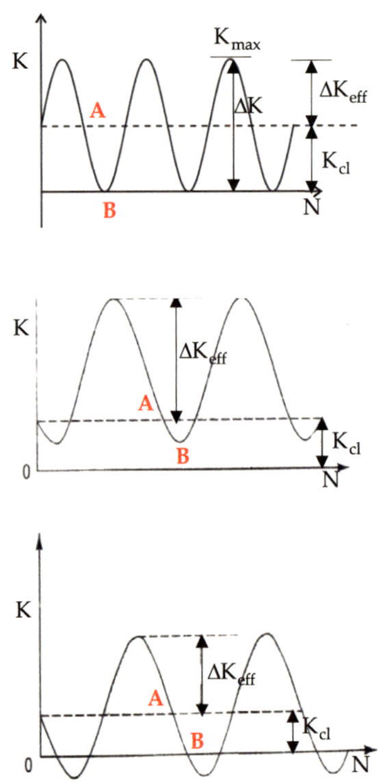

Fig. 7. Crack closure

C. *Threshold*: The condition at which the crack is just able to propagate forward by an infinitesimally small distance before it is arrested again is called the threshold state. The arrest may be due to crack tip blunting or dislocation re-arrangement. Threshold stress intensity parameter is denoted by ΔK_{th}. Experimental threshold measurement is expensive and time consuming. However, there are some empirical models to obtain ΔK_{th} which are based upon the principles of energy and dislocation dynamics. These models consider crack growth by slip at crack tip and assume grain boundaries as principal barriers to slip. Threshold prediction according to energy criterion [19] is of the form $\Delta K_{th} = \sigma_{yc}\left(\frac{2.82\pi d}{1-\nu^2}\right)^{1/2}$ where σ_{yc} is the cyclic yield strength of the metal that is equal to $2\sigma_y$, σ_y being the monotonic yield strength, d the grain size and ν the poisson's ratio of the metal. The energy model, while ignoring micro-structural effects, considers energy as controlling parameter in deciding the physical extent of the plastic zone or crack advance. Dislocation models on the other hand consider behavior of dislocations at the crack tip. Dislocations during slip pile up at the first grain boundary and limit the outward flow of further dislocations from the crack tip. A typical Hall-Petch model suggests, $\Delta K_{th} = A + B\sqrt{d}$, where A and B are material constants. The model also reduces to the concept of a slip band extending from the crack tip to a grain boundary when loaded by stress equal to σ_{yc}. With the help of linear elastic fracture mechanics (LEFM) equation and assuming the slip band to be the crack which will just be able to propagate at the threshold, a following simple relation is written [19, p. 16] as $\Delta K_{th} = \sigma_{yc}\sqrt{\pi d}$. Although this relation is somewhat invalid because of the use of principles of LEFM in plastic stress field, yet it effectively predicts the dependence of grain size and yield strength of metals on their fatigue threshold.

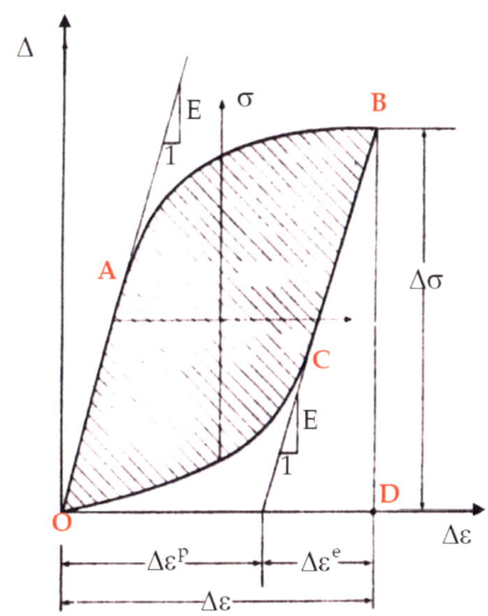

Fig. 8. Hysteresis loop for a masing material

D. *Hysteresis energy*: During cyclic loading, applied energy is dissipated in the form of plastic deformation/slip at the crack tip. Refer Fig. 8. The plastic strain energy or hysteresis energy, ΔW^p, absorbed in a load cycle per unit volume is the area of hysteresis loop (Area OABCO) and for a Masing material it is given as, $\Delta W^p = \frac{1-n'}{1+n'}\Delta\sigma\Delta\varepsilon^p$ [20], where n' is the cyclic strain-hardening exponent that is equal to zero for an elastic-perfectly plastic material. Method to obtain ΔW^p for a Non-Masing material is explained at [21]. A part of hysteresis energy is consumed in heat and vibration and the remaining part causes damage in the form of crack growth by slip.

2. Fatigue mechanisms

Nucleation and propagation of cracks constitute major fatigue mechanisms. Initiation of fatigue crack at smooth polished surface under ambient conditions may consume nearly 90% of applied cycles while crack propagation may require only remaining 10% cycles. Distribution of cycles changes in defective specimens with environment also playing a major role. The mechanisms and models for crack initiation and growth in specimens without and with the defects in ambient environment are as follows:

2.1 Specimen without notch or defect

2.1.1 Crack initiation

Initiation of a new crack in smooth polished metals under cyclic load is caused by irreversible dislocation movement leading to intrusions and extrusions. (A dislocation is the flaw in the lattice of the metal which causes slip to occur along favorable oriented crystallographic planes upon application of stress to the lattice). These dislocations agglomerate into bundles almost perpendicular to the active Burger's vector. (Burger's vector represents magnitude and direction of slip). Strain localization occurs when dislocation pattern in a few veins or bundles becomes locally unstable at a critical stress or strain thereby leading to formation of thin lamellae of persistent slip bands or PSB's. The subsequent deformation is mainly concentrated in these slip bands as they increase and fill the entire volume of the crystal. If the PSB's are removed by electro-polishing, it will be found by retesting that they reform in the same area and become persistent. That is why the slip bands are also referred to as the persistent slip bands. They are very soft as compared to hard parent metal. Mechanism of formation of PSB's is different in different metals. For example, in a single FCC crystal of copper, the strain in the matrix is accommodated by quasi-reversible to and fro bowing of screw dislocations in channels between the veins. In softer materials which experience large strains, the edge dislocations bow out of the walls and traverse the channels. In poly-crystals, the PSB's are generally found on the grains which have suitable orientations for the slip to occur. Fig. 9 shows a schematic of slip during monotonic and cyclic load. Under monotonic load, slip lines are formed in metal that are sharp and straight and are distributed evenly over each grain. Under high magnification, the individual lines appear as bands of parallel lines of various heights. On the other hand, the slip lines produced under cyclic load form in bands that do not necessarily extend right across a grain. New slip lines form beside old ones as the test proceeds. Although these bands grow wider and become more dense, there are areas between the bands where no slip takes place. Inhomogenity at the microscopic level, when the plastic strain in the PSB lamellae is at least an order of magnitude higher than the metal matrix, causes the crack to eventually form at the interface of PSB and the matrix. Also across the PSB-matrix interface, there is high strain gradient due to PSB's being softer in nature and the matrix being harder. The deformation compatibility requirement at the interface results in high shear stress along the interface leading to cracks. Crack initiation is aided by environmental effects also. Atmospheric oxygen diffuses into slip bands of PSB's thereby weakening them and accelerating initiation. On the other hand, crack initiation in an inert environment may be retarded by up to two orders of magnitude. Favorable crack initiation sites at micro-level can be stated as:- i) Slip steps between emerging extrusions of PSB's and the matrix ii) At micro notches near outer edges iii) At intrusion sites and iv) Grain boundaries in the case of high temperature and corrosive environment. Cracks once initiated can be viewed, Fig. 10, as per the following categories depending upon their location on the surface grain:- i) Trans-granular, ii) Inter-granular, iii) and iv) Surface inclusion or pore, v) Grain boundary voids and vi) Triple point grain boundary intersections. The last two are found at elevated temperatures. Models to estimate the number of cycles, N_i, for crack initiation are difficult to develop and initiation life is measured experimentally.

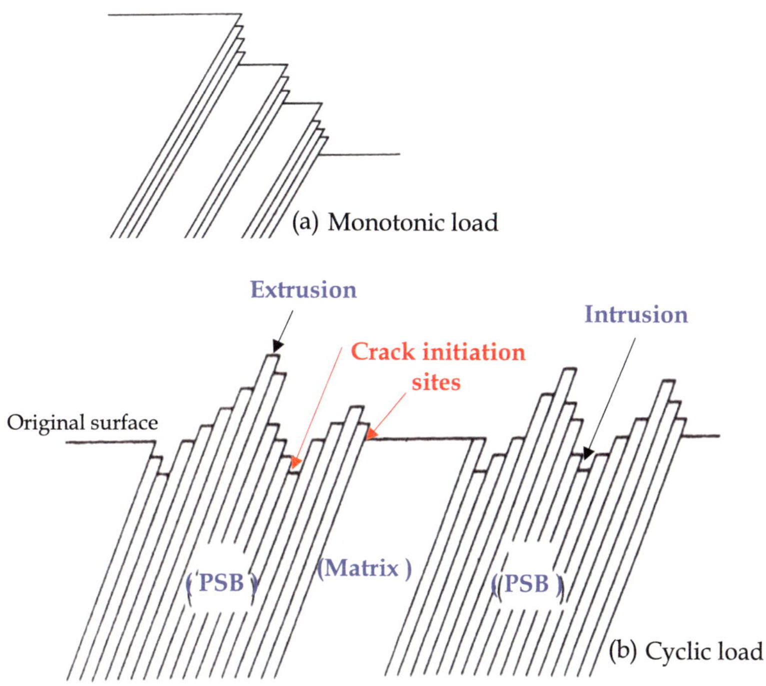

Fig. 9. Schematic of slip under (a) monotonic load and (b) cyclic load [21, p. 13]

New crack initiates from the surface. Firstly, because the surface grains are in intimate contact with the atmosphere, thus if environment is a factor in the fatigue damage process, the surface grains are more susceptible. Secondly, a surface grain is not wholly supported by adjoining grains. It can deform plastically more easily than a grain inside the body surrounded on all sides by grains. Experiments have been conducted to prove this point. If the surface of the component is hardened, either metallurgically or by surface hardening, the fatigue strength of the specimen increases as a whole. Similarly, any procedure that softens the surface decreases the fatigue strength. It has been shown that if a fatigue test is stopped after some fraction of the expected specimen life, with thin layer of metal removed from the surface of test specimen and the test restarted at same stress level, the total life of the specimen goes up. Since large number of fatigue cycles is consumed in crack initiation, removal of surface layer at frequent intervals enhances the fatigue life manifold.

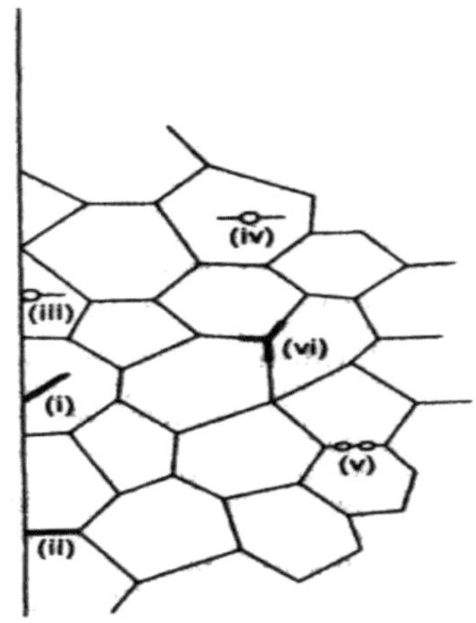

Fig. 10. Crack initiation sites [21, p. 16]

2.1.2 Crack growth

Since growth mechanism and propagation rate of a crack differs according to its size, it is essential to define the crack based on its size in increasing order, Refer Fig. 11 :- i) Metallurgically or micro-structurally small crack which is small as compared to a metallurgical variable such as the grain size. Such a crack is strongly affected by microstructure of the metal and its growth stops at micro-structural barriers if the applied stress level is below the fatigue limit of the metal. The size of this crack would generally be of the order of 1 grain. Refer Fig. 12 and Fig. 13. Its growth rate decreases with increasing length ii) Physically small crack in which the resistance to crack growth by micro-structural barriers is averaged out but it is not long enough to be called a long crack. Length of physically small crack is of the order of 3 to 4 grain size. Such a crack is also arrested at micro-structural barriers if the stress level is below the fatigue limit but it has different characteristics in comparison with long crack. It grows at threshold value of ΔK below that of a long crack and propagates at higher rate than a long crack for same value of ΔK . Like micro-structurally small crack, its growth rate also decreases with increasing length in each successive grain. Knowledge of metallurgically and physically small cracks is necessary from practical point of view as it indicates the size of the flaw or a crack which can be tolerated in the production process. Since crack closure levels are not stable in small cracks, they can not be dealt with the principles of continuum mechanics iii) Long crack that has stable closure and can be treated by continuum mechanics.

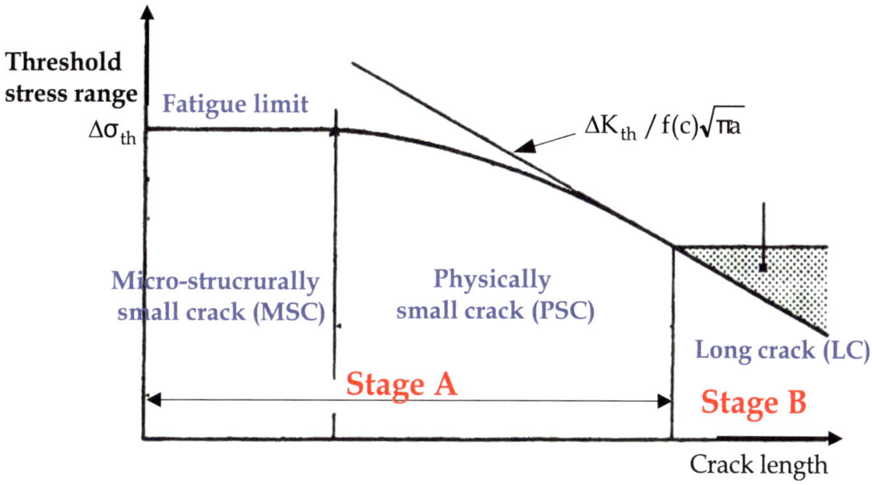

Fig. 11. Three regimes of crack size [21, p. 419]

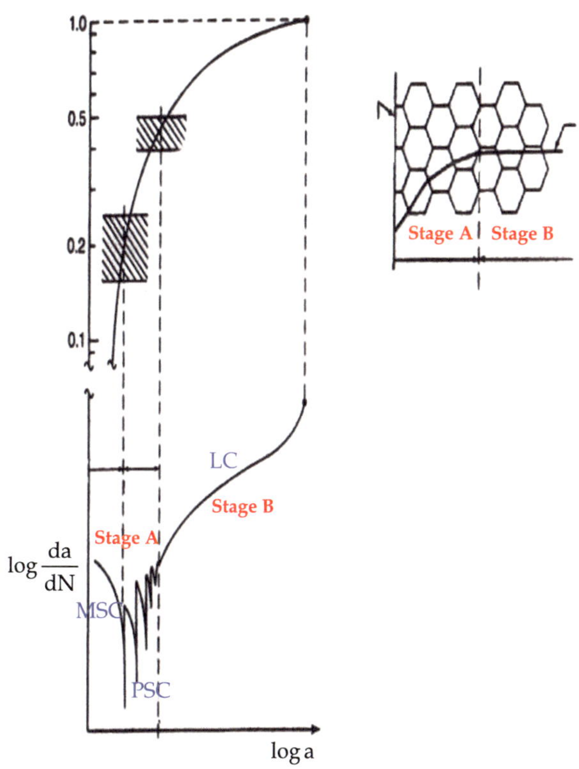

Fig. 12. Transition from small crack to long crack [21, p. 418]

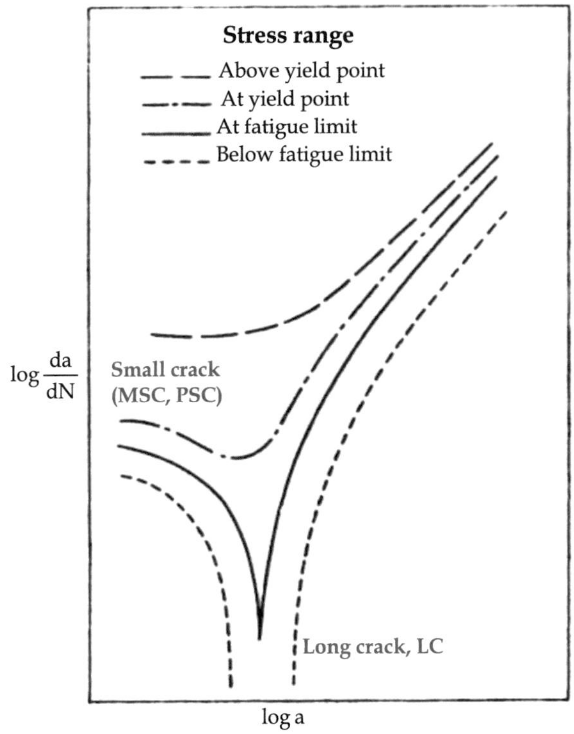

Fig. 13. Crack behavior vs applied cyclic stress [21, p. 416]

2.1.2.1 Small crack growth

The ratio of plastic zone to the overall size of a small crack is much larger than the long crack under same value of applied ΔK. Small crack is therefore strongly influenced by crack tip plasticity. It grows by irreversible plastic deformation at the tip in a slip plane along the slip direction. As shown in Fig. 14a and 14b, an intrusion forms due to the relative displacement of a slip band cc′ and the one above it due to shear stress reversal. In the next applied cycle, Fig. 14c, dislocations on cc′ plane produce a greater offset. An opposite deformation, Fig. 14d, causes the slip planes above and below cc′ to act resulting in the intrusion to eventually grow and form a crack of the order of few micro meters Crack growth is in. small integral multiples of Burger's vector. The magnitude of local cyclic plastic strain ahead of crack tip is a measure of driving force. As the crack approaches a micro-structural barrier, the primary slip becomes incompatible with adjacent grains and the micro-structural induced crack tip shielding decelerates crack growth. As a result, primary plastic zone also decreases resulting in plasticity redistribution. In the process, secondary slip system forms and the crack deviates from the original path leading to crack branching. The secondary slip system plays an important role in increasing crack opening and driving the crack across the micro-structural boundary. Once the path of the crack is altered, the roughness induced closure by mixed sliding and mismatch between crack face asperities again dips the crack driving force and growth rate. The models describing threshold and small crack behaviour are essentially micro-structure based and are as follows [4]:

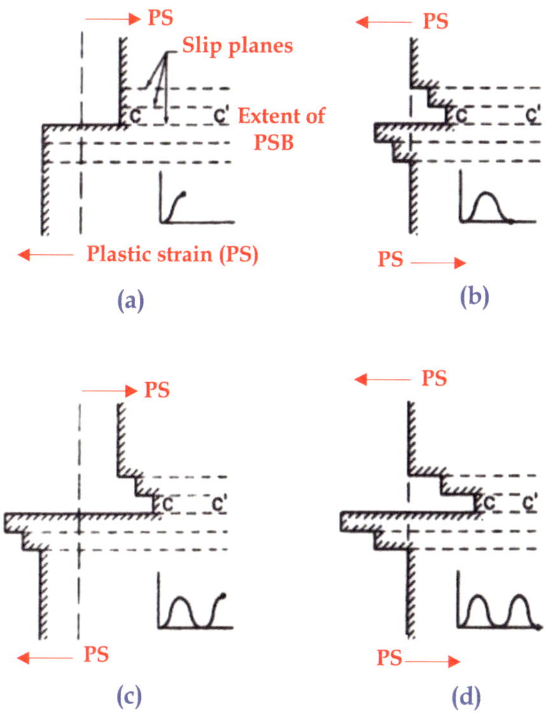

Fig. 14. Stages of small crack growth [21, p. 421]

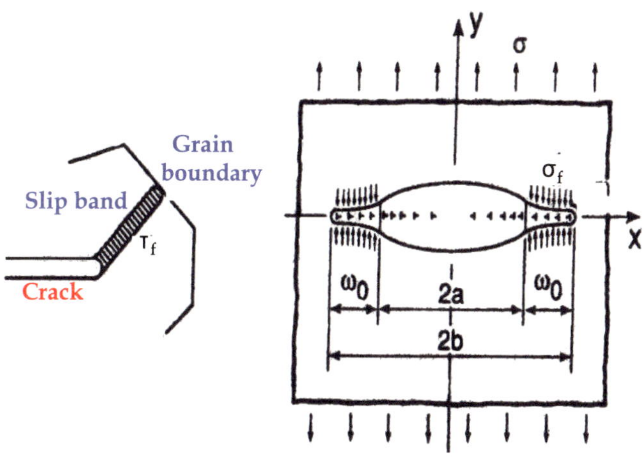

Fig. 15. Schematic of slip band model [21, p. 430]

A. Threshold models [21, pp. 430-435]

A.1 *Slip based model*: The threshold stress for short crack is defined in terms of slip band grain boundary interactions. In this model, the required condition for co-planar slip bands emanating from the crack tip when blocked by grain boundary is investigated. Fig. 15 shows a schematic of this model. An analysis of this condition yields:

$$\Delta\sigma_{th} = \frac{\Delta K_{th}}{\sqrt{\pi a}} = \frac{\Delta K_m}{\sqrt{\pi b}} + \frac{2}{\pi}\Delta\sigma_f \arccos\left\{\frac{a}{b}\right\} \tag{1}$$

where $\Delta\sigma_{th}$ is the threshold stress range, $\Delta\sigma_f$ is the frictional stress range for dislocation motion, ΔK_m is the microscopic stress intensity parameter range at the tip of slip band, 2a is the crack length, $b = a + \varpi_o$ where ϖ_o is the width of blocked slip band. The fatigue limit is obtained from above, by letting a = 0 i.e. a very small crack, in Eq. (1) as

$$\Delta\sigma_e = \frac{\Delta K_m}{\sqrt{\pi\varpi_o}} + \Delta\sigma_f \tag{2}$$

A.2 *Surface strain redistribution type model*: The model is based on non-uniformity of strains in surface layer and the development of crack closure. The surface grains oriented for easy slip are subjected to an inherent micro-structurally dependant strain concentration which decays with the depth into the material. The threshold stress range is defined as

$$\Delta\sigma_{th} = \frac{\Delta K_{th} U_{cl}}{FQ\sqrt{\pi a}} \tag{3}$$

where U_{cl} is the crack closure development parameter, F is the shape factor with a value of 0.72 for semicircular surface crack and Q is the stress concentration factor.

B. Crack growth rate model

The model is based on the principle of surface layer yield stress redistribution. In contrast to the previous model at A.2, this model is concerned with the distribution of yield stress through the surface layer. There is ample experimental evidence about the fact that the surface layer is much softer than the bulk material and the yield stress can be significantly lower in surface grains than in the bulk material. The material thickness is subdivided into a series of strips whose thickness is equal to process zone, δ^*. Refer Fig. 16. Total plastic energy density at the m-th layer is the sum of hysteresis energy per cycle and the applied energy at that layer. Hysteresis energy is provided in Section 1.3D. Applied energy given by HRR is of the form , $\Delta\sigma\Delta\varepsilon^p = \dfrac{\Delta K^2}{Er\psi(n',\theta)}$. On the crack plane, $\theta = 0$, distance of m-th layer from the crack tip, $r = \delta^* + \rho_c$ where ρ_c is critical blunting radius, $\psi(n',\theta)$ depends upon the strength of singularity field and E is the modulus of elasticity of the metal. Taking $\rho_c = 0$ for a small crack , the expression for total energy at m-th layer near the surface is obtained as

$$(\Delta\sigma\Delta\varepsilon^p)_{m,total} = \left(\frac{1-n'}{1+n'}\right)\Delta\sigma_{ym}\Delta\varepsilon_m{}^p + \frac{\Delta K^2}{E\delta^*\psi(n')} = \left(\frac{1-n'}{1+n'}\right)\Delta\sigma_{ym}\left[\Delta\varepsilon_b - \frac{\Delta\sigma_{ym}}{E}\right] + \frac{\Delta K^2}{E\delta^*\psi(n')} \tag{4}$$

Basquin and Coffin-Manson relationships when coupled give the following

$$(\Delta\sigma\Delta\varepsilon^{p})_{m,total} = 4\sigma_f^{'}\varepsilon_f^{'}(2N_f)^{b+c} \tag{5}$$

Number of cycles, N^*, required for the crack to advance through the process zone, δ^*, at m-th layer is equal to

$$N^* = \frac{1}{2}\left[\frac{(\Delta\sigma\Delta\varepsilon^{p})_{m,total}}{4\sigma_f^{'}\varepsilon_f^{'}}\right]^{\frac{1}{b+c}} \tag{6}$$

On using Eq. (4) and considering crack growth rate per cycle, $\frac{da}{dN}$, as $\frac{\delta^*}{N^*}$, the expression for $\frac{da}{dN}$ is obtained as:

$$\frac{da}{dN} = 2\delta^*\left[\frac{\left(\frac{1-n'}{1+n'}\right)\Delta\sigma_{ym}(\Delta\varepsilon_b - \frac{\Delta\sigma_{ym}}{E}) + \frac{\Delta K^2}{E\delta^*\psi(n')}}{4\sigma_f^{'}\varepsilon_f^{'}}\right]^{\frac{1}{b+c}} \tag{7}$$

Eq. (7) re-confirms that $\frac{da}{dN}$ depends upon soft surface layer properties, crack tip stress-strain range and bulk mechanical and fatigue properties of the metal.

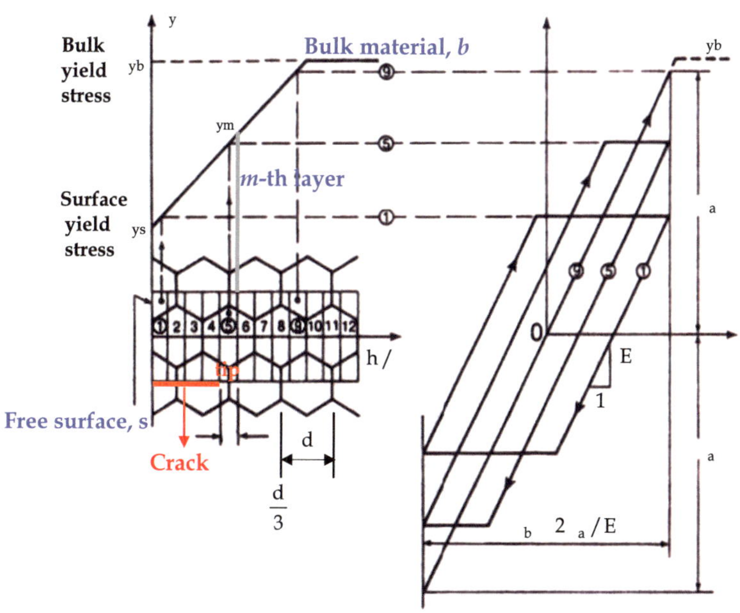

Fig. 16. Material modeling of soft surface layer and three cyclic stress-strain loops near the free surface region in short crack growth [21, p. 434]

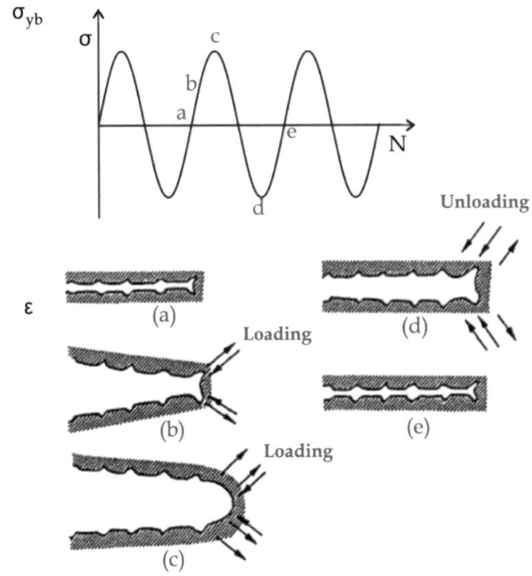

Fig. 17. Plastic blunting model for long crack growth, stages (a) to (e)

2.1.2.2 Long crack growth

Two different mechanisms namely plastic sliding and damage accumulation at the crack tip describe long crack growth in high cycle and low cycle regimes respectively. The plastic sliding mechanism in high cycle regime, proposed by Laird and Smith [22] is also known as the plastic blunting model. Fig. 17 is a schematic of crack tip opening, blunting and crack advance in this model. It can be seen that by application of tensile load, highly localized plastic deformation takes place along the slip planes of maximum shear stress. Upon further increase of load, the width of slip band increases and the crack tip blunts to semi-circular shape. As a result of blunting, the crack extends to about half the crack tip opening displacement. During compressive loading, the direction of slip is reversed and the vertical distance between crack surfaces decreases. The new surface created during tensile load is partly folded by buckling into double notch at the tip. At the maximum compressive stress, the crack tip is sharp again that facilitates further crack growth. In low cycle regime, mechanism for crack growth is based on damage accumulation in the process zone. The crack advances through the zone when sufficient amount of damage is accumulated. The models describing threshold and long crack growth characteristics are as follows:

A. Threshold model: Refer Eq. (1). Threshold condition for long crack where $a \gg \varpi_o$ is

$$\Delta\sigma_{th} = \frac{\Delta K_{th}}{\sqrt{\pi a}} = \frac{\Delta K_m}{\sqrt{\pi a}} + \frac{2}{\pi}\left(\frac{2}{a}\right)^{1/2} \Delta\sigma_f \sqrt{\varpi_o} \tag{8}$$

and

$$\Delta K_{th} = \Delta K_m + 2\left(\frac{2}{\pi}\right)^{1/2} \Delta\sigma_f \sqrt{\varpi_o} \tag{9}$$

Special case is obtained by setting $\Delta K_m = 0$ and $\Delta\sigma_f = 2\sigma_y$ in Eq. (1) as follows

$$\Delta\sigma_{th} = \frac{4}{\pi}\sigma_y \arccos\left\{\frac{a}{a+\varpi_o}\right\} \tag{10}$$

Solving for ϖ_o, one obtains

$$\varpi_o = a\left[\sec\left\{\frac{\pi\Delta\sigma_{th}}{4\sigma_y}\right\} - 1\right] \tag{11}$$

which represents Dugdale's plastic or cohesive zone size.

B. Crack growth rate models: The models in different regimes are as follows:

B.1 *High cycle regime*: Refer Fig. 18. The crack growth rate model in region B due to sliding off mechanism is of the form

$$\frac{da}{dN} \approx CTOD = \frac{\Delta K^2}{2E\sigma_y} \tag{12}$$

where CTOD is the crack tip opening displacement. Paris [23] proposed a law

$$\frac{da}{dN} = C(\Delta K)^m \tag{13}$$

where C and m are material properties. In presence of closure, the above law is modified to the form as follows:

$$\frac{da}{dN} = C(U\Delta K)^m \tag{14}$$

where U can be represented in the form, $U = C_1 + C_2R + C_3R^2 + --$, where C_1, C_2, C_3, are the constants. The log plot between ΔK and $\frac{da}{dN}$ is sigmoidal in shape and is bounded at the extremes by threshold range, ΔK_{th}, and critical ΔK_c or cyclic fracture toughness of the metal. The above law however doesn't cover crack growth rate in zones A and C i.e. at very low and very high ΔK values. In such zones, continuum mechanics is invalid and the models based on the principles of appropriate micro-fracture based like low cycle fatigue model, as discussed in Section B.2, need to be used.

Decrease in ΔK_{th} results in increased $\frac{da}{dN}$ and vice-versa. Effects of various parameters on

ΔK_{th} and $\frac{da}{dN}$ are as follows: i) Increase in grain size without changing the yield strength increases ΔK_{th} and vice-versa ii) Increase in operating temperature reduces ΔK_{th} and vice versa. But in metals that show excessive closure effects the trends may reverse. Enhancement in oxidation rates can cause more closure and may therefore increase ΔK_{th} at elevated temperature iii) Excessive corrosion results in deposits at crack surfaces that increases ΔK_{th} iv) Inert environments such as vacuum reduce ΔK_{th} by diminishing the effect of oxide induced closure v) Frequency of load cycles does not influence $\frac{da}{dN}$ in ambient and high temperatures but strongly affects it in corrosive environments vi) Increase in stress ratio (+ve value) decreases closure and reduces ΔK_{th} and vice-versa. Reduction in stress ratio (-ve value) increases ΔK_{th} and vice-versa. An overload pulse in the load cycle dips $\frac{da}{dN}$ due to the formation of a larger plastic zone during loading portion that exerts compressive stresses at the crack tip as the zone tries to regain its original shape. An underload pulse is found to increase $\frac{da}{dN}$.

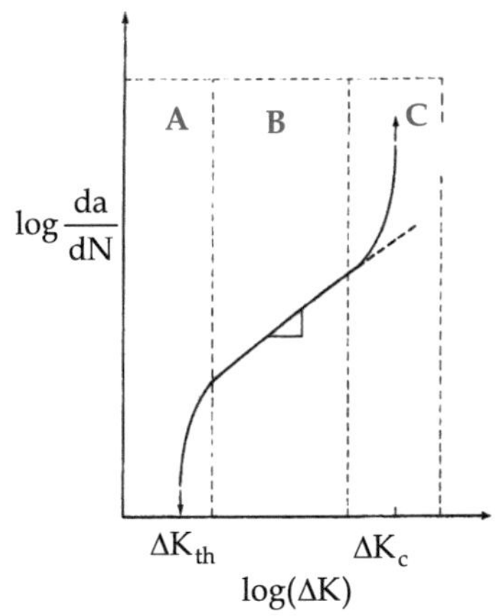

Fig. 18. Long crack growth rate vs applied stress intensity parameter

B.2 *Low cycle regime* [21, pp. 313-318]: Refer Fig. 6. Substituting HRR solution in hysteresis energy, the following expression is obtained on using $r = \delta^* + \rho_c$ and $\theta = 0$ on crack plane:

$$\Delta W^p = \frac{1-n'}{1+n'}\left[\frac{\Delta K^2}{E\psi(n')(\delta^* + \rho_c)}\right] \tag{15}$$

Basquin and Coffin-Manson relationships when coupled yield the following

$$\Delta W^p = 4\left(\frac{1-n'}{1+n'}\right)\sigma_f'\varepsilon_f'(2N_f)^{b+c} \tag{16}$$

The condition for the crack tip to advance through δ^* is obtained from Eq. (15) and Eq. (16) as

$$\frac{\Delta K^2}{E\psi(n')(\delta^* + \rho_c)} = 4\sigma_f'\varepsilon_f'(2N^*)^{-\beta} \tag{17}$$

where $\beta = -(b + c)$. Using crack growth rate, $\frac{da}{dN}$, as $\frac{\delta^*}{N^*}$, one obtains

$$\frac{da}{dN} = \frac{\Delta K^2}{4E\sigma_f{}'\varepsilon_f{}'\psi(n')} \frac{(2N^*)^\beta}{N^*} - \frac{\rho_c}{N^*} \tag{18}$$

ρ_c is obtained by using the condition, $\frac{da}{dN} \approx 0$, at $\Delta K = \Delta K_{th}$

$$\rho_c = \frac{\Delta K_{th}{}^2}{4E\sigma_f{}'\varepsilon_f{}'\psi(n')}(2N^*)^\beta \tag{19}$$

Eq. (18) and Eq. (19) result in

$$\frac{da}{dN} = 2\delta^* \left[\frac{\Delta K^2 - \Delta K_{th}{}^2}{4E\sigma_f{}'\varepsilon_f{}'\delta^*\psi(n')} \right]^{1/\beta} \tag{20}$$

In intermediate Zone B, $\Delta K_{th}{}^2$ can be ignored in comparison with ΔK^2. Therefore, $\frac{da}{dN} = 2\delta^* \left[\frac{\Delta K^2}{4E\sigma_f{}'\varepsilon_f{}'\delta^*\psi(n')} \right]^{1/\beta}$, which represents Paris law discussed earlier with coefficients $m = \frac{2}{\beta}$ and $C = 2\left[\frac{1}{4E\sigma_f{}'\varepsilon_f{}'\delta^{*(1-\beta)}\psi(n')} \right]^{1/\beta}$.

2.2 Specimen with notch or defect

In such cases, the crack initiates at the edge or corner of a stress concentration site like a notch. The word notch is used here in the generic sense to imply geometric discontinuities of all shapes, e.g. holes, grooves, shoulders (with or without fillets), keyways etc. Crack initiation is easy in notched members vis-à-vis the un-notched ones. Fundamental mechanisms of crack initiation and growth are same as presented in Section 2.1. Some aspects are discussed as follows:

A. Crack initiation life: Underlying principle in dealing with notched members is to relate the crack initiation life at the notch root with that of a smooth uni-axial specimen subjected to an equivalent cyclic load. Local stress and strain at the notch root are first found. Three approaches are then commonly employed i) Local stress or fatigue notch factor approach in which the applied far field or nominal stress range simulated over smooth specimen is K_t times the nominal stress range, $\Delta\sigma_{nom}$, over actual notched component such that $K_t\Delta\sigma_{nom} = (\Delta\sigma_{max}\Delta\varepsilon_{max}E)^{1/2}$ upon using Neuber's law, $K_t{}^2 = K_\sigma K_\varepsilon$, where $K_\sigma = \frac{\Delta\sigma_{max}}{\Delta\sigma_{nom}}$ and $K_\varepsilon = \frac{\Delta\varepsilon_{max}}{\Delta\varepsilon_{nom}}$. K_σ, K_ε and K_t are stress concentration, strain concentration and theoretical stress concentration factor determined from elastic analysis of notched element and $\Delta\sigma_{nom}$, $\Delta\sigma_{max}$, $\Delta\varepsilon_{nom}$ and $\Delta\varepsilon_{max}$ are the cyclic values of nominal stress, maximum stress, nominal strain and maximum strain respectively. K_t in the above has also been replaced by various investigators with fatigue notch factor, K_f, such that $K_f = \frac{\Delta\sigma_{nom}(\text{smooth specimen})}{\Delta\sigma_{nom}(\text{notched specimen})}$ at same fatigue life. Neuber [24] and Peterson [25] have proposed various formulae that relate K_t with K_f ii) Local strain approach in which a smooth specimen is subjected to same strain as that prevailing at the notch root of actual specimen iii) Energy approach which allows to write the following, (Refer Fig. 8 for energy under curve OABDO), $\frac{1}{2}\Delta\sigma_{max}\Delta\varepsilon_{max} = \Delta W = \frac{1}{2}\Delta W^p + \frac{1}{2}\Delta\sigma\Delta\varepsilon$ or $K_t^2\left[\frac{1}{2}\Delta\sigma_{nom}\Delta\varepsilon_{nom}\right] = \frac{1}{2}\Delta W^p + \frac{1}{2}\Delta\sigma\Delta\varepsilon$. On replacing RHS of the equation by general fatigue failure criterion based on single parameter damage representation, one obtains $K_t^2\left[\frac{1}{2}\Delta\sigma_{nom}\Delta\varepsilon_{nom}\right] = \kappa(2N_f)^{\chi} + \Delta W_o$, where ΔW_o is the hysteresis energy at the fatigue limit and is negligible, χ is the slope of plot between $\log\Delta W^p$ and $\log(2N_f)$ and $\log\kappa$ is the intercept ($2N_f = 1$) on the vertical axis. N_f is found accordingly. Initiation is significantly influenced by the radius of curvature of the notch. Fatigue crack nucleates easily at the sharp notch. N_i is small for sharp and high for blunt notches respectively.

B. Crack growth: For small crack of length l, ΔK is equal to $K_t\Delta\sigma_{nom}\sqrt{\pi(a+l)}f(c)$ where a is half notch length that is measured from load axis to the notch tip and f(c) is the configuration factor decided by the dimensions of the crack w.r.t. the body. In an infinite body, f(c) =1. For long crack, the effect of notch is not felt since the crack tip is far away form the notch tip. Therefore, $K_t = 1$ and ΔK equals $\Delta\sigma_{nom}\sqrt{\pi(a+l)}f(c)$.

3. Special cases

3.1 Fatigue under multi - axial load

Many parts in automobiles, aircrafts, pressure vessels etc. operate under multi-axial service load. Consequently, analysis of multi-axial fatigue assumes high importance although it is difficult to precisely define the fatigue behaviour of materials under such loads. When the body is subjected to stress σ_1 under uni-axial load, the state of strain in the body is three-dimensional, i.e. $\varepsilon_2 = e_3 = -\upsilon\varepsilon_1$. There are two stress components acting in the form of normal and shear stress on every plane with only one principal stress, being same as σ_1, acting over the principal plane which is the given plane itself. Multi-axial load is identified with the existence of more than one principal stress and principal plane. In addition, the principal stresses are non-proportional and their magnitude and directions change during the load cycle. Fatigue strength and ductility are found to decrease in multi-axial fatigue. Multi-axial fatigue assessment has been carried out by methods that reduce the complex multi-axial loading to an equivalent uni-axial type. The field of multi-axial fatigue theories can be classified into five categories i) Empirical formulae resulting after modification of Coffin-Manson Equation ii) Use of stress or strain invariants iii) Use of space averages of stress or strain iv) Critical plane approach and v) Use of accumulated energy. These approaches are elucidated in the succeeding section with the understanding of postulation by Dietmann et al. [26] that in order to determine the resulting fatigue strength of the metal under complex loading , the time dependence of stress wave form, the frequency, the phase difference between stress components and the number of cycles must be considered.

A. *Empirical formulae and modifications of Coffin-Manson equation*: In high cycle fatigue, Lee proposed [27] equivalent stress for a complex load case as, $\sigma_{eq} = \sigma_a\left[1+(b_{fs}C_c/2t_{fs})^{\eta}\right]^{1/\eta}$ where C_c is $2\tau_a/\sigma_a$, $\eta = 2(1+\gamma\sin\phi)$, τ_a is torsional stress amplitude, b_{fs} and t_{fs} are bending and torsional fatigue strength for a given fatigue life respectively, γ is the material constant for consideration of material hardening under out-of-phase loading and ϕ is the phase difference between bending and torsion. σ_{eq} is then used in conventional fatigue life prediction theories stated in Section 1.2.5. Lee and Chiang [28] suggested following equation to consider shear stress that were neglected by other researchers, $\left(\frac{\sigma_a}{b_{fs}}\right)^{b_{fs}(1+\gamma\sin\phi)/t_{fs}} + \left(\frac{\tau_a}{t_{fs}}\right)^{2(1+\gamma\sin\phi)} = \left\{1-\left(\frac{\sigma_m}{\sigma_f'}\right)^{n1}\right\}\left\{1-\left(\frac{\tau_m}{\tau_f'}\right)^{n2}\right\}$ where τ_f' is shear fatigue strength coefficient, τ_m is mean shear stress and n1 and n2 are empirical constants. In low cycle fatigue, Kalluri and Bonacuse [29] suggested the following Coffin-Manson type equation for von-Mises equivalent cyclic strain, $\Delta\varepsilon_{eq} = \left(\frac{\sigma_f'}{E(MF)^{b/c}}\right)(2N_f)^b + \frac{\varepsilon_f'}{MF}(2N_f)^c$ where MF is the multi-axiality factor.

B. *Use of stress or strain invariant*: Sines and Ohgi [30] described fatigue strength with the help of stress invariants in the form, $(J_2')_a^{1/2} \geq A-\alpha_1(J_1)_m-\alpha_2(J_2')_m^n$ where the first stress invariant in terms of principal stresses σ_1, σ_2 and σ_3 is $J_1=\sigma_1+\sigma_2+\sigma_3$ and $J_2' = \frac{1}{6}\left[(\sigma_1-\sigma_2)^2+(\sigma_2-\sigma_3)^2+(\sigma_3-\sigma_1)^2\right]$, $\alpha_2(J_2')_m^n$ reflects non-linearity effect in case of higher mean stress and n, A, α_1 and α_2 are the material constants. Hashin [31] generalized multi-axial fatigue failure criterion as the function $F(\sigma_{m(ij)},\sigma_{a(ij)},\phi_{ij},2N_f)=1$ where ij denotes stress component and ϕ_{ij} is the phase difference between ij stress component and a reference stress component. But the equation is limited to cases of constant load ratio fatigue only.

C. *Use of space averages of stress or strain*: Papadopoulos [32] suggested a generalized failure criterion, $\sqrt{<\tau_m>^2} + \ell\left(\max_t <\sigma_m>\right) \leq \lambda$, where <> indicates average value of the argument acting over critical plane defined by Papadopoulos with ℓ and λ being the material constants.

D. *Use of critical plane*: Fatigue analysis upon using the concept of critical plane is very effective because it is based upon the fracture mode or initiation mechanism of cracks. In the critical plane concept, the maximum shear strain or stress plane is first found. Then a parameter is defined as the combination of maximum shear strain or stress and normal strain or stress on the plane to explain multi-axial fatigue behaviour. In high cycle fatigue, McDiarmid [33] proposed equivalent shear stress amplitude as $\tau_a = t_{fs} - \frac{t_{fs} - b_{fs}}{(b_{fs}/2)^{1.5}}\sigma_a{}^{1.5} - G\sigma_m - 0.081\tau_m$ where G is an empirical constant and stress values are over the maximum shear stress plane. He further suggested [34], $\tau_a = \frac{1}{K_t}\left[t_{fs} - \frac{t_{fs} - b_{fs}}{(b_{fs}/2)^{1.5}}(K_t\sigma_a)^{1.5} - \frac{b_{fs}}{(\sigma_u/2)^2}\sigma_m\right]$ where σ_u is the ultimate tensile strength of metal. In low cycle fatigue, Brown and Miller [35] derived the relation, $\gamma_{max,pl} + \varepsilon_{n,pl} + \sigma_{no}/E = \gamma_f'(2N_f)^v$, where maximum plastic shear strain, $\gamma_{max,pl}$, and normal plastic strain, $\varepsilon_{n,pl}$, are the values of Brown and Millers parameters, σ_{no} is the component of mean stress on γ_{max} plane and v and γ_f' are shear fatigue ductility exponent and coefficient respectively. Socie [36] suggested the following for shear cracking mode, $\gamma_{max} + \varepsilon_n + \sigma_{no}/E = \gamma_f'(2N_f)^v + \frac{\tau_f'}{G}(2N_f)^w$, where w is the shear fatigue strength exponent and for tensile cracking mode, $\sigma_1^{max}\Delta\varepsilon_1/2 = \sigma_f'\varepsilon_f'(2N_f)^{b+c} + (\sigma_f')^2(2N_f)^{2b}/E$. Fatemi and Socie [37] modified Brown and Millers equation as follows:

$$\gamma_{max}\left(1 + n\frac{\sigma_n{}^{max}}{\sigma_y}\right) = (1+\nu_e)\frac{\sigma_f'}{E}(2N_f)^b + \frac{n}{2}(1+\nu_e)\frac{(\sigma_f')^2}{E\sigma_y}(2N_f)^{2b} + (1+\nu_p)\varepsilon_f'(2N_f)^c + \frac{n}{2}(1+\nu_p)\frac{\varepsilon_f'\sigma_f'}{\sigma_y}(2N_f)^{b+c}$$

where n is an empirical constant, ν_e and ν_p are poisson's ratio in elastic and plastic regions respectively.

E. *Use of energy concepts*: All aforementioned stress or strain based criteria don't consider multi-axial stress-strain response of the material. The fatigue process is generally believed to involve cyclic plastic deformations which are dependent on stress-strain path. Thus, stress or strain based criteria alone cannot reflect the path dependence on fatigue process sufficiently. After initial work by Garud [38], Ellyin and Golos [39] proposed that durability of components should be characterized with the quantity of energy which a material can contain and suggested total strain energy density, ΔW_t, as sum of elastic and plastic components as follows: $\Delta W_t = \Delta W_e + \Delta W_p$ where $\Delta W_e = \frac{1+\nu}{3E}(\bar{\sigma}^{max})^2 + \frac{1-2\nu}{6E}(\sigma_{kk}{}^{max})^2$, $\Delta W_p = \frac{2(1-n')(2\sigma_f')^{-1/n'}}{1+n'}(\Delta\bar{\sigma})^{\frac{1+n'}{n'}}$, $\bar{\sigma}$ being the von-Mises equivalent stress. However, these equations do not include the effect of strain ratio. Besides, they don't explain the additional cyclic hardening or softening of material due to interaction between loadings under out of phase multi-axial loading. Such problems were overcome by Ellyin et al. [40] who performed in-phase and out of phase multi-axial fatigue tests by using multi-axial constraint factor (MCF) to supplement the results of Ellyin and Golos.

3.2 Fatigue under aqueous and corrosive environments

Naval structures consistently work under such environments. There are various factors that influence environment assisted fracture of a metal namely its alloy chemistry, heat treatment, atmospheric humidity and temperature, salt concentration in air and work hardening of metal. It is generally agreed that presence of aggressive aqueous environment drastically reduces the fatigue life of metal. Initiation or nucleation life of fatigue crack in a polished specimen may drop to just 10% of total fatigue life in corrosive environment. Exposure to corrosive environment probably has the same effect on fatigue life as machining of sharp notch into the surface. Corrosive environment facilitates development of geometric discontinuities on the surface. These discontinuities or stress risers then become the potential sites for origination of fatigue cracks upon cyclic loading. The mechanisms of initiation and growth of crack in corrosive environment [41] are discussed below (It is important to mention here that no single mechanism can fully explain crack initiation or crack propagation behaviour of metals in corrosive environment):

A. *Crack initiation*: Several mechanisms have been proposed. They are i) Surface film rupture - In this method, nucleation is caused by mechanical damage to the protective surface film. Most metals have a thin oxide layer on their surface. This thin layer, being passive in metals, protects the metal from atmospheric moisture. In addition, both anodic and cathodic corrosion reaction sites exist on the surface of metal when exposed to aqueous solution leading to formation of a protective layer. But when the part is subjected to mechanical stressing, the film is broken resulting in exposure of fresh metal to the environment which may then act as an anodic or cathodic reaction site causing accumulation of foreign corrosive particles which act as discontinuities and potential fatigue sites ii) Pitting - Corrosion pits may be formed on and beneath the surface that assist in nucleation of fatigue cracks. High dissolution rate in the pit could unblock piled up dislocations and allow slip to occur. iii) Strain effects - Strain enhanced dissolution of emerging slip bands promotes nucleation iv) Surface adsorption - Surface energy of metal is lowered by adsorption of specific species from the environment. Reduction in surface energy enhances plasticity and early crack initiation.

B. *Crack propagation*: Two main mechanisms assist in crack growth. They are i) Anodic dissolution - When the metal is exposed to aqueous solution, it loses metallic ions close to the crack tip. The loss of ions leaves extra electrons on the surface thus making the crack tip behave as an anode of the electrolytic cell. Anodic dissolution causes growth of the crack tip within a grain till it reaches the grain boundary thus exposing the grain boundary for hydrogen embrittlement ii) Hydrogen embrittlement - Owing to tensile stresses in the vicinity of crack tip during cyclic load, the hydrogen atoms, being small in size, diffuse into grain boundaries, voids, inclusions and highly strained slip bands. The diffusion is usually slow but makes the material in the vicinity of crack tip brittle and the crack is no longer able to withstand applied ΔK. Consequently, the crack grows up to the point only where the effect of hydrogen embrittlement is diminished and the whole process then repeats itself. There are two ways in which crack grows in corrosive environment. Refer Fig. 19. i) True corrosion fatigue (TCF) in which crack growth rates are enhanced by the presence of aggressive environment through a synergistic action of corrosion and cyclic loading. TCF is characterized by environment induced change on the values for conventional Paris constants C and m. TCF is observed till ΔK is less than ΔK_{iscc} where ΔK_{iscc} represents threshold stress corrosion cracking intensity range which is material property under given environment ii) Stress corrosion fatigue (SCF) starts when ΔK exceeds ΔK_{iscc}. SCF describes static load stress concentration under fatigue conditions. Either TCF or SCF can be induced by changing the frequency and stress ratio. The kinetic processes that take place during crack growth are summarized as i) Supply of reactants and removal of products from the crack tip region ii) Reactions at the crack tip surface iii) Diffusion of atoms ahead of the crack tip iv) Partitioning of ions at various micro-structural sites v) Rupture of protective film during cyclic loading vi) Development of freshly fractured surfaces by fatigue processes and vii) Build up of corrosion products that influence crack closure and effective ΔK.

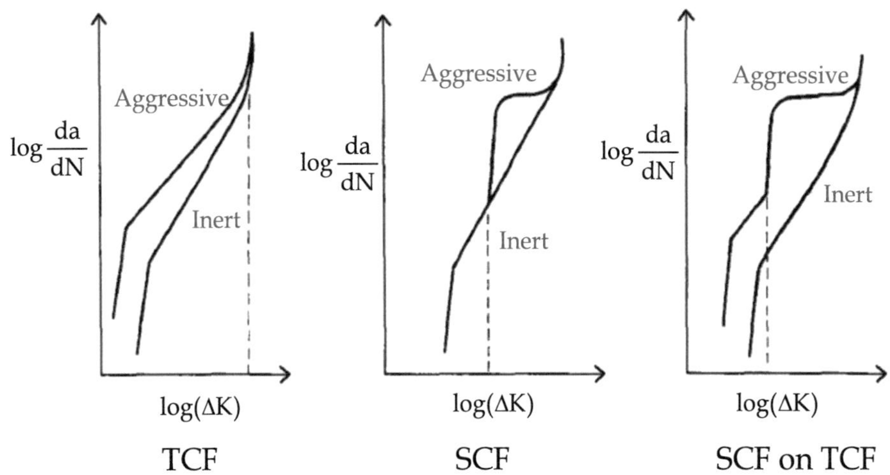

Fig. 19. Types of corrosion fatigue

B.1 *Crack propagation models*: The following models are proposed i) Process superposition model in which general form of crack growth is written by summing crack growth rates for pure mechanical fatigue and corrosion fatigue as follows: $\frac{da}{dN}=\left(\frac{da}{dN}\right)_f(1-\varphi)+\left(\frac{da}{dN}\right)_{scc}+\left(\frac{da}{dN}\right)_c\varphi$ where $\left(\frac{da}{dN}\right)_f$ represents contribution of pure mechanical fatigue (in an inert environment and independent of frequency), $\left(\frac{da}{dN}\right)_{scc}$ is contribution by corrosion at test levels above ΔK_{iscc}, $\left(\frac{da}{dN}\right)_c$ is a cycle dependant contribution that represents synergistic interaction between mechanical fatigue and stress corrosion cracking and φ stands for fractional area of crack undergoing pure corrosion fatigue ii) Process competition model in which the crack growth rate equation is written in style of Paris equation as, $C_1(\Delta K)^{m1}=C(\Delta K+d\Delta K)^m$, where constants C_1 and m_1 are constants of fatigue in corrosive environment and $d\Delta K$ represents corrosion fatigue contribution iii) Process interaction model which accounts for the fact that corrosive products at crack tip may cause closure thereby reducing crack growth rate. But the overall effect of such products is to enhance the crack growth rate. Likewise, the effect of frequency is included in this model. The crack growth rate is written as $\frac{da}{dN} = C(\Delta K)_{eff}^{m} + \int_{0}^{1/f} A\varsigma \Delta K_{eff}\wp dt$ where $(\Delta K)_{eff}$ takes into account the effects of closure, blunting and branching, the term ς includes the influence of load cycle on stress corrosion rate, f being the frequency and A and $\wp$ the constants that are experimentally obtained.

3.3 Fatigue under high temperature

Gas turbines, aircraft engines, nuclear reactors etc. operate in such an environment. The metal is said to be operating in high temperature environment when the temperature level is ≥ 0.5 times the melting point of the metal. New effects like atmospheric oxidation and creep supplement fatigue at high temperature. Metal alloys are usually designed and developed to have good creep strength at high temperatures under static loads. Because alloys with good resistance to creep generally exhibit acceptable fatigue strength up to a certain limiting temperature. But beyond that, the rate of damage in metals increases due to creep dominance coupled with fatigue resistance reduction resulting in shorter lives. The fundamental crack initiation and growth mechanisms explained earlier in Section 2 hold good at high temperature except the location of crack initiation site, subsequent path of crack propagation, crack growth rate and the quantum of damage induced with time. Since creep forms an important aspect at high temperature, corresponding fatigue models are time based. Some aspects are presented as follows:

A. Crack initiation: Most metals when subjected to cyclic load under high temperature show damage in the form of grain-boundary voids and wedge cracks, as shown in Fig. 10, because of high energy levels at the grain boundary. The latter requires grain boundary sliding which results in geometric incompatibility at 'triple points' leading to the stress concentration or crack nucleation site. Therefore, a cavity nucleates whose growth is sustained by stress assisted diffusion or grain boundary sliding or the combination of the two. Rather high tensile stresses are required to maintain the cavity. Other factors such as migration of fatigue generated vacancies, segregation of impurities and diffusion and development of internal gas pressure also contribute to the stability of the cavity.

B. Crack growth: Nucleated cavities grow along grain boundaries that are generally perpendicular to the maximum principal stress direction as shown in Fig. 20. Crack closure is more at high temperature that influences crack growth rates. Therefore, ΔK_{eff} is introduced and Paris law is modified to the form, $\frac{da}{dN} = C\left(\frac{\Delta K_{eff}}{K_o(R)}\right)^{m(R)}$, where parameters K_o and m depend upon the stress ratio, R.

C. *Life prediction methods*: Life prediction in high temperature fatigue can be grouped into five categories namely i) Linear summation of time and cycle [42,43] ii) Modifications of low-temperature fatigue relationships [44] iii) Ductility exhaustion [45] iv) Strain range partitioning [46] and v) Continuous damage parameter [47-49]. For example, the model due to i) is written as $\frac{N}{N_f}+\int\frac{dt}{t_R}=1$ where N is actual life under creep fatigue, N_f is fatigue life without creep and t_R is the creep rupture time.

The damage can be assumed to comprise two functions, namely rate independent (mechanical fatigue), Ω_f, and rate dependant (creep damage) component, Ω_c and is written as $D = D(\Omega_f, \Omega_c)$. The rate of damage accumulation per cycle is then represented as $\frac{dD}{dN}=\frac{\partial D}{\partial \Omega_f}\frac{\partial \Omega_f}{dN}+\frac{\partial D}{\partial \Omega_c}\frac{\partial \Omega_c}{dN}$.

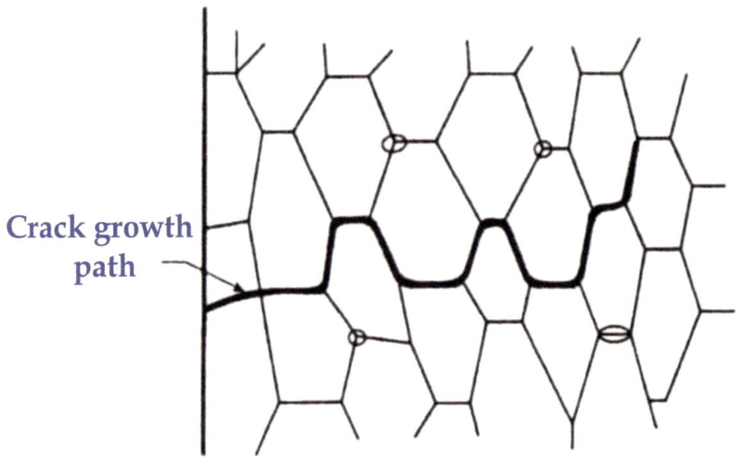

Fig. 20. Crack growth along grain boundaries at high temperature

3.4 Giga cycle fatigue

Giga or ultra high cycle fatigue has assumed importance in the design of high speed components in modern machines, the desired life of some of them ranging from 10^8 to 10^{10} cycles. This requirement is applicable to sectors like aircraft (gas turbine disks-10^{10} cycles), railway (high speed train-10^9 cycles) and automobile (car engine-10^8 cycles). Although large amount of fatigue data has been published in the form of S–N curves, the data are mostly limited to fatigue lives of 10^7 cycles. Time and cost constraints rule out the use of conventional fatigue tests for more than 10^7 cycles. Conventional test conducted at frequency of 20 Hz may require several years for generating 10^{10} cycles to test one specimen. During initial development by Wohler, gigacycle tests didn't assume importance because many industrial applications during that time, such as steam engines etc. , had a small fatigue life in comparison with modern machines. Therefore, gigacycle fatigue is more appropriate for modern technologies. The major challenge in measuring fatigue strength in gigacycle regime lies in generating very high frequencies that save time and also do not produce erroneous results by affecting basic mechanisms of crack initiation and propagation since the structure in normal operation shall have to perform under conventional frequency of smaller magnitudes.

Since the effect of frequency on fatigue life of metal is negligible in normal ambient conditions, piezoelectric based, ultrasonic fatigue machines [50] are used to generate high frequencies of the order of 20 kHz. They are reported to be reliable and capable of producing 10^{10} cycles in less than a week. But, as expected, lot of heat is generated during such a test. The temperature of the specimen is therefore continuously monitored by thermo couples for regular cooling. In some steels, the gap between fatigue strengths corresponding to 10^7 and 10^{10} cycles can reach up to 200 MPa. Therefore gigacycle machines are operated at lower stress value that should be correctly known for failure free performance of such machines.

4. Acknowledgement

Support received from the School of Mechanical and Building Sciences, VIT, Vellore during the course of this work is gratefully acknowledged.

5. References

[1] Wohler, A. (1860). Versuche uber die Festigkeit der Eisenbahnwagenachsen, *Zeitschrift fur Bauwesen*,10; English summary (1867), *Engineering*, Vol.4, pp.160-161.

[2] Campbell Glen, S. (1981). A note on fatal aircraft accidents involving metal fatigue, *International Journal of Fatigue*, Vol.3, pp. 181-185.

[3] Callister William, D. (2003). Failure, In: *Materials Science and Engineering: An Introduction*, pp. 215-217, John Wiley and sons, Inc.

[4] Basquin, O.H. (1910). The exponential law of endurance tests, *Proceedings of ASTM*, Vol. 10(II), pp. 625-630.

[5] Coffin Jr., L.F. (1954). A study of the effects of cyclic thermal stresses on a ductile metal, *Trans. ASME*, Vol. 76, pp. 931-950.

[6] Manson, S.S. (1954). Behaviour of materials under conditions of thermal stress, *NACA TN-2933*, National Advisory Committee for Aeronautics.

[7] Miner, M.A. (1945). Cumulative damage in fatigue, *Journal of Applied Mechanics*, Vol. 67, pp. A159-A164.

[8] Palmgren, A. (1924). Die Lebensdauer von Kugellagern, *Verfahrenstechinik, Berlin*, Vol. 68, pp. 339-341.

[9] Richart, F.E. and Newmark, N.M. (1948). An hypothesis for the determination of cumulative damage in fatigue, *Proceedings of ASTM*, Vol. 48, pp. 767-800.

[10] Marco, S.M. and Starkey, W.L. (1954). A concept of fatigue damage, *Trans. ASME*, Vol. 76, pp. 627-632.

[11] Kommers, J.B. (1945). The effect of overstress in fatigue on the endurance limit of steel, *Proceedings of ASTM*, Vol. 45, pp. 532-541.

[12] Bennett, J.A. (1946). A study of the damaging effect of fatigue stressing on X4130 steel, *Proceedings of ASTM*, Vol. 46, pp. 693-714.

[13] Corten, H.T. and Dolon, T.J. (1956). Cumulative fatigue damage, *Proceedings of the International Conference on Fatigue of Metals,* Institution of Mechanical Engineering and American Society of Mechanical Engineers, pp. 235-246.

[14] Freudenthal, A.M. and Heller, R.A. (1959). On stress interaction in fatigue and a cumulative damage rule, *Journal of the Aerospace Sciences,* Vol. 26, pp. 431-442.

[15] Langer, B.F. (1937). Fatigue failure from stress cycles of varying amplitude, *ASME Journal of Applied Mechanics,* Vol. 59, pp. A160-A162.

[16] Grover, H.J. (1960). An observation concerning the cycle ratio in cumulative damage, *Symposium on Fatigue of Aircraft Structures, ASTM STP 274,* ASTM, Philadelphia, PA, pp.120-124.

[17] Manson, S.S. (1966). Interfaces between fatigue, creep and fracture, *International Journal of Fracture,* Vol. 2, pp. 328-363.

[18] Ellyin, F. (1985). A strategy for periodic inspection based on defect growth, *Theoretical and Applied Fracture Mechanics,* Vol. 4, pp. 83-96.

[19] Taylor, D. (1989). Theories and mechanisms, In: *Fatigue Thresholds,* p. 13. Butterworths.

[20] Morrow, J.D. (1965). Cyclic plastic strain energy and fatigue of metals, In: *Internal Friction, Damping and Cyclic Plasticity, ASTM STP 378,* Philadelphia, PA, pp. 45-84.

[21] Ellyin, F. (1997). Phenomenological approach to fatigue life prediction under uniaxial loading, In: *Fatigue Damage, Crack Growth and Life Prediction,* pp. 88-90, Chapman and Hall.

[22] Laird, C. and Smith, G.C. (1963). Initial stages of damage in high stress fatigue in some pure metals, *Phil. Mag.,* Vol. 8, pp. 1945-1963.

[23] Paris, P.C., Gomez, M.P. and Anderson, W.E. (1961). A rational analytic theory of fatigue, *The Trend in Engineering,* Vol. 13, pp. 9-14, University of Washington, Seatle (WA).

[24] Neuber, H. (1958). Theory of notch stress, *Kerbspannungslehre,* Springer, Berlin.

[25] Peterson, R.E. (1974). *Stress Concentration Factors,* John Wiley, New York.

[26] Dietmann, H., Bhongbhidhat, T. and Schmid, A. (1991). Multi-axial fatigue behaviour of steels under in-phase and out-of-phase loading including different waveforms and frequencies, In: *Fatigue under bi-axial and multi-axial loading,* p. 449, ESIS10, Mechanical Engineering Publications, London.

[27] Lee, S.B. (1980). Evaluation of theories on multi-axial fatigue with discriminating specimens, Ph.D. Thesis, Stanford University, Stanford.

[28] Lee, Y.L. and Chiang, Y.J. (1991). Fatigue predictions for components under bi-axial reversed loading, *Journal of Testing and Evaluation,* Vol. 19, p. 359.

[29] Kalluri, S. and Bonacuse, P.J. (1993). In-phase and out-of-phase axial-torsional fatigue behaviour of Haynes 188 superalloy at 760 deg. C., In: *Advances in multi-axial fatigue,* ASTM STP 1191, p.133.

[30] Sines, G. and Ohgi, G. (1981). Fatigue criteria under combined stresses or strains, *Journal of Engineering Materials and Technology,* Trans. ASME, Vol. 103, p.82.

[31] Hashin, Z. (1981). Fatigue failure criteria for combined cyclic stress, *International Journal of Fracture,* Vol. 17, p. 101.

[32] Papadopoulos, I.V. (1994). A new criterion of fatigue strength for out-of-phase bending and torsion of hard metals, *International Journal of Fatigue,* Vol. 16, p. 377.

[33] McDiarmid, D.L. (1985). The effects of mean stress and stress concentration on fatigue under combined bending and twisting, *Fatigue and Fracture of Engineering Materials and Structures*, Vol. 8, p.1.
[34] McDiarmid, D.L. (1987). Fatigue under out-of-phase bending and torsion, *Fatigue and Fracture of Engineering Materials and Structures*, Vol.9, p. 457.
[35] Brown, M.W. and Miller, K.J. (1982). Two decades of progress in the assessment of multi-axial low cycle fatigue life, In: *Low cycle fatigue and life prediction*, ASTM STP 770, p. 482.
[36] Socie, D. (1987). Multi-axial fatigue damage models, *Journal of Engineering Materials and Technology*, Trans. ASME, Vol. 109, p. 293.
[37] Fatemi, A. and Socie, D.F. (1988). A critical plane approach to multi-axial fatigue damage including out-of-phase loading, *Fatigue and Fracture of Engineering Materials and Structures*, Vol. 11, p. 149.
[38] Garud, Y.S. (1979). A new approach to the evaluation of fatigue under multi-axial loading, *Proceedings of Symposium on Methods for Predicting Material Life in Fatigue*, p. 247, ASME.
[39] Ellyin, F. and Golos, K. (1988). Multi-axial fatigue damage criterion, *Journal of Engineering Materials and Technology*, Trans. ASME, Vol. 110, p. 63.
[40] Ellyin, F., Golos, K. and Xia Z. (1991). In-phase and out-of-phase multi-axial fatigue, *Journal of Engineering Materials and Technology*, Trans. ASME, Vol. 113, p. 112.
[41] Sudarshan, T.S., Srivatsan, T.S. and Harvey II, D.P. (1990). Fatigue processes in metals-Role of aqueous environments, *Engineering Fracture Mechanics*, Vol. 36, pp. 827-852.
[42] Robinson, E.L. (1952). Effect of temperature variation on the long time rupture strength of steels, *Trans. ASME*, Vol. 74, pp. 777-781.
[43] Taira, S. (1962). Lifetime of structures subjected to varying load and temperature, In: *Creep in structures*, Academic Press, pp. 96-124.
[44] Coffin, L.F. (1973). Fatigue at high temperatures, In: *Fatigue at high temperatures*, ASTM STP 520, pp. 5-43.
[45] Polhemus, J.F., Spaeth, C.E. and Vogel, W.H. (1973). Ductility exhaustion model for prediction of thermal fatigue and creep interaction, In: *Fatigue at elevated temperatures*, ASTM STP 520, pp. 625-636
[46] Manson, S.S., Halford, G.R., and Hirschberg, M.H. (1971). Creep-fatigue analysis by strain range partioning, In: *Design for elevated temperature environment*, ASME, pp. 12-24.
[47] Kachanov, L.M. (1999). Rupture time under creep conditions, *International journal of fracture*, Vol. 97, pp. 1-4.
[48] Majumdar, S. and Maiya, P.S. (1980). A mechanistic model for time-dependent fatigue, *Journal of Engineering Materials and Technology*, Trans. ASME, Vol. 102, pp. 159-167.
[49] Majumdar, S. (1964). Relationships of creep, creep-fatigue and cavitation damage in Type 304 austenitic stainless steels, *Journal of Engineering Materials and Technology*, Trans. ASME, Vol. 111, pp. 123-131.

[50] Marines, I., Bin, X. and Bathias, C. (2003). An understanding of very high cycle fatigue of metals, *International Journal of Fatigue*, Vol. 25, pp. 1101-1107.

10

A New Systemic Study Regarding the Behaviour of Some Alloy Steels During Low Cycles Fatigue Process

Macuta Silviu
Dunarea de Jos University of Galati
Romania

1. Introduction

The research of the metallic materials used in machine manufacturing to which high stress and a small number of cycles is applied have been increasingly gained interest in the last 40 years; this is because during cyclical stress at critical points in terms of resistance repeated strain occurs in many major constructions.

Fatigue breaking to a small number of cycles and high strains is encountered in the operation of various types of machines: power machine, elements of heating boilers or heat exchangers in electro nuclear industry, pressure vessels, steam and hydro turbines, turbo compressors, the plane landing trains and other transport means and mechanisms.

The process of cracking and breaking by fatigue is characterized by different mechanisms at different levels of strain. Thus Wöhler's curve can be studied in 3 areas:

The first area called the quasi-static field, where $N=0\div10^4$ cycles, when there are large plastic strains;

The second area called the area of limited durability, where $N=10^4\div10\cdot10^6$ cycles, breaking in this situation implies elastoplastic strain and calculations will be carried out to limited durability.

The third area - the area of unlimited durability, or the fatigue resistance range, when $N>10\cdot10^6$ cycles. Breaking is characterized by elastic strain only.

Regarding the first area, a more careful research led to the conclusion that there are two distinct zones:

The quasi-static area itself, for $N<10^3$ cycles - area where high strains close to the material yielding point occur, where breaking is characterized by large plastic strain, similar to static fracture.

The low-cycle fatigue zone also called the *low cycle fatigue* where $N=5\cdot10^2\div10^4$ cycles, where this time breaking is characterized by elastoplastic strain. A deeper insight into this area has revealed the existence of anomalies (discontinuities) of the fatigue curve. The problem was first studied by R. Moore in 1923, then by Sabolin, Finney and Mann .

The phenomenon of fatigue to a small number of cycles has three specific features.

1. High level of strains
2. Low testing time (10^3-10^4, max. 10^5 cycles)
3. Reduced testing frequency(up to50 cycles/min)

Investigating the low cycle fatigue damage, it can be distinguished between a rigid and a soft loading regime.

The studies carried out have a general character, the fatigue behaviour being determined from the fatigue curves, the curves of mechanical hysteresis and the cyclic cold-hardening curves. Based on the investigations, a set of rules and criteria to predict the material behaviour to *low cycle* strains has been established.

2. A new systemic approach of the fatigued surface layer behavior

The concept of a *structural cybernetic pattern* has been introduced in order to obtain an as complete as possible approach of the surface layer behavior and also with a view to analyzing and emphasizing the main factors and parameters which determine the fatigue process. This concept, introduced by professor I. Crudu from Dunarea de Jos University of Galati-Romania in order to characterize a tribosystem, was extended to the characterization of the fatigued surface layer and it represents a totally new approach of the fatigue process.

In this chapter a new research methodology has been developed along with a way to approach the fatigue behaviour of the surface layer to high strains and small number of cycles by extending the concept of structural cybernetic tribo-system.

For the research purpose, special equipment was needed consisting of a patented Universal Machine for fatigue testings and related facilities. The development of certain surface layer parameters was monitored (1st and 2nd order strains, dislocation density, crystalline lattice parameter, texture, microstructure, and micro-hardness of the surface layer) during the fatigue process according to the control parameter (number cycles, strain, frequency, type of material required).

In order to have an as extensively as possible approach to the surface layer behavior and to track and highlight the main factors and parameters determining the process of destruction by fatigue, the concept of *structural cybernetic model* was introduced. This concept was introduced to characterize a tribosystem and allowed its extension to the characterization of the surface layer of the material subject to fatigue, which is an entirely new approach to fatigue processes.

Fig. 1 shows a structural cybernetic model by means of which the changes of the input parameters can be systematically monitored by measuring the output parameters of the surface layer undergoing the action of destruction by fatigue.

The input-output parameters of the cybernetic model include the parameters of the surface layer ((S_s-S'_s) and the control parameters (U). Surface layer parameters can be grouped into:

geometric parameters (macro-geometry and micro-geometry- X_1- X'_1)

mechanical parameters (hardness and micro hardness – X_2 -X'_2 and the strain state – X_3- X'_3);

physical and metallurgical parameters (chemical composition – X_4- X'_4, structure – X_5- X'_5, purity – X_6- X'_6)..

Some of these parameters, such as micro hardness - X_2, strain - and structure X_3 - X_5, may be modified from outside so that the life time of a material under fatigue process can be modified within certain limits, as desired.

The control parameters (U) also called external factors are those parameters which by their action on the functioning of the fatigue testing machine may modify some parameters of the surface layer of the material the test-piece is made of. The control parameters can be grouped into:

Constructive parameters (nature of the material - U_1, test-piece shape - U_2, the test-piece dimensions - U_3);

Operating parameters (working environment - U4, cinematic - U5 and energy parameters - U6).

Under the experimental program, out of all surface layer parameters, the evolution with respect to the initial state (input parameters) of the following parameters (output parameters) has been investigated: mechanical parameters (micro hardness - X2; state of strain - X3 namely: 1st order strain (σ_I) - X_3^1,) -2nd order strain (σ_{II}) - X_3^2; -,3rd order strain ((ρ) - X_3^3, - and from the physical metallurgical parameters, the structure changes were monitored - X5 (network parameters, the texture).

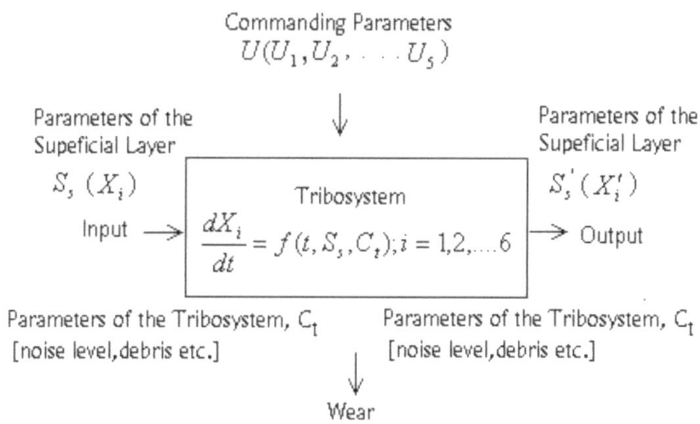

Fig. 1. A cybernetic model used in study of friction process adapted in study of the low cycle fatigue process.

From among the control parameters (U), for the purpose of the experiment, the constructive parameter was acted upon by the nature of the material U1, using two grades of steel OL52 and 10TiNiCr180, both shape and dimensions of the test pieces remaining unchanged throughout the experiment. From among the operating parameters, it was maintained the same working environment - air at ambient temperature (U4), acting upon the cinematic parameters U_5, namely the testing frequency ν_1= 20 cycles / min or ν_2= 40 cycles / min

It was also acted upon the energy parameters (U6),in an attempt to investigate the evolution of the surface layer parameters by changing the imposed strains (ε_1, ε_2, ε_3)) and the number of strain cycles (N1, N2, N3, N4, N5) under an experimental program.

Knowing at any time the parameters of the surface layer, this is one of the safest procedures in assessing and forecasting the degree of degradation of metallic materials under fatigue processes .

Determination of the structural changes in the surface layers may allow for the optimization of the metal components manufacturing technology. In practice the control of the surface layer parameters often requires the use of physical methods of investigation which do not affect its structure and physical-chemical condition.

3. Experimental researches

This chapter presents only the experimental research carried out for the steel OL52. The experimental research was performed in steps from 2000 to 2000 cycles up to 10^4 cycles.

At each step of the number of cycles (2000 cycles) investigations were carried out on the evolution of the crystalline network parameter, trap, density, texture, analysis of the micro hardness and the evolution of the layer micro-structure. For all these investigations use was made of a diffractometer of X radiation, DRON-3, micro hardness meter PMT3, and optical microscope Olympus BX60M of Japanese construction.

Mention must be made that there has not been made a systematic research of the surface layer for the following reasons: wide variety of materials, wide range of physical, chemical, mechanical and metallurgical factors influencing on the surface layer, various deficient physical methods of non-destructive control. The relatively long time taken for some of the analyzing methods as compared to the stress relaxation period/time of certain structure modifications, gets in the way of finding a common point of view on and general methods for analyzing and controlling the different processes occurred in the surface layer.

The experimental program consists of:

Experiment 1 - carbon steel sample OL52 under stress at frequency ν_1=20 cycles/min;

Experiment 2 - carbon steel sample OL52 under stress at frequency ν_2=40 cycles/min;

Experiment 3 - alloyed steel sample 10TiNiCr180 under stress at frequency ν_1=20 cycles/min;

Experiment 4 -alloyed steel sample 10TiNiCr180 under stress at frequency ν_2=40 cycles/min;

All the 4 experiments were focused on the modifications of the surface layer parameters depending on the number of cycles N_1, N_2, N_3, N_4, N_5 and the prescribed deformations ε_1, ε_2, ε_3.

Tables 1, 2, 3, 4 illustrates the experimental program, highlighting all the parameters involved in the experiment.

TEST PROGRAM (Tables 1,2)

OAL_{ji}, $1AL_{ji}$ - A(OL52) steel samples at two stress frequencies $\sigma_{I\ i,j,k}$ – Ist order stress $a_{\ i,j,k}$ - lattice parameter ε_j – induced deformations, (j=1...3). $\sigma_{II\ i,j,k}$ – IInd order stress $texture_{i,j,k}$ N_k – number of stress cycles, (k=1...5).$\rho_{i,j,k}$ – displacement density, $HV_{i,j,k}$ – micro-hardness

TEST 1					
OL52 steel - frequency ν_1=20 cycles/min					
OAL_{11}	ε_1 [μm/m] N_1 N_2 N_3 N_4 N_5 $\sigma_{I\,i,j,k}$; $\sigma_{II\,i,j,k}$; $\rho_{i,j,k}$; $a_{i,j,k}$; texture, micro-hardness $HV_{i,j,k}$	OAL_{21}	ε_2 [μm/m] N_1 N_2 N_3 N_4 N_5 $\sigma_{I\,i,j,k}$; $\sigma_{II\,i,j,k}$; $\rho_{i,j,k}$; $a_{i,j,k}$; texture, micro-hardness $HV_{i,j,k}$	OAL_{31}	ε_3 [μm/m] N_1 N_2 N_3 N_4 N_5 $\sigma_{I\,i,j,k}$; $\sigma_{II\,i,j,k}$; $\rho_{i,j,k}$; $a_{i,j,k}$; texture, micro-hardness $HV_{i,j,k}$
OAL_{12}	ε_1 [μm/m] N_1 N_2 N_3 N_4 N_5 surface layer properties $\sigma_{I\,i,j,k}$; $\sigma_{II\,i,j,k}$; $\rho_{i,j,k}$; $a_{i,j,k}$; texture, micro-hardness $HV_{i,j,k}$	OAL_{22}	ε_2 [μm/m] N_1 N_2 N_3 N_4 N_5 surface layer properties $\sigma_{I\,i,j,k}$; $\sigma_{II\,i,j,k}$; $\rho_{i,j,k}$; $a_{i,j,k}$; texture, micro-hardness $HV_{i,j,k}$	OAL_{32}	ε_3 [μm/m] N_1 N_2 N_3 N_4 N_5 surface layer properties $\sigma_{I\,i,j,k}$; $\sigma_{II\,i,j,k}$; $\rho_{i,j,k}$; $a_{i,j,k}$; texture, micro-hardness $HV_{i,j,k}$
OAL_{13}	ε_1 [μm/m] N_1 N_2 N_3 N_4 N_5 $\sigma_{I\,i,j,k}$; $\sigma_{II\,i,j,k}$; $\rho_{i,j,k}$; $a_{i,j,k}$; texture, micro-hardness $HV_{i,j,k}$	OAL_{23}	ε_2 [μm/m] N_1 N_2 N_3 N_4 N_5 $\sigma_{I\,i,j,k}$; $\sigma_{II\,i,j,k}$; $\rho_{i,j,k}$; $a_{i,j,k}$; texture, micro-hardness $HV_{i,j,k}$	OAL_{33}	ε_3 [μm/m] N_1 N_2 N_3 N_4 N_5 $\sigma_{I\,i,j,k}$; $\sigma_{II\,i,j,k}$; $\rho_{i,j,k}$; $a_{i,j,k}$; texture, micro-hardness $HV_{i,j,k}$

Table 1. Test program for OL52 steel - frequency ν_1=20 cycles/min

TEST 2					
OL52 steel - frequency ν_1=40 cycles/min					
$1AL\ 11$	ε_1 [μm/m] N_1 N_2 N_3 N_4 N_5 $\sigma_{I\,i,j,k}$; $\sigma_{II\,i,j,k}$; $\rho_{i,j,k}$; $a_{i,j,k}$; texture, micro-hardness $HV_{i,j,k}$	AL_{21}	ε_2 [μm/m] N_1 N_2 N_3 N_4 N_5 $\sigma_{I\,i,j,k}$; $\sigma_{II\,i,j,k}$; $\rho_{i,j,k}$; $a_{i,j,k}$; texture, micro-hardness $HV_{i,j,k}$	$1AL_{31}$	ε_3 [μm/m] N_1 N_2 N_3 N_4 N_5 $\sigma_{I\,i,j,k}$; $\sigma_{II\,i,j,k}$; $\rho_{i,j,k}$; $a_{i,j,k}$; texture, micro-hardness $HV_{i,j,k}$
$1AL_{12}$	ε_1 [μm/m] N_1 N_2 N_3 N_4 N_5 $\sigma_{I\,i,j,k}$; $\sigma_{II\,i,j,k}$; $\rho_{i,j,k}$; $a_{i,j,k}$; texture, micro-hardness $HV_{i,j,k}$	$1AL_{22}$	ε_2 [μm/m] N_1 N_2 N_3 N_4 N_5 $\sigma_{I\,i,j,k}$; $\sigma_{II\,i,j,k}$; $\rho_{i,j,k}$; $a_{i,j,k}$; texture, micro-hardness $HV_{i,j,k}$	$1AL_{32}$	ε_3 [μm/m] N_1 N_2 N_3 N_4 N_5 $\sigma_{I\,i,j,k}$; $\sigma_{II\,i,j,k}$; $\rho_{i,j,k}$; $a_{i,j,k}$; texture, micro-hardness $HV_{i,j,k}$
$1AL_{13}$	ε_1 [μm/m] N_1 N_2 N_3 N_4 N_5 $\sigma_{I\,i,j,k}$; $\sigma_{II\,i,j,k}$; $\rho_{i,j,k}$; $a_{i,j,k}$; texture, micro-hardness $HV_{i,j,k}$	$1AL_{23}$	ε_2 [μm/m] N_1 N_2 N_3 N_4 N_5 $\sigma_{I\,i,j,k}$; $\sigma_{II\,i,j,k}$; $\rho_{i,j,k}$; $a_{i,j,k}$; texture, micro-hardness $HV_{i,j,k}$	$1AL_{33}$	ε_3 [μm/m] N_1 N_2 N_3 N_4 N_5 $\sigma_{I\,i,j,k}$; $\sigma_{II\,i,j,k}$; $\rho_{i,j,k}$; $a_{i,j,k}$; texture, micro-hardness $HV_{i,j,k}$

Table 2. Test program for OL52 steel - frequency ν1=40 cycles/min

TEST 3					
10TiNiCr180 alloy steel - frequency ν_1=20 cycles/min					
OBL_{11}	ε_1 [μm/m] N_1 N_2 N_3 N_4 N_5 $\sigma_{I\,i,j,k}$; $\sigma_{II\,i,j,k}$; $\rho_{i,j,k}$; $a_{i,j,k}$; texture, micro-hardness $HV_{i,j,k}$	OBL_{21}	ε_2 [μm/m] N_1 N_2 N_3 N_4 N_5 $\sigma_{I\,i,j,k}$; $\sigma_{II\,i,j,k}$; $\rho_{i,j,k}$; $a_{i,j,k}$; texture, micro-hardness $HV_{i,j,k}$	OBL_{31}	ε_3 [μm/m] N_1 N_2 N_3 N_4 N_5 $\sigma_{I\,i,j,k}$; $\sigma_{II\,i,j,k}$; $\rho_{i,j,k}$; $a_{i,j,k}$; texture, micro-hardness $HV_{i,j,k}$
OBL_{12}	ε_1 [μm/m] N_1 N_2 N_3 N_4 N_5 surface layer properties $\sigma_{I\,i,j,k}$; $\sigma_{II\,i,j,k}$; $\rho_{i,j,k}$; $a_{i,j,k}$; texture, micro-hardness $HV_{i,j,k}$	OBL_{22}	ε_2 [μm/m] N_1 N_2 N_3 N_4 N_5 surface layer properties $\sigma_{I\,i,j,k}$; $\sigma_{II\,i,j,k}$; $\rho_{i,j,k}$; $a_{i,j,k}$; texture, micro-hardness $HV_{i,j,k}$	OBL_{32}	ε_3 [μm/m] N_1 N_2 N_3 N_4 N_5 surface layer properties $\sigma_{I\,i,j,k}$; $\sigma_{II\,i,j,k}$; $\rho_{i,j,k}$; $a_{i,j,k}$; texture, micro-hardness $HV_{i,j,k}$
OBL_{13}	ε_1 [μm/m] N_1 N_2 N_3 N_4 N_5 $\sigma_{I\,i,j,k}$; $\sigma_{II\,i,j,k}$; $\rho_{i,j,k}$; $a_{i,j,k}$; texture, micro-hardness $HV_{i,j,k}$	OBL_{23}	ε_2 [μm/m] N_1 N_2 N_3 N_4 N_5 $\sigma_{I\,i,j,k}$; $\sigma_{II\,i,j,k}$; $\rho_{i,j,k}$; $a_{i,j,k}$; texture, micro-hardness $HV_{i,j,k}$	OBL_{33}	ε_3 [μm/m] N_1 N_2 N_3 N_4 N_5 $\sigma_{I\,i,j,k}$; $\sigma_{II\,i,j,k}$; $\rho_{i,j,k}$; $a_{i,j,k}$; texture, micro-hardness $HV_{i,j,k}$

Table 3. Test program for 10TiNiCr180 alloy steel - frequency ν_1=20 cycles/min

TEST 4					
10TiNiCr180 alloy steel - frequency ν_2=40 cycles/min					
$1BL_{11}$	ε_1 [μm/m] N_1 N_2 N_3 N_4 N_5 $\sigma_{I\,i,j,k}$; $\sigma_{II\,i,j,k}$; $\rho_{i,j,k}$; $a_{i,j,k}$; texture, micro-hardness $HV_{i,j,k}$	$1BL_{21}$	ε_2 [μm/m] N_1 N_2 N_3 N_4 N_5 $\sigma_{I\,i,j,k}$; $\sigma_{II\,i,j,k}$; $\rho_{i,j,k}$; $a_{i,j,k}$; texture, micro-hardness $HV_{i,j,k}$	$1BL_{31}$	ε_3 [μm/m] N_1 N_2 N_3 N_4 N_5 $\sigma_{I\,i,j,k}$; $\sigma_{II\,i,j,k}$; $\rho_{i,j,k}$; $a_{i,j,k}$; texture, micro-hardness $HV_{i,j,k}$
$1BL_{12}$	ε_1 [μm/m] N_1 N_2 N_3 N_4 N_5 $\sigma_{I\,i,j,k}$; $\sigma_{II\,i,j,k}$; $\rho_{i,j,k}$; $a_{i,j,k}$; texture, micro-hardness $HV_{i,j,k}$	$1BL_{22}$	ε_2 [μm/m] N_1 N_2 N_3 N_4 N_5 $\sigma_{I\,i,j,k}$; $\sigma_{II\,i,j,k}$; $\rho_{i,j,k}$; $a_{i,j,k}$; texture, micro-hardness $HV_{i,j,k}$	$1BL_{32}$	ε_3 [μm/m] N_1 N_2 N_3 N_4 N_5 $\sigma_{I\,i,j,k}$; $\sigma_{II\,i,j,k}$; $\rho_{i,j,k}$; $a_{i,j,k}$; texture, micro-hardness $HV_{i,j,k}$
$1BL_{13}$	ε_1 [μm/m] N_1 N_2 N_3 N_4 N_5 $\sigma_{I\,i,j,k}$; $\sigma_{II\,i,j,k}$; $\rho_{i,j,k}$; $a_{i,j,k}$; texture, micro-hardness $HV_{i,j,k}$	$1BL_{23}$	ε_2 [μm/m] N_1 N_2 N_3 N_4 N_5 $\sigma_{I\,i,j,k}$; $\sigma_{II\,i,j,k}$; $\rho_{i,j,k}$; $a_{i,j,k}$; texture, micro-hardness $HV_{i,j,k}$	$1BL_{33}$	ε_3 [μm/m] N_1 N_2 N_3 N_4 N_5 $\sigma_{I\,i,j,k}$; $\sigma_{II\,i,j,k}$; $\rho_{i,j,k}$; $a_{i,j,k}$; texture, micro-hardness $HV_{i,j,k}$

Table 4. Test program for 10TiNiCr180 alloy steel - frequency ν_2=40 cycles/min

TEST PROGRAM

$0BL_{ji}$, $1BL_{ji}$ - B (10TiNiCr180) steel samples at two stress frequencies $\sigma_{I\,i,j,k}$ – Ist order stress a $_{i,j,k}$ - lattice parameter ε_j – induced deformations, (j=1...3).$\sigma_{II\,i,j,k}$ – IInd order stress texture$_{i,j,k}$

N_k – number of stress cycles, (k=1...5).$\rho_{i,j,k}$ – displacement density $HV_{i,j,k}$ – micro-hardness

3.1 Sample preparation

The samples have been prepared metallographically according to the standards in force. The metallographic analysis has been made on samples from the two steel grades investigated. Samples have been taken longitudinally and investigated at a size of (x100). For purity purpose samples have been prepared and analyzed acc to [121], and for microstructure and grain size acc. to STAS 7626-79, STAS 5490-80 and SR ISO 643-93 [120].

The attack for the sample made from OL 52 was achieved by means of the natal reactant , 2% and for the sample made from TiNiCr180 with nitrogen acid reactant, 50% under electrolytic attack regime. Results are given in table 5 and 6.

Material	Purity STAS 5949-80	Microstructures STAS 7626-79 and SR ISO 643-93	Figures
OL52	Silicates + punctilious oxides score>5	Ferrite + perlite Grain size = 9 Ratio Pe/Fe=15/85	2, 3
10TiNiCr180	Titanium nitrure score=2,5	Austenitic structure with **maclați** grains and chrome carbides distributed in rows; M:G:=4-5	4 5

Table 5. Initial result for samples

In order to closely watch any microstructure modifications with the samples coded OAL_{ji}, $1AL_{ji}$, OBL_{ji} and $1BL_{ji}$, surface micro-photos were taken at size (x200, x500, x1000) directly on the samples subject to strain.

Due to the big sample sizes, vs an optimum size of a metallographic slif , it was rather difficult to prepare the surface being investigated .

The samples not chemically attacked by reactants were photographed at size (x100), and those attacked at (x200, x500, x1000).

Since the samples made from alloyed steel (austenitic stainless steel strongly anti-corrosive) are of bigger size, they have been immersed into the reactant for 30 min. The chemical composition of the reactant was: nitrogen acid (1,4) 5 ml, fluoride acid 1 ml, distilled water 44 ml, and for the carbon steel , the reactant was inital 2%.

In order to adjust the machine to 3 prescribed deformations, acc to another experimental program, captor-samples were prepared.

sample	Structure	Notifications
Şlif 1 Carbon steel	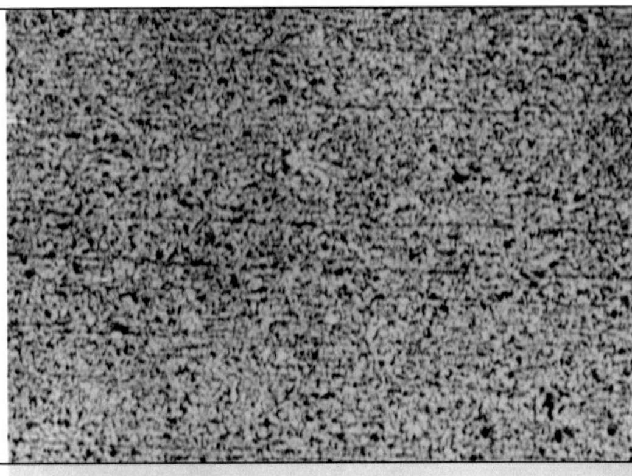	Fig. 2. Ferrite-perlitic microstructure OL52x100 grain size = 9 ratio Pe/Fe=15/85 natal attack 2%
Şlif 2 oțel carbon	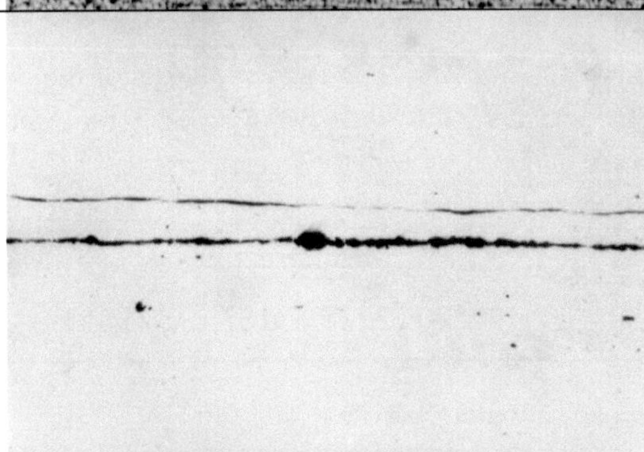	Fig. 3. OL52x100 purity – nonmetallic inclusions fragile silicates - punctilios oxides- score > 5
Şlif 3 Alloyed steel	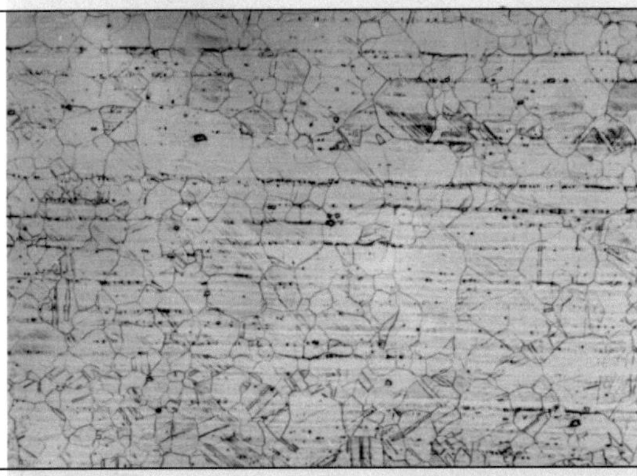	Fig. 4. Austenitic microstructure 10TiNiCr180x100
Şlif 4 Alloyed steel		Fig. 5. 10TiNiCr180x100

3.2 Determination of the prescribed deformations

In order to determine the upper limit of the conventional elastic range, and to assess the prescribed deformations used for the experiment purpose an experimental program was designed consisting of :

1. preparation of the captor-samples
2. analysis of the captor -samples operation
3. partial plotting of the characteristic curves for the two materials specifying the prescribed deformations
4. experimental determination of the longitudinal elastic module and max strains.

Preparing the captor-samples

The samples were marked to facilitate identification:

the samples from material A (OL52) was marked 2.1, 2.2, 2.3 acc to the three prescribed deformations to be used in the experiment, and those from material B (10TiNiCr180) marked 1.1, 1.2,1.3. In the central measuring zone, on both sides a mechanical processing of the surfaces with abrasive paper to achieve the desired roughness : 1.5 ... 2 μm.

After this processing the cross sections were measured and the results given in Table 6 (fig 6).

sample Dimensions [mm]	1.1	1.2	1.3	2.1	2.2	2.3
b	10.16	10.17	10.17	10.13	9.35	10.05
h	4.78	4.85	4.60	4.25	4.35	4.60

Table 6. Sample dimensions

The processed surfaces were marked (the middle of the surface was plotted on both directions) and subsequently chemically prepared .The chemical preparation involves degreasing with **dicloretan** and carbon **tetraclorură**, with a number of flushes in **isopropilic** alcohol. The adhesive used to stick the marks was "Z70" - Hottinger, a fast-hardening cyanoacrylate. For an even as possible hardness over the entire surface of the mark (without polimerisation poles) a neutralizing solution "NZ70" Hottinger was resorted to

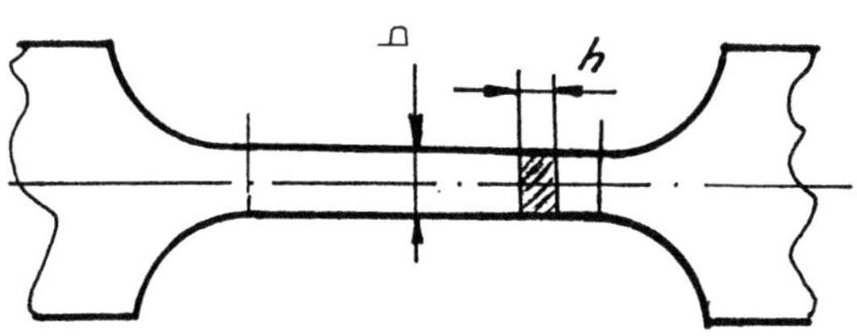

Fig. 6. Captor sample

The tensometric marks were of the type: 3P/120LY11 made by Hottinger with constant k=2.04 ±1%. The complex mark-adhesive and coating was chosen depending on the degree of compliance with the rule of the dissipated power over tensometric mark. The dissipated power on the grill surface is calculated

Two grades of rolled steels mainly used in pressure boilers and vessels industry, namely, OL52k and austenite alloyed steel 10TiNiCr180, were considered. Samples made from the above mentioned steels were tested on the universal machine at 50 tf hydraulically-driven pull at the department of material strength, the Faculty of Mechanical Engineering Galati.

On the same machine pure bending testings were conducted by making use of a construction illustrated in figure 7.

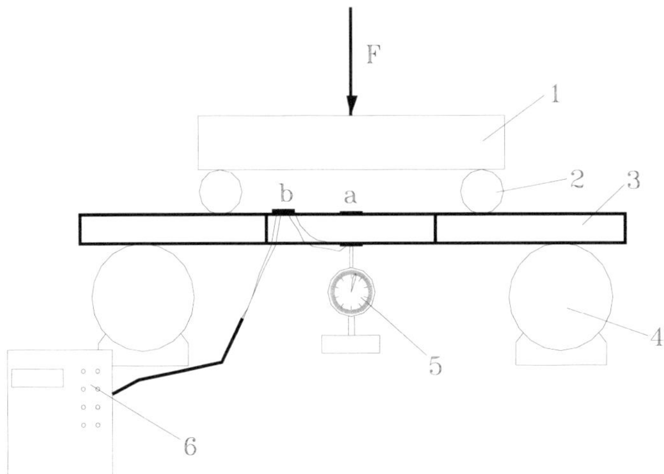

Fig. 7. The construction used for pure bending testings, were: 1 - plate, 2 - roller, 3 - sample, 4 - base roller, 5 - watch, 6 - digital tensometric amplifier, a - tensometric mark, b - connector.

The tensometric marks used are of the type 3P/120LY11 manufactured by Hottinger, constant k=2.04±1%. The two tensometric marks were connected in half-bridge and the connector was tied to the tensometric amplifier by a 6-thread cable. The arrow **f** was easily measured in the section a by means of a comparator (5) at all testing stages.

Using the relation Mohr-Maxwell procedure Veresceaghin, the arrow expression becomes:

$$f_A = \frac{Fl_0}{2EI_y}\left[\frac{l_0^2}{3} + \frac{l_1}{2}(2l_0 + l_1)\right] \tag{1}$$

Relation (1) indicates the proportionality between the force **A** and the applied force **F**, or between **fA** (the arrow) and the bending moment $M = \frac{Fl_0}{2}$. Since the beam is subject to pure bending at the middles zone, the tensions σ are proportional to the bending moment (Navier relation).

The proportionality zone on the characteristics curve is highlighted by plotting an arrow curve (f) depending on the specific deformation (ε). From the experimental results, the diagram of the austenite alloyed steel 10TiNiCr180 was plotted in Figure 8.

At the Tensometry Laboratory of ICEPRONAV bending testing were conducted at variable moment. Each sample was installed on a device to the diagram in Figure 9. The samples were built in at one end while forces of known values were attached to the other end, gradually.

The point for forces application is 100 mm from the central reference of the tensometric mark according to the diagram. The applied forces were obtained with calibrated weigts (order 4) of 0.5 and 1 kgf respectively, which allow for calibrations higher than the accuracy class 0.5. After the experiments and calculations performed the curve in Figure 10 was plotted.

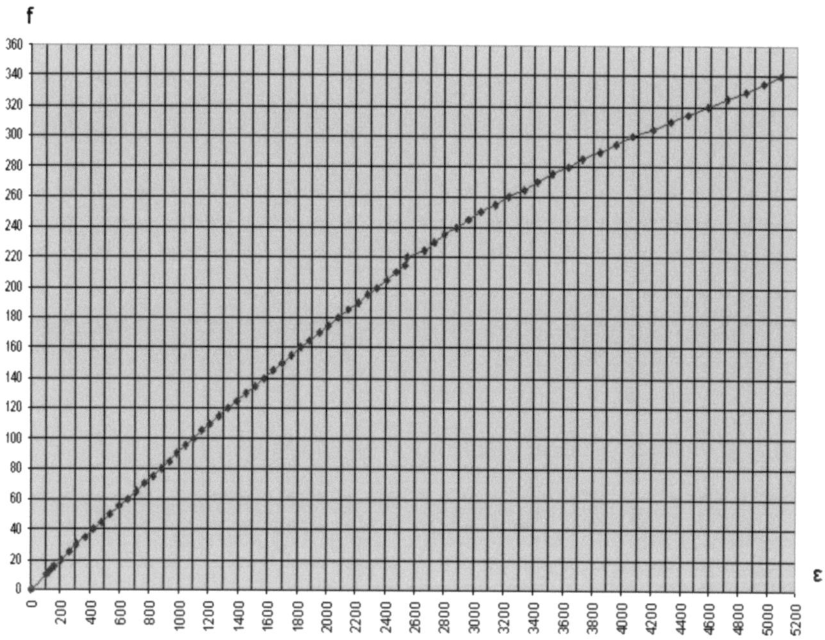

Fig. 8. Pure bending diagram for 10TiNiCr180

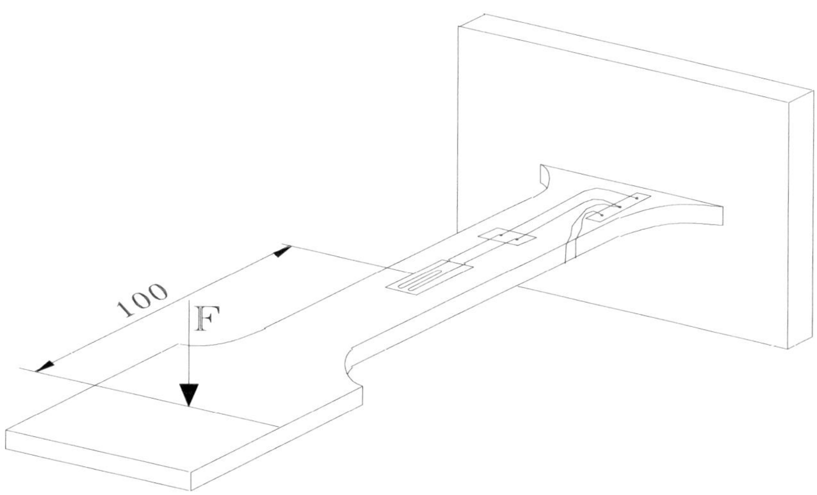

Fig. 9. The sample installed on a device.

It should be underlined that the specific deformations in Figures 9 and 10 are those read on the digital tensometric amplifier N2313; the tensometric marks being connected in half-bridge , $\varepsilon_{real} = \varepsilon_{citit}/2$.

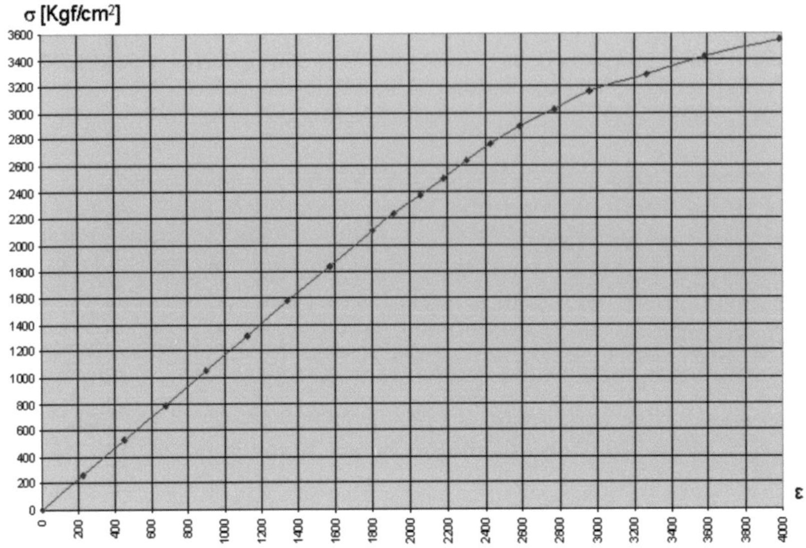

Fig. 10. Bending at variable moment diagram for 10TiNiCr180

From both the experiments and the characterictic curves, three deformations were obtained for the alloyed steel in the transition from the elastic-plastic domain.

The same methodology was used for the carbon steel OL52 and other three deformations were obtained for the above transition domain (according to table 7).

Steel type	Imposed deformations		
	ε_1 [μm/m]	ε_2 [μm/m]	ε_3 [μm/m]
10TiNiCr180	1500	2000	2500
OL52K	2000	2500	3500

Table 7. Imposed deformations in elasto-plastic area.

The values in the table are those recorded while the real ones are half due to the half-bridge arrangement

3.3 The evolution of certain parameters in the surface layer during low cycle fatigue proces

In this chapter the evolution of certain parameters in the surface layer during low cycle fatigue process are presented: evolution of lattice parameter, evolution network parameter, evolution of texture level, variations of microhardness, evolution of microstructures in the surface layer only for ol52 samples tested to pure bending fatigue.

3.3.1 The internal 3rd order strain. The trap density

Figure 11 presents the dependence of $\left(I_f / I_{\max}\right)$ on the number of strain cycles for 3 imposed strains ε_1, ε_2, ε_3 at frequency ν_1= 20 cycles / min for steel OL52.

In general, it is found out the existence of a process of decrease in the trap density in case of small strains (ε_1, ε_2) relative to the original state (level 1 and level 2) and an increase in the trap density (ρ) in case of higher strains ε_3. With increased strain ,a lower degree of deformation of the crystalline lattice around atoms or groups of atoms (dislocation density) is visible.

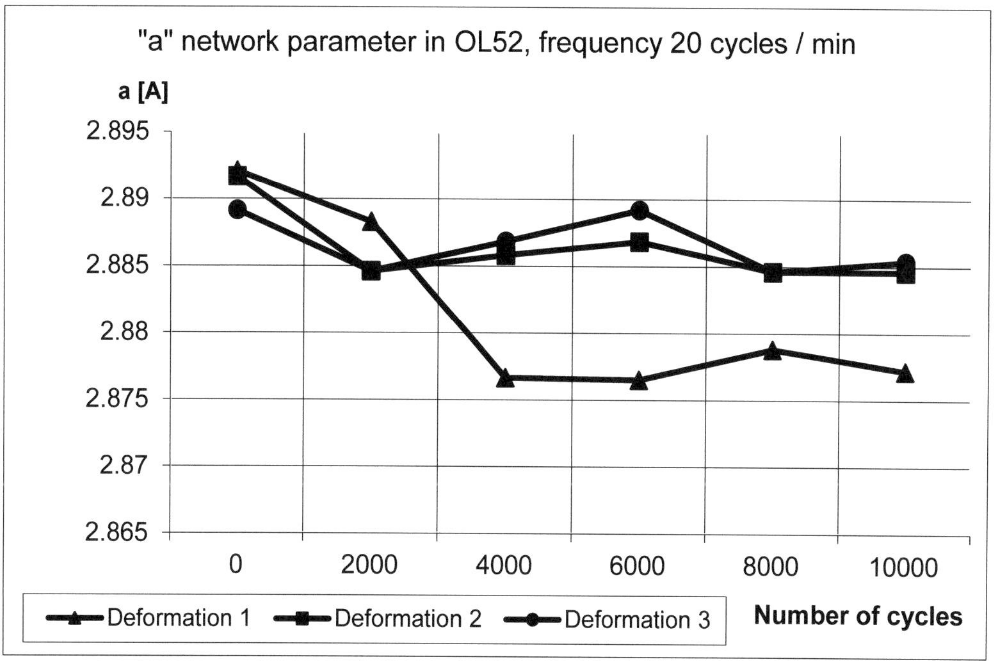

Fig. 11. Evolution of lattice parameter for f = 20 cycles/min

For the average strain ε_2=2500 μm/m it is found a tendency to increase the trap density with the number of strain cycles, said increase taking place in jumps. This increase in the trap density can lead to their agglomeration and subsequently to the generation of micro-cracks. For a strain of ε_1=2000 μm/m the trap density decreases with increasing number of cycles, which can be accounted for by a greater durability of the sample put to strain, when compared with the strain ε_2=2500 μm/m.

As regards the development of the trap densities in the case of strains ε_1= 2000 μm/m and ε_2 = 2500 μm/m, two opposite trends of its variation are found. This may be related to a process of transition from the elastic to elasto-plastic.

With ε_3=3500 μm/m there is a general tendency to increase the trap density, which will hasten the destruction by fatigue as compared with the strain ε_2 = 2500 m / m.

Figure 12 shows the dependence of $\left(I_f / I_{\max}\right)$ on the number of strains for the 3 strain imposed at the frequency of 40 cycles / min for carbon steel OL52.

It is found that for the 3 imposed strains there is an overall decrease in the trap density relative to the initial state.

There is a tendency to increase the trap density with the number of cycles, for the strains of ε_2= 2500 μm / m and ε_3 = 3500 μm/m, and a slight downward tendency of the trap density for a small strain of ε_1=2000 μm/m.

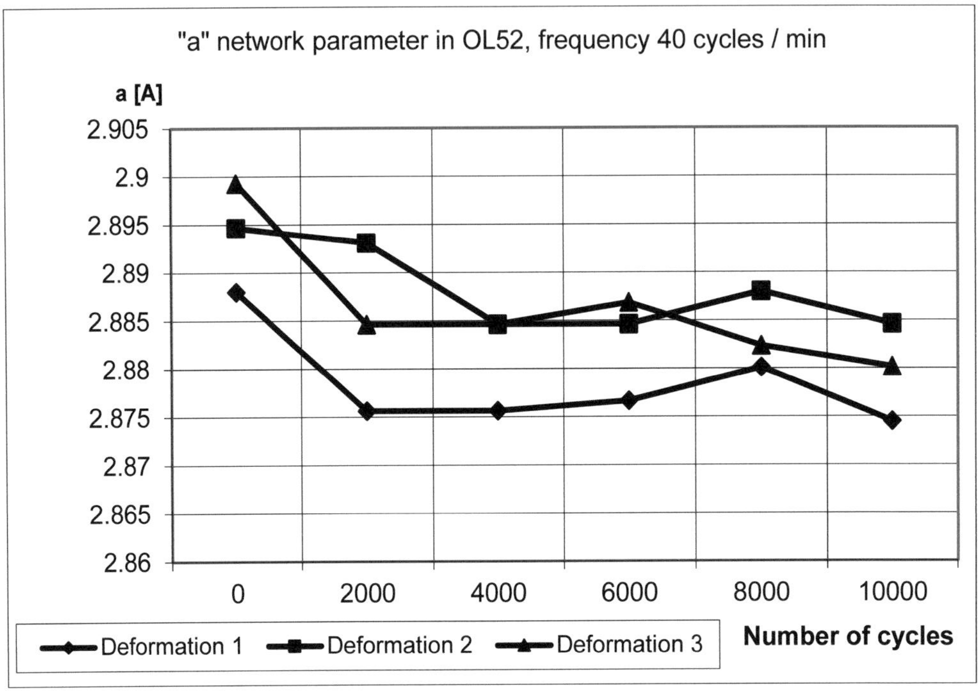

Fig. 12. Evolution of the lattice parameter for f = 40 cycles/min

The downward and upward slopes of the trap density (ε_1=2000 μm/m and ε_2=2500 μm/m) are lower as in the case of frequency of 20 cycles / min above. In addition, the slopes are reversed when compared with the case of frequency of 20 cycles/min.

The fact that there is a change of sign in the slopes at ε_1=2000 μm/m and ε_2=2500 μm/m, again justifies the existence, between the two strains, of a strain of transition from elastic to elasto-plastic.

The trap density variation with the number of cycles has a minimum value the position of which appears increasingly later, as the degree of strain decreases.

From the analysis of the two diagrams it is found that, in order to compensate for the fatigue damage to large strains, high-frequency strains should be applied.

Therefore, at low frequency and large strain, the occurrence of cracks through the process of trap agglomeration is much higher.

3.3.2 Analysis of metallurgical characteristics of the surface layer. Structure analysis. Network parameter

From the analysis of Figures 13 and 14 it is found that the network parameter tends to decrease with increasing number of strain cycles for the 3 strains imposed. The decrease is stronger at small strain and low frequencies and at high frequencies and large strains, respectively. Modification of the network parameter may be associated with the existence of a migration process of the atoms in the alloying elements from the network to the material surface.

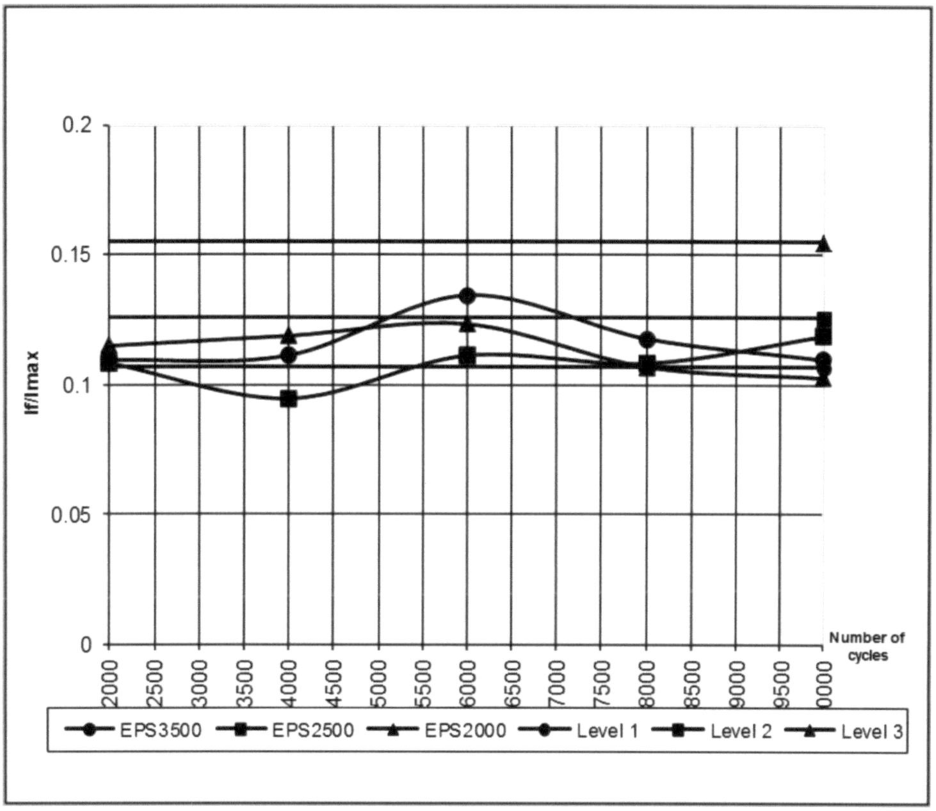

Fig. 13. Variation of the $\left(I_f / I_{\max}\right)_{220}$ on number of testing cycles for frequency v1=20 cycles/min

This is also supported by the slight increase in the network parameter for a given number of cycles. This increase occurs earlier when the strain is greater. The process of atom migration in and from the elementary cell of the ferritic phase indicates a high atom kinetics in the surface layer during the fatigue process.

The high kinetics may have adverse effects if the material would be put to strain in corrosive environments.

Analyzing the two figures it can be seen that the process of atom migration in and from the elementary cell occurs more slowly for low frequencies and faster for high frequencies and large strain; at small strains the process is more pronounced at lower frequencies.

It follows that if the strains were applied in aggressive environments, the life time would be much shorter for low frequencies/low strain and high frequency/ large strains.

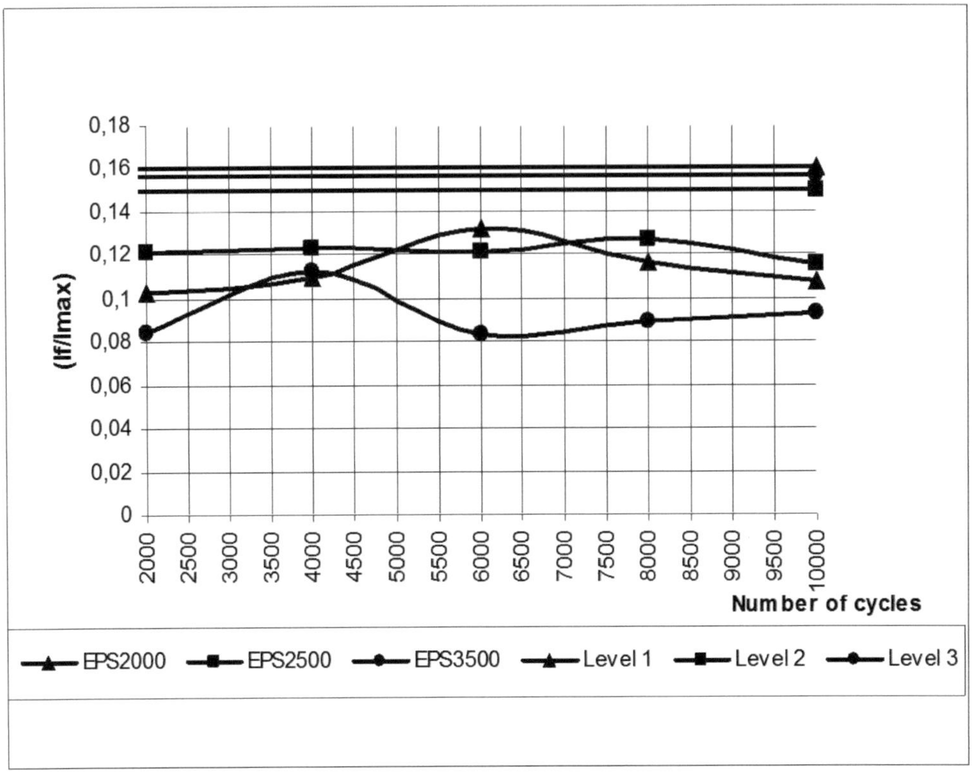

Fig. 14. Variation of the $\left(I_f / I_{max}\right)_{220}$ on number of testing cycles for frequency ν_1=40

There is therefore the possibility to "manage" from outside the life time by appropriate adjustments of the relationship between the strains imposed and the frequency applied.

3.3.3 Evolution of texture level

Diffractometry investigations with X-rays have highlighted the degree of texturing of OL52 steel sample subjected to high fatigue at the limit of the elastic range and to low frequencies.

As in previous cases, the texture analysis was made in increments of 2,000 cycles to 10,000 cycles and 3 imposed strains. The histogram in Figure 15 shows the dependence of I_{max}/I_0 on the number of cycles N and strain ε, at a frequency of 20 cycles / min.

Analyzing the resulting graphical representation, it is found the predominance of a retexturing process of the material to crystallographic direction (220).

The highest degree of retexturing becomes apparent at the largest strain value, ε_3. This retexturing is associated with the mechanical micro-processes leading to the loss of preferential orientation of crystal planes according to the direction (220) with respect to the state reached after rolling (I_{max}/I_0=1).

With the first strain ε_1, it is found a slight tendency of texturing which increases with increasing number of cycles. With the intermediate strain it is found that there is a stronger texturing which can be associated with the forced orientation of the crystalline planes (220). This material behaviour can be explained by analyzing the first micro processes of elastoplastic strain although, according to the characteristic curve of the material, we found ourselves in the elastic range. Figure 16 provides the histogram of the dependence of I_{max}/I_0 on the number of cycles N and the required strains ε_1, ε_2, ε_3 at the frequency of 40 cycles / min

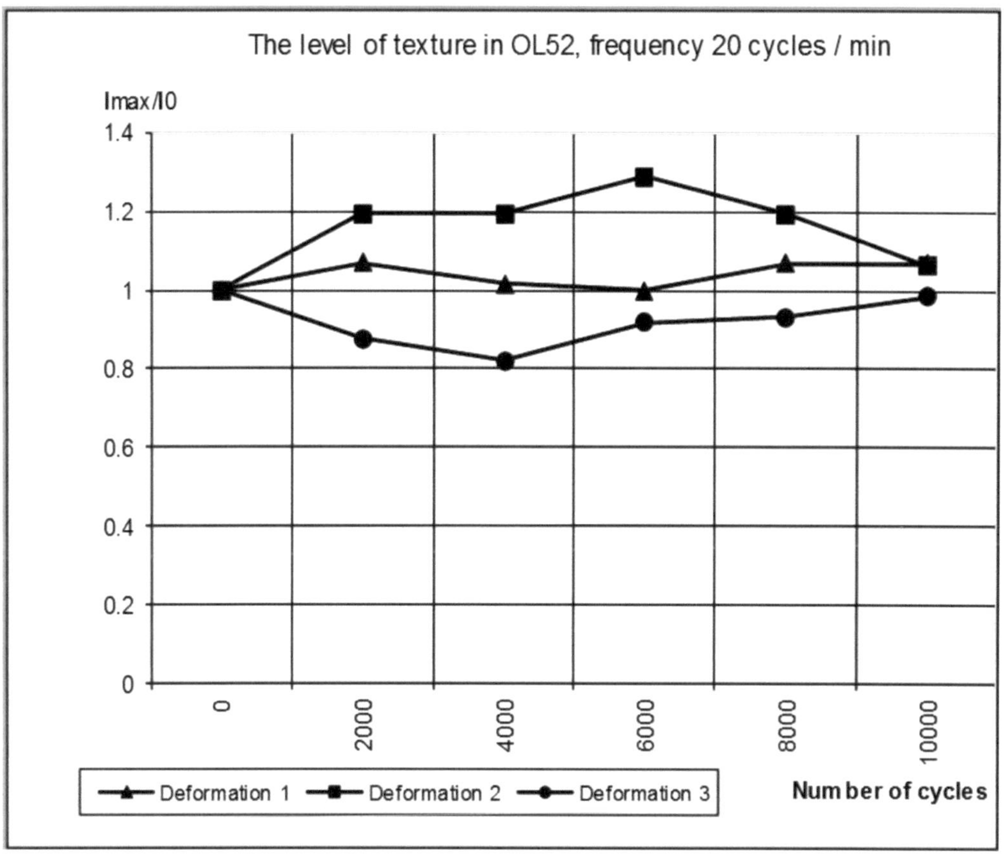

Fig. 15. Evolution of texture level for f = 20 cycles/min

Compared with the previous case, the graphical representation analysis shows that there is a general tendency of increased texture of the crystalline network both with increased number of cycles and the strain required.

With a frequency of 40 cycles / min. no tendency of retexturing was revealed for any amount of strain or number of strain cycles, as in the case described above at a frequency of 20 cycles / min. The fact that at low frequency (ν_1= 20 cycles / min) there are texturing and retexturing processes leads us to the conclusion that the material does not present inertia to changes in structure (hysteresis functions normally), while with high frequency (40 cycles /

min), the material loses part of its elastic properties responding to external factors - inertia to structure changes being much lower.

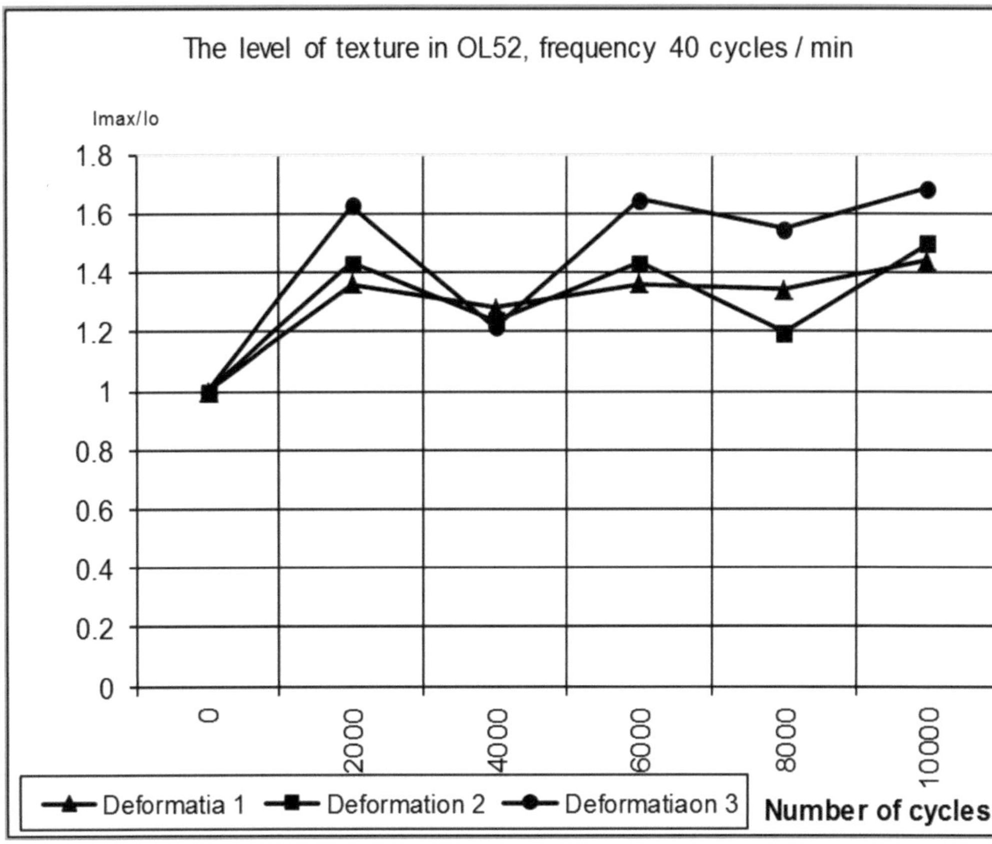

Fig. 16. Evolution of texture level for f = 40 cycles/min

3.3.4 Analysis of the surface layer microhardness

Figure 17 and 18 show the micro hardness HV variation for sample OL52 HV depending on the number of cycles to the strain ε_3=3500 μm/m for the two frequencies. The analysis of the experimental data and curves shows that micro hardness decreases in jumps. The decrease in the micro hardness occurs through processes of hardening and softening. With low frequencies (ν_1=20 cycles/min) decreases in micro hardness is less than the initial state, while at high frequency, the decrease in micro hardness is higher than the initial state. The amplitude of the softening and hardening processes is much higher for low frequencies than for high frequency (ν_2= 40 cycles / min). The period of the hardening and softening processes is lower at low frequencies and higher at higher frequencies. It can be said, therefore, that the velocities of the hardening and softening processes are higher at low frequencies than at higher frequencies for the same strain imposed.

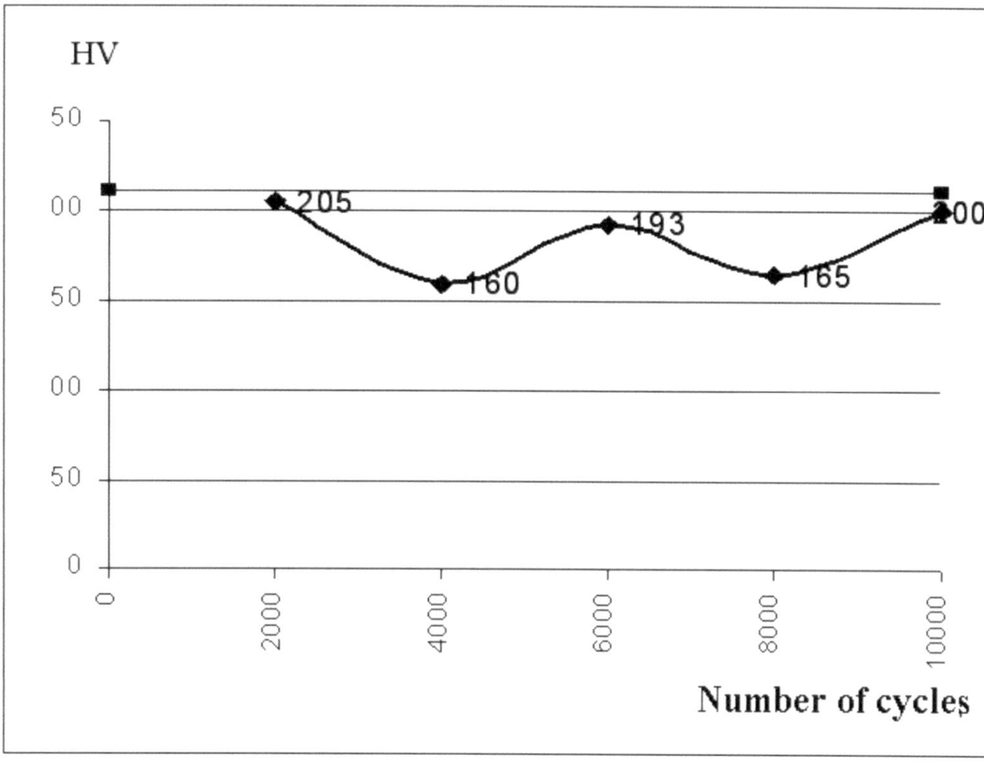

Fig. 17. Variation of microhardness vs testing cycles for v = 20 cycles/min

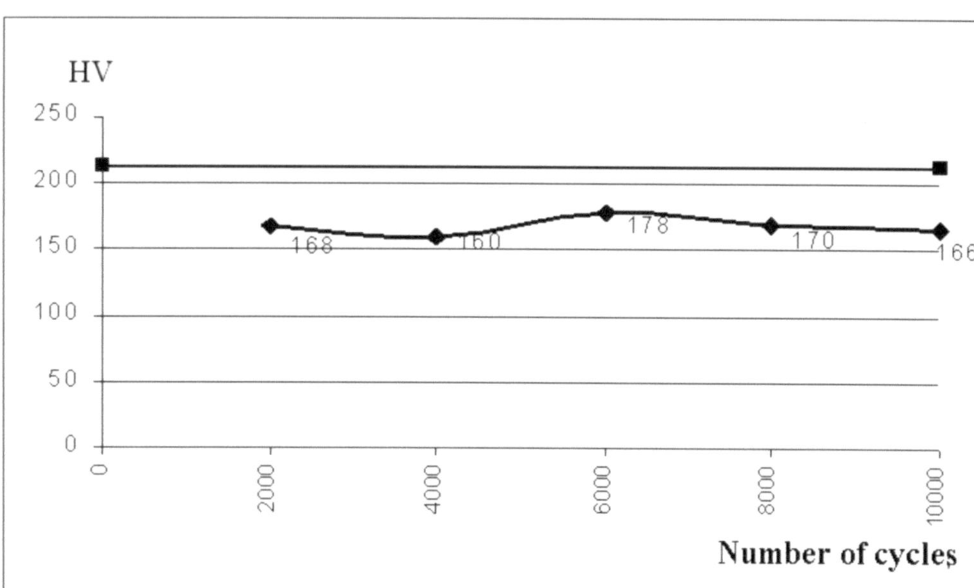

Fig. 18. Variation of microhardness vs testing cycles for v = 40 cycles/min

Number of cycles	x500 v_1=20 cycles/min ε=3500μm/m	x500 v_1=40 cycles/min ε=3500μm/m
Initial state	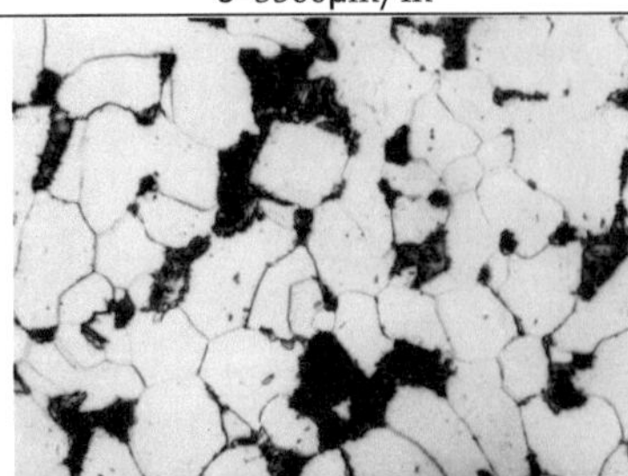	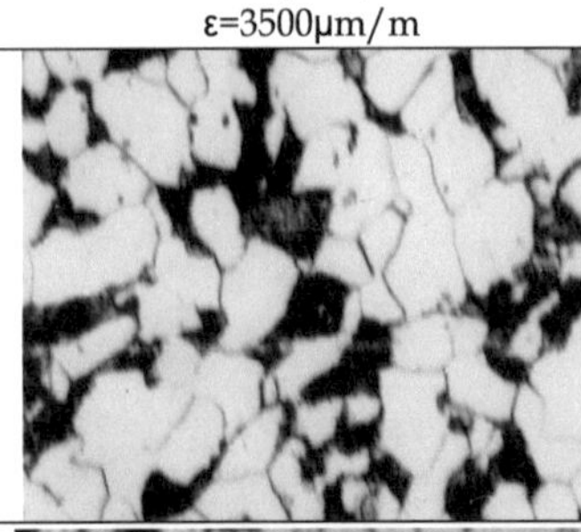
2000	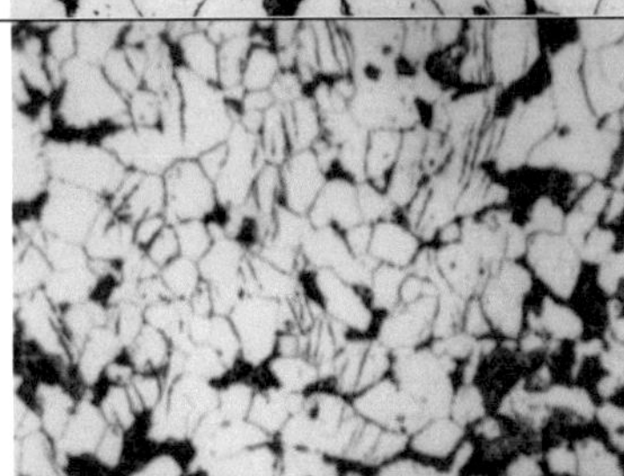	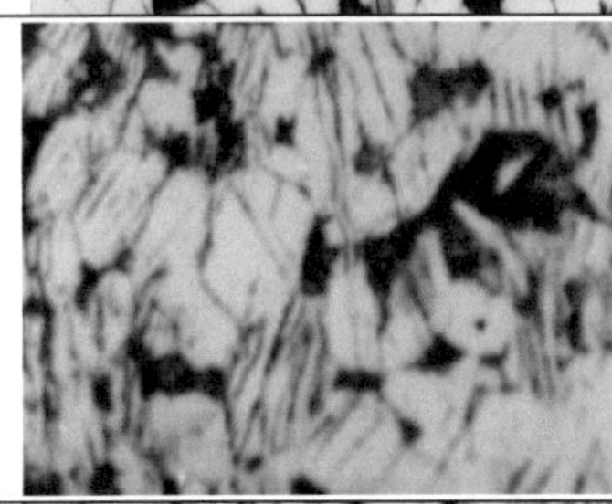
4000	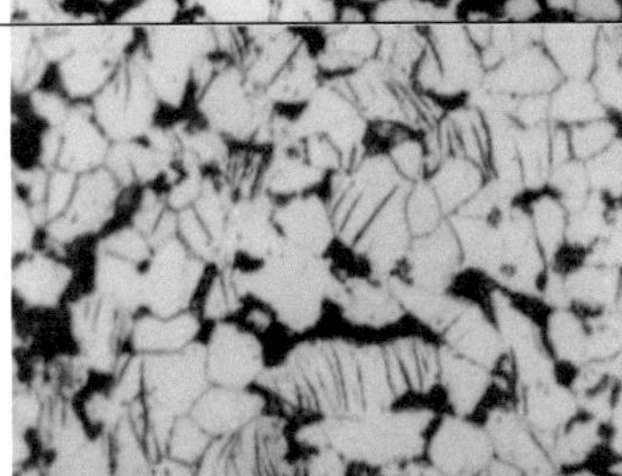	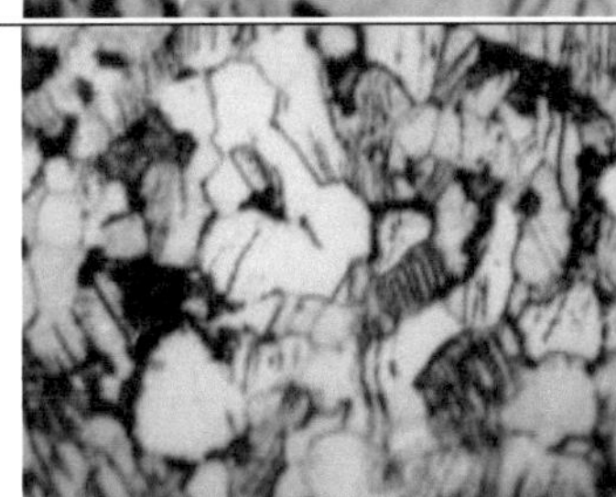
6000	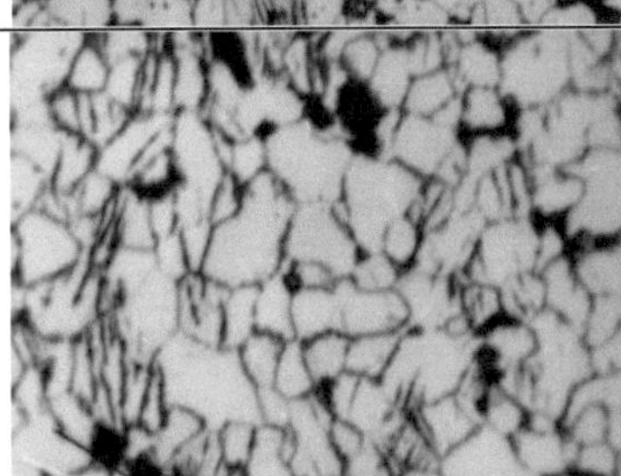	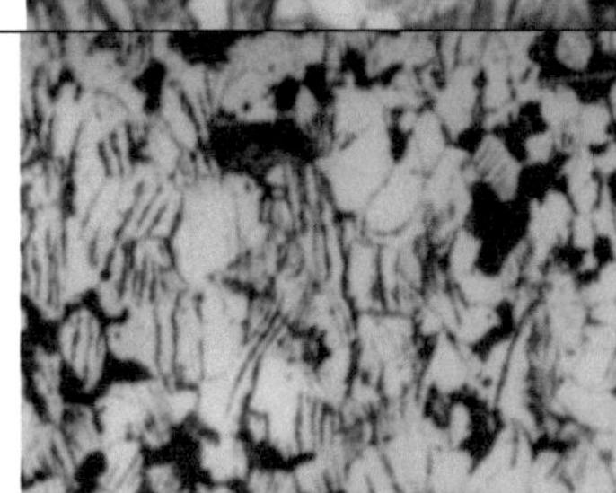
10000	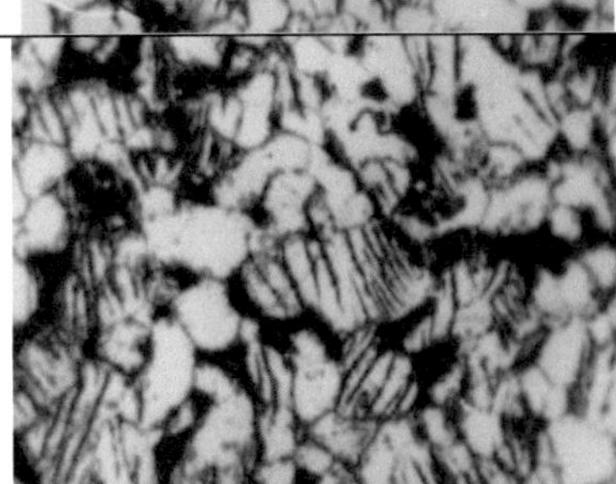	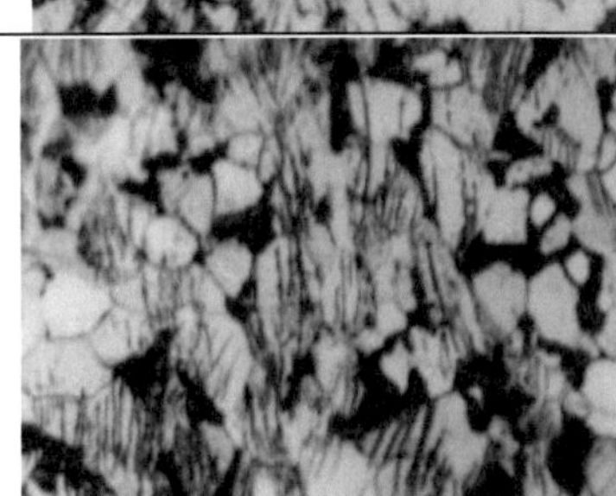
	Fig. 19. v_1=20 cycles/min	Fig. 20. v_1=40 cycles/min

3.3.5 Analysis of microstructure

The microstructure analyses were also carried out in increments of 2,000 cycles up to 10,000 cycles for 3 imposed strains and at the two frequencies (ν_1 şi ν_2).

The Figures 19 and 20 present the microstructures in the surface layer of the OL52 samples tested to pure bending fatigue to the strain ε_3=3500 μm/m, at the two frequencies.

The analysis of the microstructures reveals that with increasing number of cycles there is an increase in the density of the sliding bands in the ferrite grains for a given frequency. With the same strain and number of strain cycles it is observed that with higher frequency the sliding bands density is lower compared with that at lower frequencies.

3.4 Macroscopic aspects of the fractures

Generally, in all cases (Figure 21, 22, 23) the fatigue fracture process is initiated from the sample surface from spots featuring microscopic surface defects (roughness, more intense local hardening because of previous processing, surface defects of the material structure, such as inclusions, intermetallic phases, intersecting the processing surface).

In the section damaged by fatigue process, the samples present a characteristic shiny area and the sudden breaking zone under the strain applied to them.

In case of the sample shown in Figure 23a, on the polished surface near the fracture zone, sliding bands are clearly visible due to the relatively high speed of the strain propagation onto the crystalline grains favorably oriented and of relatively low total hardening intensity. The weight of the plastic strains under elasto-plastic regime is relatively high in a relatively small period of time ((N_{1r} = 30,065 cycles until breaking) to the strain ε_3=3500 μm/m.

The weight of the fatigue fracture surface is relatively small and located near the originator (concentrator) of the breaking/ fracture process (Figure 23b, c).

In Figure 21a, the polished surface reveals sliding bands specific to a very large number of cycles (N_{3r} = 106,488 cycles up to breaking) at a relatively large distance from the break/fracture zone. This indicates that for strains with small strains ((ε_1=2000 μm/m), the elasto-plastic strain zone before fracture is more extensive. The explanation is that the rate of hardening of the material is relatively small and therefore we believe that plastic strains will be taken, at the next cycles, by less hardened neighboring areas which feature lower strain resistance. Extension of the plastic strain area in the vicinity of the fracture zone is accounted for by the propagation of plastic strains, progressively to grains from the neighboring hardened areas.

In Figure 22a, with the sample put to strain ε_2=2500 μm/m, on the polished surface in the vicinity of the breaking zone it is highlighted the presence of sliding bands of highly fine granulation due to the extension of the elasto-plastic range to cover a larger period of time and a larger number of cycles until breaking (N_{2r} = 62,635 cycles). We believe that the weight of plastic strains under elasto-plastic regime is smaller than the previous case.

In Figure 22 b,c the fatigue fracture surface (shiny area) has a relatively greater expansion in the vicinity of the fracture originator and is developed over the entire width of the sample. Both on the previous sample and the sample mentioned above, the fatigue fracture surface is unilateral (on one side).

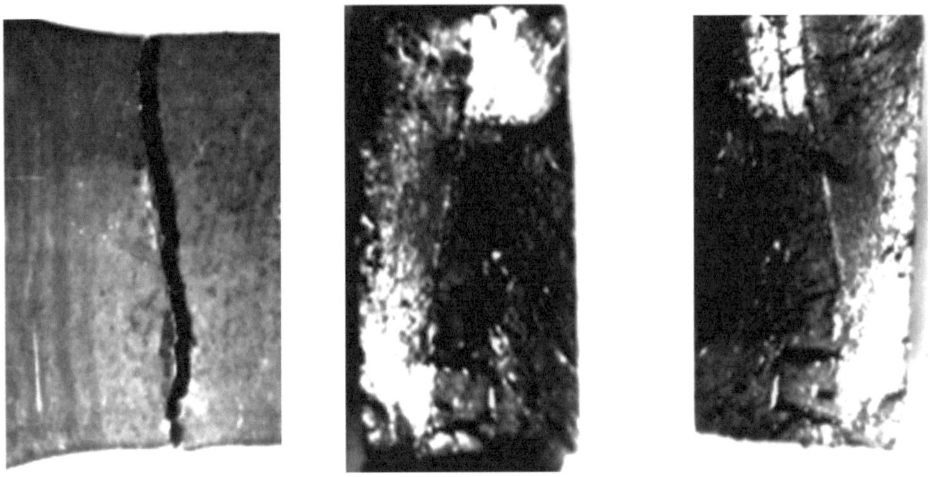

(a) (b) (c)

Fig. 21. Macroscopic aspects of crack for ε_3 = 3500 μm/m

(a) (b) (c)

Fig. 22. Macroscopic aspects of crack for ε_2 =2500 μm/m

(a) (b) (c)

Fig. 23. Macroscopic aspects of crack for ε_1=2000 μm/m

Sliding bands have greater width, this being possible by the accumulation of plastic strain in a relatively large time, i.e. for a larger number of cycles.

In Figure 21b, c, as far as the fracture section is concerned, the fatigue damaged area is much larger than the sudden fracture area, developed over the entire width of the sample, and of bilateral aspect. This can be explained by the fact that the development speed of the fatigue fracture surface from a concentrator is small which allows the initiation of the fatigue fracture from a concentrator on the opposite side.

In the previous cases, since strains are higher, it is sufficient to initiate the fatigue fracture from a stress concentrator because the growth rate of the fatigue fracture is much higher which makes no longer possible the initiation of a fatigue fracture from another concentrator.

4. Conclusions

1. By extension of the tribolayer and tribosystem concepts to the study low cycle fatigue process of the steel the structural changes in the superficial layer are shown. This allows to establish a relationship between structural parameters of superficial layer and damage degree during fatigue tests. It was evinced a microfatigue process which is strong influenced of: frequency testing, strain level, and number of the fatigue tests.
2. Our results can be used to account for the damage mechanism of the tested samples subjected to low frequency fatigue test and high tensions

5. References

Buzdugan, Gh., & Blumenfield, I. (1979), *Calculul de rezistenta al organelor de masini ,* Editura Tehnica Bucuresti -Romania, Bucuresti

Constantinescu, I. & Stefanescu, D.M. & Sandu, M., (1989), *Masurarea marimilor mecanice cu ajutorul tensometriei,* Editura Tehnica Bucuresti, ISBN 973-31-0127-3, Bucuresti

Crudu, I.& Macuta, S. & Palaghian, L. & Fazekas L.(1991) "*Masina universala de incercat materiale*", Patent nr. 102714/1991 ,Bucuresti

Gheorghes, C. (1990), *Controlul structurii fine a metalelor cu radiatii X,* Editura Tehnica Bucuresti, ISBN 973-31-0151-6, Bucuresti

Lieurade, H.P. (1982) *La Pratique des Essais de Fatigue,* PYC Editon , ISBN 2-85330-053-6, Paris

Macuta, S. - "*Evolution of some structural fine parameter in the superficial layer during low cycle fatigue process*", Tome I of International Conference on Advanced in Materials and Processing Technologies AMPT'01, vol1.ISBN 84-95821-06-0, Leganes, Madrid - Spania. September 2001

Macuta, S. & Rusu, E., (2008)., Experimental researches regarding the evolution of some parameters of the superficial layer in low cycle fatigue processes ,In: *Maritime Industry Ocean Engineering and Coastal Resources,* Guedes Soares & Kolev, pp. 219 - 223, Taylor and Francis Group , ISBN 978-0-415-45523-7, London

Macuta, S. & Rusu, E., (2009), Experimental researches regarding the evolution of some parameters of the superficial layer in low cycle fatigue processes, *Proceedings of*

13th International Congress International Maritime Association of Mediterranean, tom. 1 - ISBN 978-975-561-356-7, Istanbul Turkey, October 2009

Macuta, S. & Rusu, L., (2009), Modelling by finite element method of stress state establshing and experimental research regarding the elasto-plastic deformations of some steels alloys , *Proceedings of 13th International Congress International Maritime Association of Mediterranean,* tom. 3 - ISBN 978-975-561-358-1, Istanbul Turkey, October 2009

Macuta, S. (2007),. *Oboseala oligociclica a materialelor,* Editura Academiei Romane, ISBN 978-973-27-1382-2, Bucuresti

Macuta, S. (2010) - The Evolution of Certain Parameters In The Surface Layer During Low Cycle Fatigue Process-*Metalugia International Journal* vol.XV Special Issue no.8 , (augusut 2010), pp .20-25 ,ISSN 1582-2214

Macuta, S., (2004)- *"Establishing the elasto-plastic deformations of some steel alloys"*, Proceeding of The 29-th Annual Congress of the American Romanian Academy of Arts and Sciences, ISBN 973-632-140-1 Bochum,Germany september 2004.

Mocanu, Ds. (1982), *Incercaera materialelor Vol. 1 & 2,* Editura Tehnica Bucuresti - Romania , Bucuresti